POWER CONVERTERS WITH DIGITAL FILTER FEEDBACK CONTROL

POWER CONVERTERS WITH DIGITAL FILTER FEEDBACK CONTROL

KENG WU 吳耿志
Switching Power, Inc
Ronkonkoma, New York

Amsterdam • Boston • Heidelberg • London
New York • Oxford • Paris • San Diego
San Francisco • Singapore • Sydney • Tokyo
Academic Press is an imprint of Elsevier

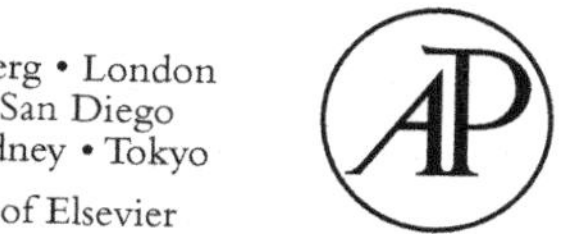

Academic Press is an imprint of Elsevier
125 London Wall, London EC2Y 5AS, UK
525 B Street, Suite 1800, San Diego, CA 92101-4495, USA
50 Hampshire Street, 5th Floor, Cambridge, MA 02139, USA
The Boulevard, Langford Lane, Kidlington, Oxford OX5 1GB, UK

Notices
Knowledge and best practice in this field are constantly changing. As new research and experience broaden our understanding, changes in research methods, professional practices, or medical treatment may become necessary.

Practitioners and researchers must always rely on their own experience and knowledge in evaluating and using any information, methods, compounds, or experiments described herein. In using such information or methods they should be mindful of their own safety and the safety of others, including parties for whom they have a professional responsibility.

To the fullest extent of the law, neither the Publisher nor the authors, contributors, or editors, assume any liability for any injury and/or damage to persons or property as a matter of products liability, negligence or otherwise, or from any use or operation of any methods, products, instructions, or ideas contained in the material herein.

British Library Cataloguing-in-Publication Data
A catalogue record for this book is available from the British Library

Library of Congress Cataloging-in-Publication Data
A catalog record for this book is available from the Library of Congress

ISBN: 978-0-12-804298-4

For information on all Academic Press publications
visit our website at http://store.elsevier.com/

To
Shwu
Stephanie
戴雪月

CONTENTS

BIOGRAPHY

Keng C. Wu, a native of Chiayi, Dalin, Taiwan, received a BS degree from Chiaotung University, Taiwan, in 1969 and an MS degree from Northwestern University, Evanston, Illinois in 1973.

He was a lead member of the technical staff of Lockheed Martin, Moorestown, New Jersey; a well recognized expert in high reliability power supply, power systems, and power electronics product design, including all component selection, board layout, modeling, large scale system dynamic study, prototype, testing and specification verification; and an author of four books: "*Pulse Width Modulated DC-DC Converters*" January 1997; "*Transistor Circuits for Spacecraft Power System*" November 2002; "*Switch-mode Power Converters: Design and Analysis*" Elsevier, Academic Press, November 2005; and "*Power Rectifiers, Inverters, and Converter*" November 2008. He also holds a dozen U.S. patents, was awarded "Author of the Year" twice (2003 and 2006, Lockheed Martin), and presented a three-hour educational seminar at IEEE APEC–2007.

PREFACE

There are two sayings; neither depicts the disheartening anguish that power supply industries had encountered in the face of digital advances. One goes like this, "Everything is going digital, but power supply," and the other, "Everything is going digital, power supply must follow." Taking it at its face value, the former is downright dispirited, while the latter a bit condescending. But both also carry some truth.

Here is the fact. Analog, audio, magnetic cassette tapes had been buried by digital, optical CDs. Analog, video, magnetic VHS tapes had been wiped out by digital, optical DVDs. Analog, landline telephones had been replaced by digital, wireless cell phones. Similarly, motor control has witnessed the inroad of digital techniques, such as space vector modulation and *d–q* decomposition. Therefore, at least 10–15 years ago, the expectation was that the switch-mode power supply (SMPS) will also be taken in by the digital tide.

However, it did not happen.

What could have been causing such a disappointment?

It is not that components suitable for the task were not available. It is not because of the lack of professionals well versed in the trade of SMPS design. It is not due to shrinking market. And, of course, it is not the lack of support from academics in digital signal processing.

What is missing is a scaffold linking them all.

Modern power supply in general, and SMPS in particular, are nonlinear feedback systems. Conventional, analog feedback systems with a single loop had been well studied and understood. Analytic tools and techniques for ensuring loop stability were readily available. And, thanks to late Prof. Robert Middlebrook and the power electronics group at the California Institute of Technology in the 1980s, deriving gain function for the nonlinear power stage was made feasible. A fact shared by all those advances is that they are effective only in the environment of analog domain and not all need to be, or can be, translated to the digital realm.

It turns out that only the error amplifier, which always resides in a feedback loop, and perhaps part of a modulator, which follows, needs to be converted into digital form. The rest, including the power stage, the switch driver, and many filters remains in analog form.

The saying that ONLY error amplifier needs to be moved to the digital form actually masks the degree of difficulties in disguise. This overly simplistic view ends up costing the power supply industry more than a decade in an attempt to transition to digital control.

Moving analog controlling amplifiers to digital entails more actions than what one would anticipate. It is not a one-man show and requires at least four sets of skill.

First, the analog controller ensuring feedback stability must be designed and its transfer function identified. Extracting and expressing the function in *s*-parameter takes skill.

Next, the analog function is converted to the digital z-transform plane. Digital signal processing (DSP) insights abound for treating digitized data streams obtained from low-level analog signals. Performing similar tasks for high-level power processing does not, however, enjoy the ease of harvesting the low-hanging fruits.

Then, all designs shall be simulated to verify or confirm the performance of the analog system and its digital equivalent. Newer tools capable of performing mixed signal simulation and accepting the functional model, rather than physical model, are required, as is experienced staff.

The last, results of the second step must be coded and implemented with a selected microcontroller. Professionals well trained in the conventional analog system may not master the new skill. New generation of digital experts are needed at this step.

Therefore, it is the attempt, better the goal, of this writing to expound the four steps.

Given the extreme challenges, and the utmost purpose of serving the industrial sector, it is considered better to proceed based on example.

Part I employs a forward converter and follows all four steps in sequence. Both voltage-mode and current-mode control are covered. Part II presents the flyback converter. Part III gives precision linear regulators and current regulators intended for driving LED array or charging battery. Part IV covers boost topology. Part V treats special converters, including resonant.

For all parts, the presentation is geared toward those who are already experienced in analog power processing. Therefore, minimal time will be spent in topics considered basic in that subject, for instance filters, operational amplifiers, pulse-width modulators, solid-state switch drivers, and basic transfer functions.

Ideally, a wholesome digital loop shall include both the digital filter/amplifier and the digital PWM. However, this writing for the time being does not vigorously cover the latter since, in terms of criticality, the digital filter occupies higher priority.

As mentioned before, a single individual, the principal writer included, simply cannot master all skill sets required for digital power supply design. Alex Krasner, a young, brilliant engineer helps cover MATLAB SIMULINK simulations and last chapter on digital implementation. In addition, Rizwan Ahmad, a Technology VP, reviewed the manuscript. Their efforts are gratefully appreciated.

Last, but not the least, heartfelt gratitude are also extended to Elsevier editorial team, Lisa Reading and Peter Jardim.

Keng Wu
Princeton, NJ, September 2015

NOTE TO THE READER

In this writing, simulations based on difference equations (MathCAD Professional 2000) and SIMULINK (MATLAB 2007a) are extensively invoked. The former requires complex key entries that are prone to typographical errors while the latter yields drawings that are short in meeting high-quality print requirement.

In order to mitigate both shortcoming and to serve reader, simulation files and source codes are collected and posted on the publisher's website (http://booksite.elsevier.com/9780128042984). MathCAD Professional 2000 and MATLAB 2007a are required to view these files.

PART I

Forward Converter

For power level higher than 100 W, this converter type is preferred in contrast to the Flyback that will be discussed in Part II.

This part begins by giving directly a design schematic in its entirety with local, housekeeping supply implied, but not shown explicitly, and with switch driver represented merely by a block. No introductory materials are given in either this part or other parts considering print page limitation and to avoid duplication of basic materials that are available in many other texts.

Based on the schematic, appropriate procedures are employed to derive the transfer function for all building blocks except the compensation error amplifier; the controller. All transfer functions are grouped and named "the modulator." Given a desired close-loop stability requirement, in terms of gain and phase margin, and by evaluating the modulator performance, a compensator that is able to meet the loop gain is identified and its analytical function extracted. By applying bilinear transformation, the controller in analog s-domain is moved to digital z-domain. Simulations are then performed for both the original analog version and its digital equivalent to verify the design.

Both voltage-mode and current-mode controls are covered and both follow the same sequence.

CHAPTER 1

Forward Converter with Voltage-Mode Control

1.1 SCHEMATIC WITH ANALOG CONTROLLER AND SAWTOOTH

A typical forward, pulse-width-modulated (PWM) DC–DC converter with voltage-mode control is shown in Figure 1.1.

At the moment, a type-III error amplifier, with C_{3a} and R_{3a} included, is shown. In later analysis, if it becomes apparent that just a type-II is sufficient, both components can then be removed. A type-III amplifier is known to provide three poles and two zeros; a type-II two poles and one zero. If double pole and double zero, $(s + z_2)^2/(s + p_2)^2$ term, are intentionally formed, the former can provide phase boost, ideally, up to 180°. The latter with $(s + z_1)/(s + p_1)$ term can boost phase up to only 90°. In both cases, the pole at zero frequency, which is an integrator $-1/(k \cdot s)$ term, is accounted for in phase boost computation.

The output filter is shown to include shunt damping, r_d and C_d, such that filter peaking, in frequency, can be minimized at light load. A ×1 buffer is also provided to isolate the controller's zero/pole setting from the output feedback divider in case the feedback factor needs to be tweaked.

The sawtooth clock is shown externally. In actual implementation, it is generally an internal block of an integrated PWM chip.

1.2 DERIVATION OF MODULATOR GAIN

The modulator is defined in this manner: circling around the control loop, the converter has a loop gain. With the error amplifier's function excluded, the rest is named the modulator.

In early 1970s, Prof. Robert D. Middlebrook and his then students G.W. Wester and Slobodan Ćuk initiated and developed the now famed technique, "State Space Averaging" [1,2], for the nonlinear power stage of switch-mode power converter. Appendix A briefly outlines the process.

Given Appendix A, we begin the arduous task of developing the state equations associated with the power stage or the power train. The goal is to come up with multiple transfer functions that will lead to the modulator gain.

Power Converters with Digital Filter Feedback Control
http://dx.doi.org/10.1016/B978-0-12-804298-4.00001-0

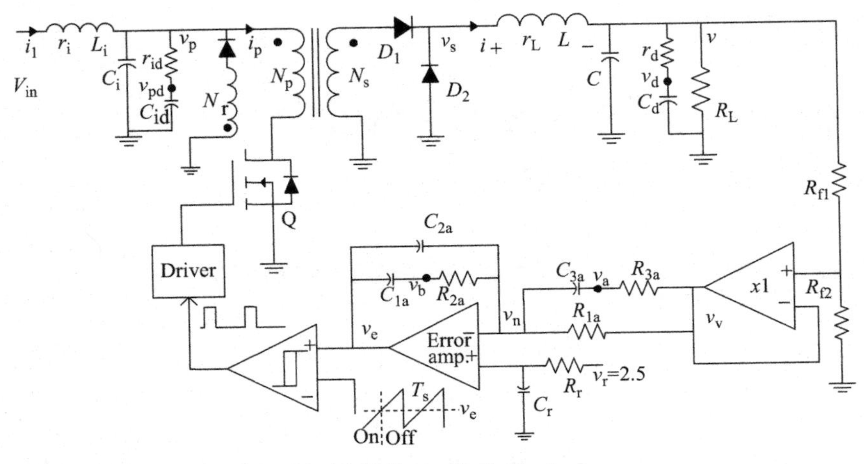

Figure 1.1 ***Forward Converter with Voltage-Mode Control.***

First, six variables are identified: output inductor current i, damping capacitor voltage v_d, output voltage v, input current i_1, primary voltage v_p, and input damping capacitor voltage v_{pd}.

Next, when the power switch Q is turned on (D_1 on, D_2 off) and considering the transformer action, the power stage is in ON configuration, Figure 1.2. There are six first-order differential equations to be derived.

The output inductor L and its parasitic series resistance r_L combination sustain a potential and is described by

$$\frac{di}{dt} = \frac{-r_L}{L} i + 0 + \frac{-1}{L} v + 0 + \frac{N_s}{N_p L} v_p + 0 \tag{1.1}$$

It is arranged in the order of six state variables explained earlier. Zero terms are intentionally inserted to remind us the absence of effect from that corresponding variable and to make matrix formulation less prone to error later.

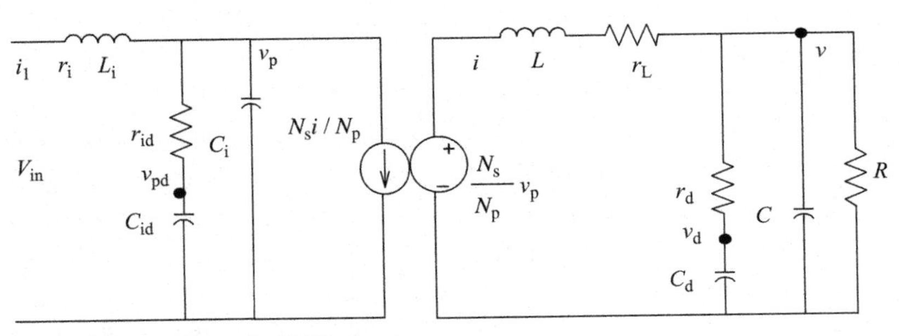

Figure 1.2 ***Power Stage in ON State.***

We than establish equations for the rest state variables according to the sequence given. At the output damping capacitor node,

$$\frac{dv_{\mathrm{d}}}{dt} = 0 + \frac{-1}{r_{\mathrm{d}}C_{\mathrm{d}}}v_{\mathrm{d}} + \frac{1}{r_{\mathrm{d}}C_{\mathrm{d}}}v + 0 + 0 + 0 \tag{1.2}$$

at the output node

$$\frac{dv}{dt} = \frac{1}{C}i + \frac{1}{r_{\mathrm{d}}C}v_{\mathrm{d}} + \frac{-1}{R_{\mathrm{p}}C}v + 0 + 0 + 0 \tag{1.3}$$

where $R_{\mathrm{p}} = (r_{\mathrm{d}}R_{\mathrm{L}})/(r_{\mathrm{d}} + R_{\mathrm{L}})$, at the primary input loop

$$\frac{di_1}{dt} = 0 + 0 + 0 + \frac{-r_{\mathrm{i}}}{L_{\mathrm{i}}}i_1 + \frac{-1}{L_{\mathrm{i}}}v_{\mathrm{p}} + 0 + \frac{1}{L_{\mathrm{i}}}V_{\mathrm{in}} \tag{1.4}$$

at the primary voltage node

$$\frac{dv_{\mathrm{p}}}{dt} = \frac{-N_{\mathrm{s}}}{N_{\mathrm{p}}C_{\mathrm{i}}}i + 0 + 0 + \frac{1}{C_{\mathrm{i}}}i_1 + \frac{-1}{r_{\mathrm{id}}C_{\mathrm{i}}}v_{\mathrm{p}} + \frac{1}{r_{\mathrm{id}}C_{\mathrm{i}}}v_{\mathrm{pd}} \tag{1.5}$$

at the input damping capacitor node

$$\frac{dv_{\mathrm{pd}}}{dt} = 0 + 0 + 0 + 0 + \frac{1}{r_{\mathrm{id}}C_{\mathrm{id}}}v_{\mathrm{p}} + \frac{-1}{r_{\mathrm{id}}C_{\mathrm{id}}}v_{\mathrm{pd}} \tag{1.6}$$

When the switch is turned off, the free-wheeling diode D_2 commences conduction. If an ideal case is assumed, the diode forward voltage may be considered zero. With the assumption, a new set of equations can be written for the switch-off configuration, Figure 1.3. However, only two equations

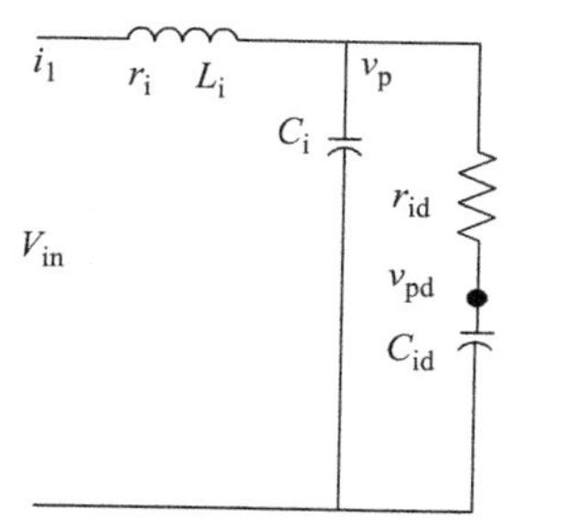

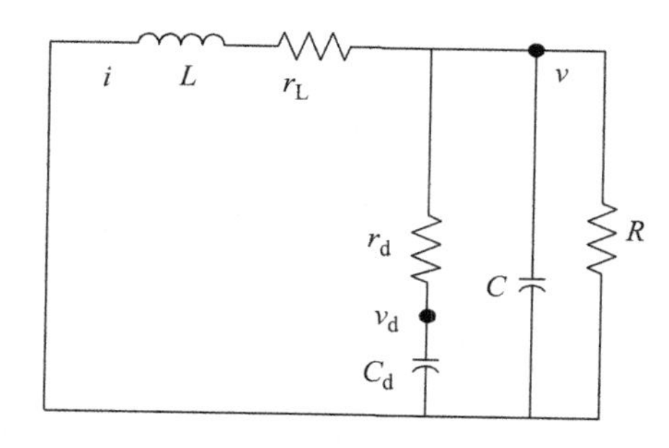

Figure 1.3 ***Power Stage in OFF State.***

(1.1) and (1.5), need to be modified, since transformer coupling no longer exists.

$$\frac{di}{dt} = \frac{-r_L}{L} i + 0 + \frac{-1}{L} v + 0 + 0 + 0 \quad (1.7)$$

$$\frac{dv_p}{dt} = 0 + 0 + 0 + \frac{1}{C_i} i_1 + \frac{-1}{r_{id} C_i} v_p + \frac{1}{r_{id} C_i} v_{pd} \quad (1.8)$$

Given (1.1–1.6), and with the understanding of the state vector $x = [i \ v_d \ v \ i_1 \ v_p \ v_{pd}]^T$, we are able to identify matrix A_1, B_1, and C_1 for the switch-on configuration.

$$A_1 = \begin{bmatrix} \frac{-r_L}{L} & 0 & \frac{-1}{L} & 0 & \frac{N_s}{N_p L} & 0 \\ 0 & \frac{-1}{r_d C_d} & \frac{1}{r_d C_d} & 0 & 0 & 0 \\ \frac{1}{C} & \frac{1}{r_d C} & \frac{-1}{R_p C} & 0 & 0 & 0 \\ 0 & 0 & 0 & \frac{-r_i}{L_i} & \frac{-1}{L_i} & 0 \\ \frac{-N_s}{N_p C_i} & 0 & 0 & \frac{1}{C_i} & \frac{-1}{r_{id} C_i} & \frac{1}{r_{id} C_i} \\ 0 & 0 & 0 & 0 & \frac{1}{r_{id} C_{id}} & \frac{-1}{r_{id} C_{id}} \end{bmatrix}, B_1 = \begin{bmatrix} 0 \\ 0 \\ 0 \\ \frac{1}{L_i} \\ 0 \\ 0 \end{bmatrix}, C_1 = \begin{bmatrix} 0 \\ 0 \\ 1 \\ 0 \\ 0 \\ 0 \end{bmatrix} \quad (1.9a)$$

By the same token, matrix A_2, B_2, and C_2 for the switch-off configuration are also identified.

$$A_2 = \begin{bmatrix} \frac{-r_L}{L} & 0 & \frac{-1}{L} & 0 & 0 & 0 \\ 0 & \frac{-1}{r_d C_d} & \frac{1}{r_d C_d} & 0 & 0 & 0 \\ \frac{1}{C} & \frac{1}{r_d C} & \frac{-1}{R_p C} & 0 & 0 & 0 \\ 0 & 0 & 0 & \frac{-r_i}{L_i} & \frac{-1}{L_i} & 0 \\ 0 & 0 & 0 & \frac{1}{C_i} & \frac{-1}{r_{id} C_i} & \frac{1}{r_{id} C_i} \\ 0 & 0 & 0 & 0 & \frac{1}{r_{id} C_{id}} & \frac{-1}{r_{id} C_{id}} \end{bmatrix}, B_2 = B_1, C_2 = C_1 \quad (1.9b)$$

Next, with (A.12), we shall find the closed-loop, steady-state duty cycle D_x. To that end, we define symbolically $A(D_x) = D_x A_1 + (1 - D_x) A_2$, $B(D_x) = D_x B_1 + (1 - D_x) B_2$, $C(D_x) = D_x C_1 + (1 - D_x) C_2$, error amplifier

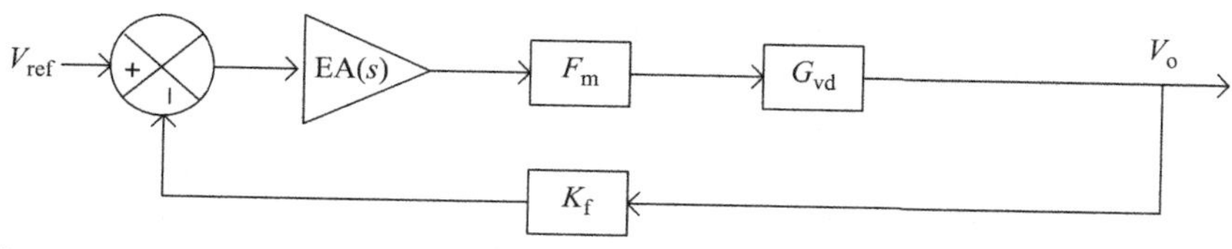

Figure 1.4 ***AC Small Signal Block Diagram.***

open-loop gain G_A, and triangular carrier voltage swings, V_L to V_H. The duty cycle is then determined by

$$G_A\left[v_r - [-C(D_x)^T A(D_x)^{-1} B(D_x) V_{in}]\frac{R_{f2}}{R_{f1}+R_{f2}}\right] = \\ V_L + \frac{V_H - V_L}{0.98} D_x \quad \text{or} \quad V_e = V_L + \frac{1}{F_m} D_x \tag{1.10}$$

Equation (1.10) denotes the intercept of the triangular carrier ramp against the error voltage, which is generated by comparing the feedback with a reference under high gain condition. This is the mechanism under which the duty cycle is determined, assuming 0.98 maximum duty cycle corresponding to the full error voltage range V_L to V_H.

Once the closed-loop, steady-state duty cycle is found, (A.7) and (A.16) yield the duty cycle-to-output transfer function, $G_{yd}(s)$, and we are done with the heavy lifting. Two simple blocks remain. The PWM gain is easily identified, $F_m = 0.98/(V_H - V_L)$. While the feedback factor, K_f, equals $R_{f2}/(R_{f1} + R_{f2})$. Another form of (1.10), and together with K_f and $G_{yd}(s)$, also implies a small signal AC block diagram (Figure 1.4). Ref. [3] gives more in detail.

Furthermore, the error amplifier gain is identified too.

$$E_A(s) = -\frac{[(R_{2a} + \frac{1}{C_{1a}s})^{-1} + C_{2a}s]^{-1}}{[(R_{3a} + \frac{1}{C_{3a}s})^{-1} + \frac{1}{R_{1a}}]^{-1}}$$

At this point, transfer functions of contributing blocks for constructing the modulator have all been identified. They are consolidated to give the modulator gain.

$$M(s) = \left(\frac{R_{f2}}{R_{f1}+R_{f2}}\right)\left(\frac{0.98}{V_H - V_L}\right) G_{yd}(s) \tag{1.11}$$

1.3 IDENTIFY CONTROLLER AND EXTRACT TRANSFER FUNCTION

We are now proceeding to the next phase, which is to identify the proper error amplifier needed to make the overall loop meet the stability requirement and to extract its analytical expression in s-domain.

This is how it is done. The modulator gain (1.11) is plot for its magnitude, in db, and phase, in degree against frequency. Furthermore, assume the total loop gain crossover frequency, f_c, is specified. The modulator plot may show a gain or a deficit, M_{db} and a phase α_M at the crossover. Further assume a phase margin α_m is desired. (1.12) dictates which amplifier, type-II or type-III, is selected. If α_b is less than 90°, type-II compensator is selected. If it is more than 90°, but less than 180°, type-III is then required.

$$\alpha_b = \alpha_m - (\alpha_M + 90) \tag{1.12}$$

With known phase boost and controller type selected, pole/zero frequency separation factor, k, is obtained

$$\begin{aligned} k &= \sqrt{\frac{f_{p1}}{f_{z1}}} = \tan\left[\tfrac{\pi}{180}\left(\tfrac{\alpha_b}{2} + 45\right)\right], \text{ type- II} \\ k &= \frac{f_{p2}}{f_{z2}} = \left\{\tan\left[\tfrac{\pi}{180}\left(\tfrac{\alpha_b}{4} + 45\right)\right]\right\}^2, \text{ type- III} \end{aligned} \tag{1.13}$$

The separation factor and modulator gain/deficit together yields the component value for either type. Figure 1.5 gives both types, and their component values are given below with R_1 often preselected, about 2 kΩ, for op-amp offset and input bias current considerations. Readers are reminded again that f_{p1}/f_{z1} is a single pole–zero pair, while f_{p2}/f_{z2} a double pair, and they are also referred to [3] for more details.

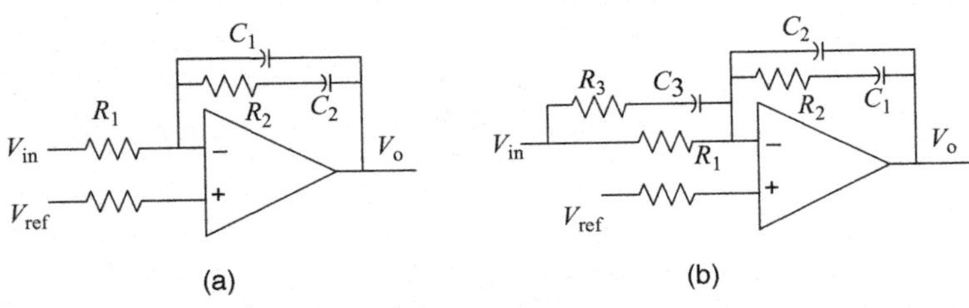

Figure 1.5 ***Error Amplifier.*** (a) Type-II and (b) Type-III.

For type-II, the following shall be computed in the order given,

$$\begin{aligned}
R_1 &\approx 2\,\mathrm{K}\Omega, \text{preselected} \\
C_1 &= \left(2\pi R_1 k^2 f_c \times 10^{\left[\frac{-M_{db}}{20} - \log(k)\right]}\right)^{-1} \\
C_2 &= (k^2 - 1)C_1 \\
R_2 &= \left(2\pi \frac{f_c}{k} C_2\right)^{-1}
\end{aligned} \tag{1.14}$$

and for type-III

$$\begin{aligned}
R_1 &\approx 2\,\mathrm{K}\Omega, \text{preselected} \\
C_2 &= \left(2\pi R_1 \sqrt{k} f_c \times 10^{\left[\frac{-M_{db} - 10\cdot\log(k)}{20}\right]}\right)^{-1} \\
C_1 &= (k-1)C_2 \\
R_2 &= \sqrt{k}\left(2\pi f_c C_1\right)^{-1} \\
R_3 &= R_1 (k-1)^{-1} \\
C_3 &= \left(2\pi R_3 \sqrt{k} f_c\right)^{-1}
\end{aligned} \tag{1.15}$$

In addition, transfer function for both are given

$$\begin{aligned}
\mathrm{EA}_{\mathrm{II}}(s) &= -\frac{R_2 C_2 s + 1}{R_1 (C_1 + C_2) s \left(R_2 \frac{C_1 C_2}{C_1 + C_2} s + 1\right)} \\
\mathrm{EA}_{\mathrm{III}}(s) &= -\frac{(R_2 C_1 s + 1)[(R_1 + R_3) C_3 s + 1]}{R_1 (C_1 + C_2) s (R_3 C_3 s + 1)\left(R_2 \frac{C_1 C_2}{C_1 + C_2} s + 1\right)}
\end{aligned} \tag{1.16}$$

Once the error amplifier type is selected and components value calculated, (1.16) is combined with (1.11) to yield the total loop gain $T(s) = \mathrm{EA}(s)M(s)$.

1.4 DERIVATION OF DIGITAL TRANSFER FUNCTION

We are now ready to transform the analog error amplifier to the digital realm. Readers are forewarned that the procedure is not straightforward and requires both patience and skill. Luckily, modern computer algebra with symbolic processing power does help.

First, (1.16) shall be placed in rational function form with both the numerator and the denominator in polynomial. We proceed first with type-II; inverting sign omitted.

$$\mathrm{EA_{II}}(s) = \frac{R_2C_2s+1}{R_1(C_1+C_2)s\left(R_2[C_1C_2/(C_1+C_2)]s+1\right)} = \frac{A_1s+A_0}{B_2s^2+B_1s+B_0} \tag{1.17}$$

where
$A_1 = R_2C_2, A_0 = 1, B_2 = R_1R_2C_1C_2, B_1 = R_1(C_1+C_2),$
$B_0 = 0$

Next, s operator is replaced by the bilinear transformation that maps the complex s-plane to the z-plane: $s = C(1 - z^{-1})/(1 + z^{-1})$. C is an unknown constant yet to be determined, considering sampling rate.

$$\mathrm{EA_{II}}(z) = \frac{A_1C\left((1-z^{-1})/(1+z^{-1})\right)+A_0}{B_2C^2\left((1-z^{-1})/(1+z^{-1})\right)^2+B_1C\left((1-z^{-1})/(1+z^{-1})\right)} \tag{1.18}$$

With the help of a tool, for instance MathCAD, with algebraic symbolic processing capability, the mapping (Appendix B) process is done. The analog error amplifier is now becoming a digital filter.

$$\begin{aligned} H_{\mathrm{II}}(z) &= \frac{a_0+a_1z^{-1}+a_2z^{-2}}{1+b_1z^{-1}+b_2z^{-2}} \\ a_0 &= \frac{A_0+A_1C}{B_2C^2+B_1C+B_0}, a_1 = \frac{2A_0}{B_2C^2+B_1C+B_0}, \\ a_2 &= \frac{A_0-A_1C}{B_2C^2+B_1C+B_0}, b_1 = \frac{-2B_2C^2+2B_0}{B_2C^2+B_1C+B_0}, \\ b_2 &= \frac{B_2C^2-B_1C+B_0}{B_2C^2+B_1C+B_0} \end{aligned} \tag{1.19}$$

However, it shall be noted that all lowercase coefficients given in (1.19) are extremely complicated functions of the actual circuit components, R_1, R_2, C_1, and C_2, since all capital coefficients, A_n and B_n, given in (1.17) are functions of the same. They are all left as they are, and no attempt is made to express explicitly each in component function form. Only numerical computation is advised. The leading minus sign of $\mathrm{EA_{II}}(s)$ may be included if needed. For now, the inverting sign is not included.

At this point, it is also advisable to address the significance of bilinear transformation $s = C(1 - z^{-1})/(1 + z^{-1})$ or $z = (C + s)/(C - s)$.

In treating the discrete sampled signal, a mapping between the continuous s-plane and the periodic z-plane exists because of sampling at a selected clock of high frequency. It is defined by $z = e^{sT}$, or $s = \ln(z)/T$, with T being the sampling period. This mapping function has the property that the entire left half side, including the imaginary axis, of the s-plane is mapped into the interior and the circumference of a unit circle in z-plane. Since variable s is complex, exponential of s is periodic. As a result, z variable and function of z are also periodic.

However, if one attempts to ideally transform (1.17) to z-domain by plugging $s = \ln(z)/T$, one immediately sees the impossibility because of terms like $\ln(z)$, $\ln(z)^2$, etc.

It turns out, there is a way out. That is the inverse $z^{-1} = e^{-sT}$. The following holds:

$$e^{-sT} = 1 - sT + \tfrac{1}{2}(sT)^2 - \tfrac{1}{6}(sT)^3 + \tfrac{1}{24}(sT)^4 - +..... \quad (1.20)$$

Interestingly, a long division shows the following also holds

$$\frac{1-(sT/2)}{1+(sT/2)} = 1 - sT + \tfrac{1}{2}(sT)^2 - \tfrac{1}{4}(sT)^3 + \tfrac{1}{8}(sT)^4 - +..... \quad (1.21)$$

In other words, form (1.21) is an acceptable approximation of the exact transform (1.20), which is $z^{-1} = (1 - sT/2)/(1 + sT/2)$, or $s = (2/T)(1 - z^{-1})/(1 + z^{-1}) = C(1 - z^{-1})/(1 + z^{-1})$; the bilinear transform with constant C equals twice, at least, the sampling frequency.

Next, we shall repeat the same transform process for type-III.

$$\begin{aligned}
\mathrm{EA}_{\mathrm{III}}(s) &= \frac{(R_2C_1s+1)[(R_1+R_3)C_3s+1]}{R_1(C_1+C_2)s(R_3C_3s+1)\left(R_2[C_1C_2/(C_1+C_2)]s+1\right)} \\
&= \frac{A_2s^2+A_1s+A_0}{B_3s^3+B_2s^2+B_1s+B_0} \\
A_2 &= R_2C_1C_3(R_1+R_3), A_1 = R_1C_3+R_2C_1+R_3C_3, A_0 = 1 \\
B_3 &= R_1R_2R_3C_1C_2C_3, B_2 = R_1(R_3C_2C_3+R_2C_1C_2+R_3C_1C_3), \\
B_1 &= R_1(C_1+C_2), B_0 = 0
\end{aligned} \quad (1.22)$$

$$
\begin{aligned}
H_{III}(z) &= \frac{a_0 + a_1 z^{-1} + a_2 z^{-2} + a_3 z^{-3}}{1 + b_1 z^{-1} + b_2 z^{-2} + b_3 z^{-3}} \\
a_0 &= \frac{A_2 C^2 + A_1 C + A_0}{B_3 C^3 + B_2 C^2 + B_1 C + B_0}, \\
a_1 &= \frac{-A_2 C^2 + A_1 C + 3A_0}{B_3 C^3 + B_2 C^2 + B_1 C + B_0}, \\
a_2 &= \frac{-A_2 C^2 - A_1 C + 3A_0}{B_3 C^3 + B_2 C^2 + B_1 C + B_0}, \\
a_3 &= \frac{A_2 C^2 - A_1 C + A_0}{B_3 C^3 + B_2 C^2 + B_1 C + B_0}, \\
b_1 &= \frac{-3B_3 C^3 - B_2 C^2 + B_1 C + 3B_0}{B_3 C^3 + B_2 C^2 + B_1 C + B_0}, \\
b_2 &= \frac{3B_3 C^3 - B_2 C^2 - B_1 C + 3B_0}{B_3 C^3 + B_2 C^2 + B_1 C + B_0}, \\
b_3 &= \frac{-B_3 C^3 + B_2 C^2 - B_1 C + B_0}{B_3 C^3 + B_2 C^2 + B_1 C + B_0}
\end{aligned}
\tag{1.23}
$$

1.5 REALIZATION OF DIGITAL TRANSFER FUNCTION

Digital transfer functions $H_{II}(z)$ in (1.19) and $H_{III}(z)$ in (1.23) are recognized to take the following rational, output-to-input form:

$$
H(z) = \frac{Y(z)}{X(z)} = \frac{\sum_{i=0}^{k} a_i z^{-i}}{1 + \sum_{i=1}^{k} b_i z^{-i}}
\tag{1.24}
$$

In other words, the output can be expressed as

$$
Y(z) = \frac{X(z)}{1 + \sum_{i=1}^{k} b_i z^{-i}} \sum_{i=0}^{k} a_i z^{-i} = W(z) \sum_{i=0}^{k} a_i z^{-i}
\tag{1.25}
$$

That is,

$$
\begin{aligned}
&W(z) = \frac{X(z)}{1 + \sum_{i=1}^{k} b_i z^{-i}}, W(z) = X(z) + \sum_{i=1}^{k} (-b_i) W(z) z^{-i} \\
&Y(z) = W(z) \sum_{i=0}^{k} a_i z^{-i} = \sum_{i=0}^{k} a_i W(z) z^{-i}
\end{aligned}
\tag{1.26}
$$

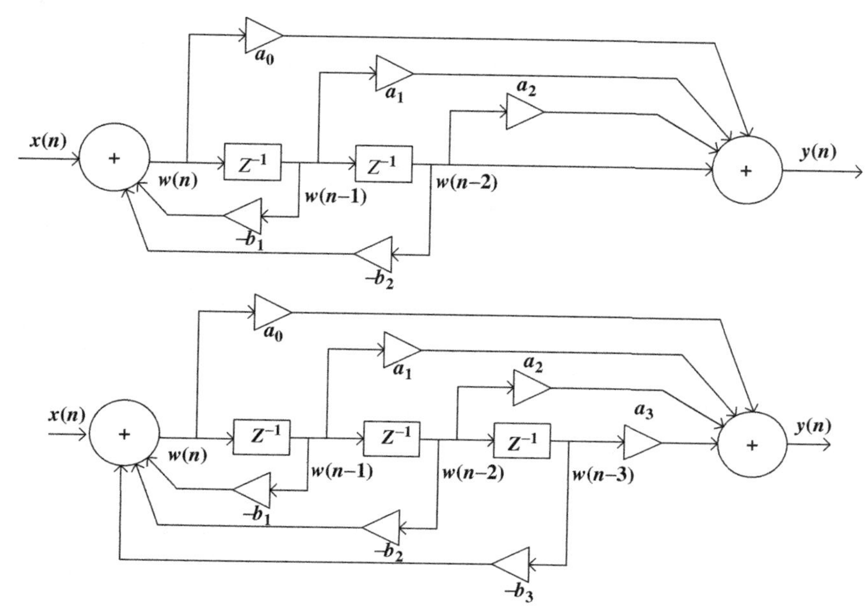

Figure 1.6 ***Direct Form*** **H(z)** ***for Type-II and Type-III.***

By taking the inverse transform, (1.26) implies

$$\begin{aligned} w(n) &= x(n) + \sum_{i=1}^{k}(-b_i)w(n-i) \\ y(n) &= \sum_{i=0}^{k} a_i w(n-i), \end{aligned} \tag{1.27}$$

(1.27) yields the direct form realization, Figure 1.6, for general digital filters (1.24).

1.6 IMPLEMENTATION IN CIRCUIT FORM

In the previous process, efforts were focused in small signal, AC frequency domain. As such, DC bias condition, for example, reference voltage V_r, was not included. It must be reinstated to set the DC condition correct.

The circuit surrounding the error amplifier is represented by Figure 1.7; Z_i and Z_f, respectively, the input and feedback blocks.

Conventional theory for operational amplifier circuits says that the error output is given by

$$V_e(s) = \left(1 + \tfrac{Z_f(s)}{Z_i(s)}\right)V_r - \tfrac{Z_f(s)}{Z_i(s)}V_f \tag{1.28}$$

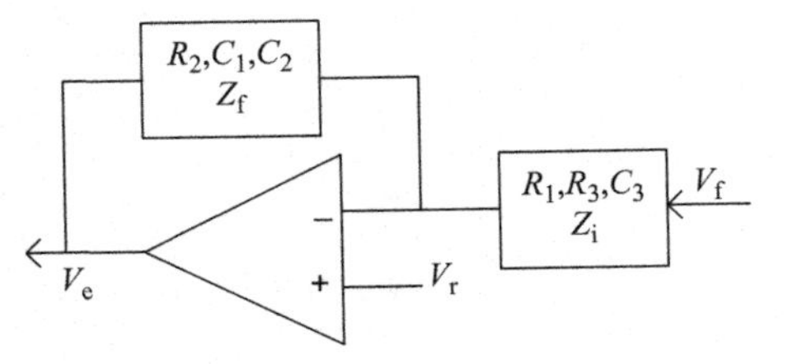

Figure 1.7 ***Operational Amplifier.***

where the reference voltage enters at the noninverting terminal, while the feedback does at the inverting terminal. (1.28) indicates that both have different gains.

(1.28) can be rewritten as

$$V_e(s) = V_r + (V_r - V_f)\frac{Z_f(s)}{Z_i(s)} \tag{1.29}$$

The rewritten form conveys an extremely important insight that is somewhat clouded in (1.28). What (1.29) declares is that only the differential part between the reference and the feedback is subjected to filtering. Whereas, the reference merely sets the DC state. This observation is critical in that it makes the implementation of digital filter quite unique in signal flow form.

That is, in actual circuit environment, Figure 1.6 shall be modified to account for the DC operating point. The modification results in Figure 1.8.

Readers are reminded that A-to-D (analog-to-digital) and D-to-A shall be included in actual, practical circuits, and with that in mind, the controller in digital form is ready to replace the analog amplifier in Figure 1.1.

In Figure 1.9, the digital filter block is identified also with a s-to-z mapping constant C = 2 MHz, (1.18 OR 1.23), to remind us that it plays a sensitive role in setting the magnitude of coefficients for the numerator and the denominator polynomial of $H(z)$. In addition, a sample and hold

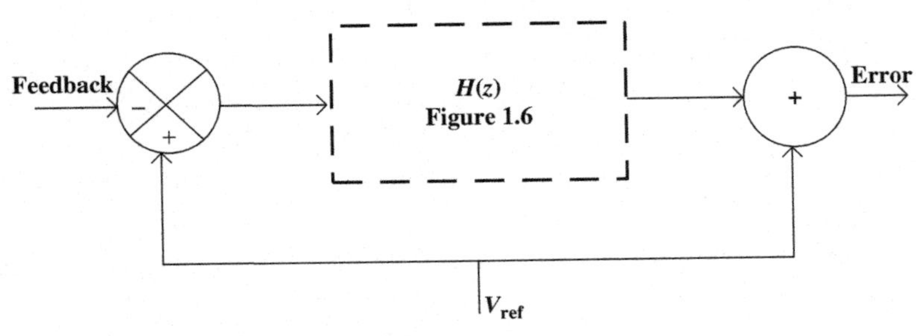

Figure 1.8 ***Digital Filter With Proper DC Setting.***

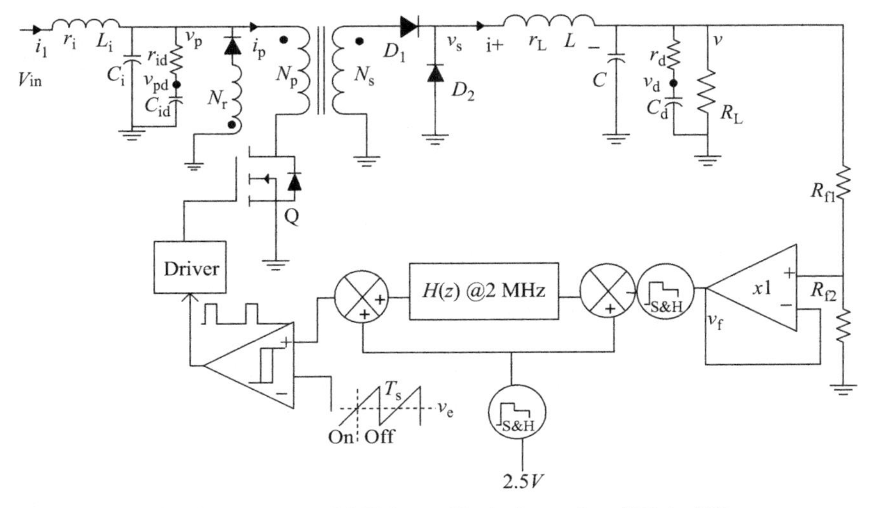

Figure 1.9 ***Forward Converter With Voltage-Mode Control and Digital Filter.***

is inserted symbolically to represent some sort of analog-to-digital (A/D) action. In order to utilize the full scale of A/D resolution more effectively, it may be better to sample the difference ($V_r - V_f$), a smaller quantity than V_f alone. It is also recognized that the PWM block and its action can be quite easily carried out in digital form. That means that the digital filter and the PWM can both be integrated into a single block. With that, the majority of the feedback circuit is all in digital. This is a better arrangement, actually more desirable, since the power switch driver and the power train cannot be digitized and might be located remotely.

1.7 OTHER APPROACHES AND CONSIDERATIONS

In the process of converting an analog error amplifier to digital in Section 1.4, the bilinear transform was invoked. The transform is considered valid, based on the fact of acceptable mathematical approximation alone. It does not take into consideration its physical significance. Briefly, in particular the choice of constant C, there were many choices, none perfect. One approach attempts to match both at a single, chosen frequency. In another, the responses of both versions across a low frequency band are made almost equal. Both efforts set an eye on the performance in the frequency domain. There are alternatives, of course, and that is changing the focus to the time domain. One is called impulse invariance method. It implies that the impulse response of a digital filter is forced to be identical to the impulse of its analog counterpart. It is basically a procedural matter that does not require cumbersome theoretical support. We outline only the process here and will give a demonstration later in an example.

Three steps are called for in the impulse invariance method. Step one takes the inverse Laplace transform of the analog compensator function (inverting sign excluded), identified by way of Section 1.3. This yields the time-domain impulse response of the corresponding analog amplifier. Step two takes z-transform of the impulse response function in the time domain. Then the last step follows by arranging the z-domain function in the form of (1.24).

In the analog world, circuit operations and performances are sensitive to the value of components. Digital filters obtained through the above fare even worse by the fact of (1.17), (1.19), (1.20), and (1.23). The performance of a digital filter is extremely sensitive to the coefficients of its numerator and denominator polynomial. Readers are strongly advised to retain coefficients' numerical precision to at the least 10 decimal places.

The last, but not the least, concern is the local stability of the compensator. Poles of $H_{II}(z)$ in (1.19), $H_{III}(z)$ in (1.23), or in general, $H(z)$ of (1.24), must lie within the unit circle in the z-plane. We will see to it in the example to follow.

1.8 EXAMPLE

In Figure 1.1, all components were shown without a corresponding value. In order to proceed with an actual numerical evaluation, component values are assigned at this time, starting from the input port to the output and circling around the feedback loop. In addition, several operation parameters shall also be given. Here is the listing: input voltage ranging from 43 V to 53 V with nominal V_{in} = 48 V, switching frequency f_s = 125 kHz, transformer primary-to-secondary turn ratio N_p/N_s = 7/2, output voltage = 5.2 V, reference voltage = 2.5 V, feedback factor or ratio, k_f = 2.5/5.2, sawtooth swing from 1 V to 5.7 V, dead time = 2%, maximum duty cycle = 98%, input filter L_i = 40 μH, r_i = 0.001 Ω, C_i = 60 μF, C_{id} = 400 μF, r_{id} = 0.4 Ω, switch on resistance = 0.01 Ω, transformer magnetizing inductance = 1 mH, output filter inductor L = 170 μH, inductor resistance = 0.001 Ω, output filter capacitor C = 37 μF, output filter shunt damping capacitor C_d = 86 μF, damping resistor r_d = 1.52 Ω, output power = 100 W. Components for the error amplifier will be selected later.

Starting from the output filter, here is how component values given earlier are derived:

Assume a requirement of 1% peak-to-peak, output voltage ripple, δv, and a desired output filter inductor ripple current, δi, at 10% of the

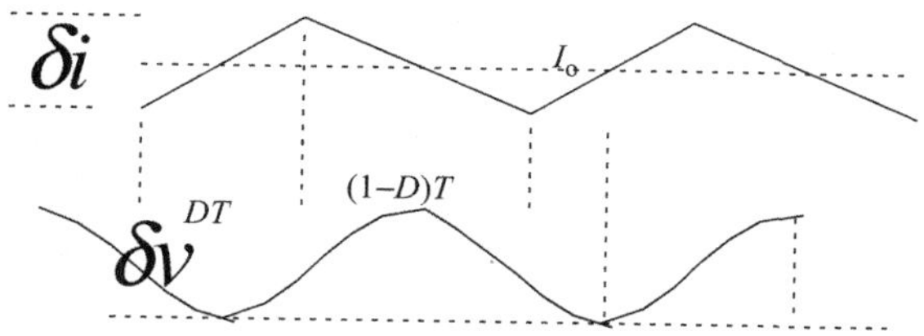

Figure 1.10 ***Output Filter Inductor Current and Output AC Part.***

maximum load, the current through the filter inductor and the output alternating part are understood to resemble Figure 1.10.

Since only the alternating current flows through the main output filter capacitor, both waveforms yield $C = \delta i/(8f_s\delta v)$. By choosing a 2 KHz filter bandwidth, F_{bw}, the filter inductor is given by $1/(4CF_{bw}^{2}\pi^2)$. As a result and for controlled, critical damping, instead of load damping, the filter damping network is derived: $r_d = [\sqrt{(L/C)}]/(2\times 0.707)$; $C_d = 100/(2\pi f_s r_d)$.

In the above, the output filter is designed from the viewpoint of output quality. On contrast, the input filter is designed from the need of ensuring control loop stability, for which the interaction between the input and output filters plays a key role. Basically, the output impedance magnitude of the input filter must be much lower than the input impedance magnitude of the reflected output filter. From the design already given and the output loading, the reflected input impedance is expressed as

$$Z_{ip}(s) = \left(\frac{N_s}{N_p}D\right)^{-2}\left\{Ls + r_L + \left[R_L^{-1} + Cs + \frac{1}{r_d + (C_d s)^{-1}}\right]^{-1}\right\} \tag{1.30}$$

The input filter's output impedance is expressed as

$$Z_g(s) = \left[\frac{1}{r_i + L_i s} + C_i s + \frac{1}{r_{id} + (C_{id} s)^{-1}}\right]^{-1} \tag{1.31}$$

All four components for (1.31) are selected such that its magnitude, in db, is way less than the reflected input impedance of the output filter across a wide band as shown in Figure 1.11. In physical terms, this means that the input source has a low source impedance, such that it almost represents an ideal voltage source, one that does not degrade the voltage it delivers when loaded.

Next, with both the input and the output circuits ready, (1.9a), (1.9b), (A.12), (A.7), (A.16), and (1.10) can now be invoked to compute the modulator gain (1.11), in which the leading element equals the feedback factor $k_f = 0.48$; $V_H = 5.7$, and $V_L = 1$ yield PWM gain $F_m = 0.98/4.7$; (A.16)

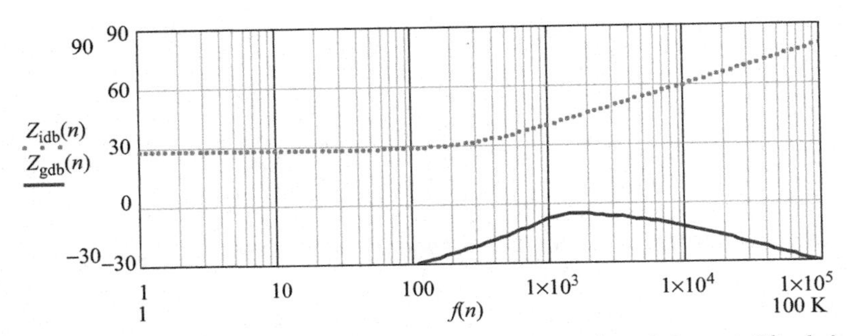

Figure 1.11 ***Input Filter's Output Impedance (Solid Trace) and Output Filter's Input Impedance Reflected to the Primary Side.***

the power stage's duty cycle-to-output gain. Together, for this example, the modulator gain is identified (Figure 1.12).

If a desired phase margin, $\alpha_m = 60°$, is chosen at a close-loop cross over frequency of 2 kHz, (1.12) indicates that a phase boost $\alpha_b = 64°$ is required. In this case, a type-II is sufficient to do the job. But we choose a type-III just to demonstrate the procedure. Anyway, with type-III amplifier, (1.13) gives pole–zero separation factor $k = 3.275$, and, (1.15) provides amplifier component values in the order given: $R_{1a} = 2\text{K}$, $C_{2a} = 6\times10^{-9}$, $C_{1a} = 14\times10^{-9}$, $R_{2a} = 10\text{K}$, $R_{3a} = 879$, $C_{3a} = 50\times10^{-9}$. However, those capacitor values as

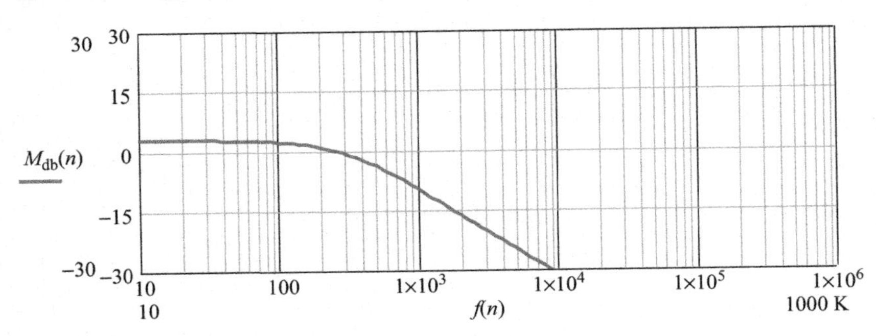

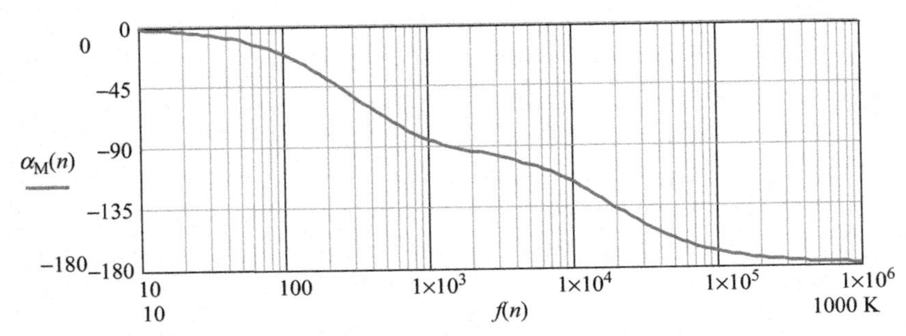

Figure 1.12 ***Modulator Gain and Phase at 2 kHz, $M_{db} = -16$, $\alpha_M = -94°$.***

computed are not standard values. Actual parts selected from the commercial market and with values close to the theoretical estimate may be the compromised choice. If necessary, multiple parts may be parallel to reach the estimate. This being said, and considering the educational purpose of this writing, we retain those computational values as the basis of next step. With this understanding, we have reached the type-III analog amplifier, $EA_{III}(s)$, expected to properly compensate the modulator and make the control loop stable with the desired phase margin at the selected crossover frequency. The resulting open loop gain is therefore $T(s) = M(s)EA_{III}(s)$. Figure 1.13 gives the loop gain magnitude in db and phase in degree.

Clearly, the control loop crosses at 2 kHz with the desired phase margin at 60°, and it is almost ready for conversion to digital, except for selecting the bilinear transformation constant C. This shall be done by first examining the resulting $EA_{III}(s)$ derived earlier. In the frequency domain, it gives the following gain in db, Figure 1.14.

The plot, which is gain over frequency, indicates a bandwidth of 50 kHz. But some activity, about 25 db down or 1/17, which is not insignificant, at 1 MHz still presents. A decision was then made to employ sampling at 2 MHz. That means the z-transform constant C shall be 4×10^6. Moving through (1.22) and (1.23) and assuring components' designator is correctly named, polynomial coefficients for the corresponding $H_{III}(z)$ are deduced.

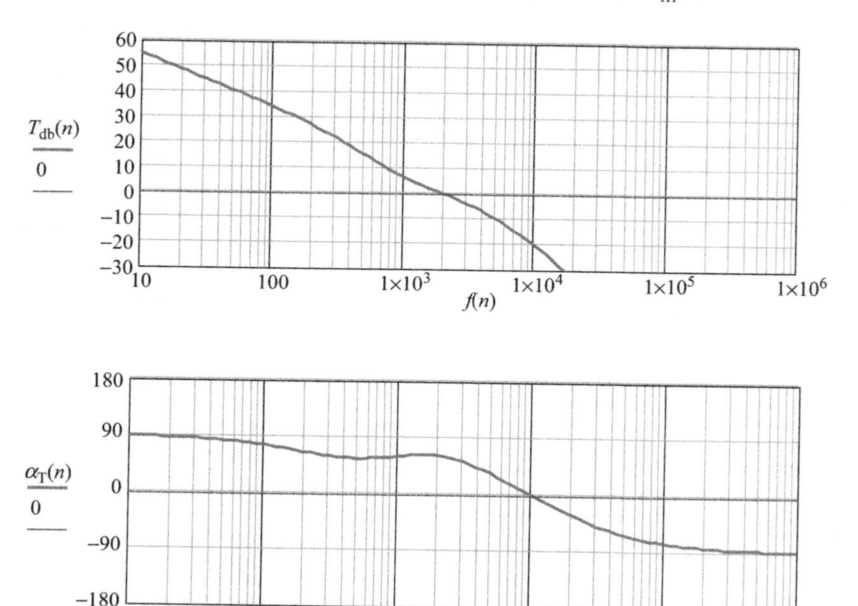

Figure 1.13 ***Loop Gain Magnitude and Phase.***

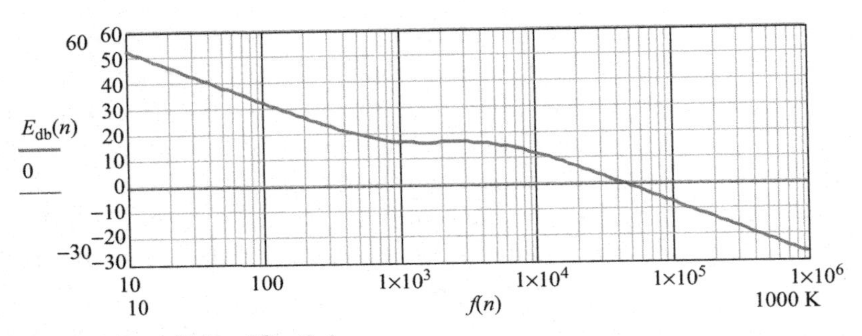

Figure 1.14 ***Type-III Amplifier Gain.***

$$a_0 = 66.66718 \times 10^{-3} \qquad a_1 = -66.205034 \times 10^{-3}$$

$$a_2 = -66.666379 \times 10^{-3} \qquad a_3 = 66.205835 \times 10^{-3}$$

$$b_1 = -2.977388 \times 10 \qquad b_2 = 2.954904 \times 10$$

$$b_3 = -977.515893 \times 10^{-3}$$

An important step shall also be taken here before going further. It is the frequency response of the corresponding digital filter obtained so far. With $z = e^{sT} = e^{j\omega T}$, the digital filter's response is superimposed and compared with the analog amplifier's response, Figure 1.15. The plot shows that a very minor magnitude difference exists at the high frequency section, while both phase responses match. Next, we shall inspect the stability of the digital filter. This is done by rewriting the $H_{III}(z)$'s polynomial in a positive power form.

$$H_{III}(z) = \frac{a_0 z^3 + a_1 z^2 + a_2 z + a_3}{z^3 + b_1 z^2 + b_2 z + b_3} \tag{1.32}$$

The three roots of denominator are found to be within the unit circle in the z-plane, Figure 1.16. Therefore, local stability is established.

Another way for ensuring stability is by looking at the impulse response, i.e., the inverse z-transform of the digital filter. With the help of modern software tools, the impulse discrete time response is obtained, Figure 1.17, and it does show the decaying behavior of a stable circuit.

1.9 SIMULATION AND PERFORMANCE VERIFICATION

In this section, we shall evaluate the time-domain performance of the converter designed so far. Cycle-by-cycle switching waveforms, for both the analog version Figure 1.1 and the digital version Figure 1.9, under transient turn-on and steady state are generated.

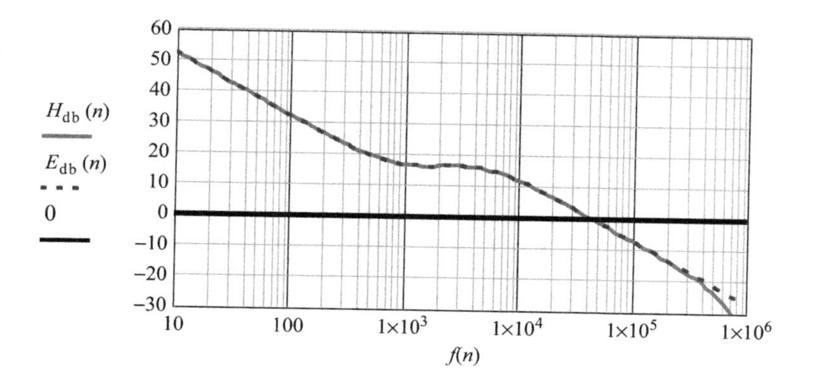

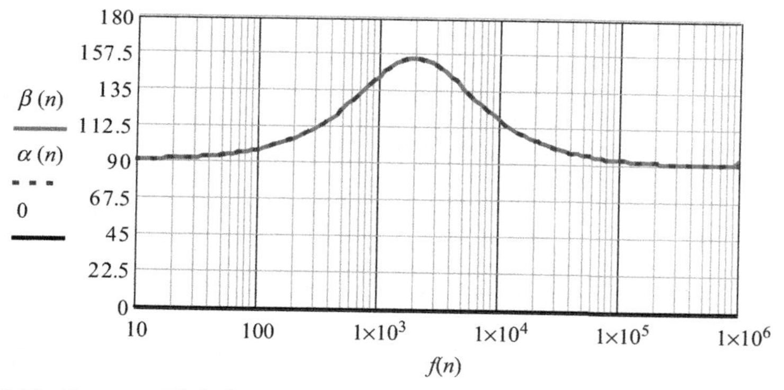

Figure 1.15 ***Compare Digital Response (Solid) with Analog (Dash).***

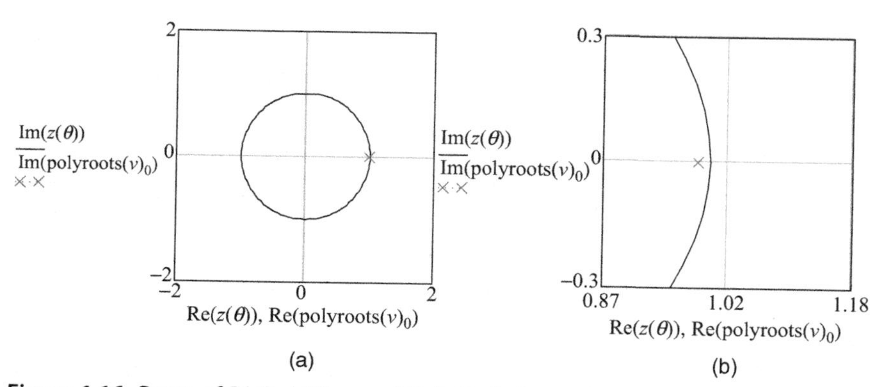

Figure 1.16 ***Roots of Digital Filter and Unit Circle in z-Plane.*** (a) Normal view and (b) zoom in view.

For the analog version, an additional feature, a slow start by ramping up the reference voltage, is included. The task begins by placing (1.1) through (1.8) in iterative (MathCAD, Mathsoft Inc.) discrete difference form (Appendix C) with zero starting state and a preset sawtooth, sw, clock.

$$\begin{bmatrix} i_{1_{j+1}} \\ v_{p_{j+1}} \\ i_{j+1} \\ v_{o_{j+1}} \\ v_{cd_{j+1}} \\ v_{a_{j+1}} \\ v_{b_{j+1}} \\ v_{e_{j+1}} \\ i_{m_{j+1}} \\ v_{id_{j+1}} \\ i_{r_{j+1}} \end{bmatrix} = \begin{bmatrix} i_{1_j} + \left[V_{in} - \left(r_1 i_{1_j} + v_{p_j}\right)\right]\dfrac{\delta t}{L_1} \\ v_{p_j} + \left[i_{1_j} - \left(\dfrac{v_{p_j} - v_{id_j}}{R_{d2}} + \dfrac{N_2}{N_1}\,\text{if}\left(v_{e_j} > s_{w_j}, i_j, 0\right) + i_{m_j}\right)\right]\dfrac{\delta t}{C_1} \\ \text{if}\left[i_j < 0, 0, i_j + \dfrac{\delta t}{L}\left[\text{if}\left[v_{e_j} > s_{w_j}, \dfrac{N_2}{N_1}\left[v_{p_j} - R_{on}\left(\dfrac{N_2}{N_1} i_j + i_{m_j}\right)\right] - V_d, -V_d\right] - \left(r i_j + v_{o_j}\right)\right]\right] \\ v_{o_j} + \dfrac{\delta t}{C}\left[i_j - \left(\dfrac{v_{o_j} - v_{cd_j}}{R_d} + \dfrac{v_{o_j}}{R_L}\right)\right] \\ v_{cd_j} + \dfrac{\delta t}{C_d}\dfrac{v_{o_j} - v_{cd_j}}{R_d} \\ v_{a_j} + \dfrac{\delta t}{C3}\dfrac{k_f v_{o_j} - v_{a_j}}{R29} \\ v_{b_j} + \delta t\left(\dfrac{v_{r_j} - v_{b_j}}{R28C16} - \dfrac{k_f v_{o_j} - v_{r_j}}{R30C16} - \dfrac{k_f v_{o_j} - v_{a_j}}{R29C16} - \dfrac{v_{r_j} - v_{b_j}}{R28C15}\right) \\ \text{if}\left[v_{e_j} < 0, 0, \text{if}\left[v_{e_j} > 15, 15, v_{e_j} + \dfrac{\delta t}{C16}\left(\dfrac{v_{r_j} - v_{b_j}}{R28} - \dfrac{k_f v_{o_j} - v_{r_j}}{R30} - \dfrac{k_f v_{o_j} - v_{a_j}}{R29}\right)\right]\right] \\ \text{if}\left[v_{e_j} > s_{w_j}, i_{m_j} + \dfrac{\delta t}{L_m}\left[v_{p_j} - \left(\dfrac{N_2}{N_1} i_j + i_{m_j}\right) R_{on}\right], 0\right] \\ v_{id_j} + \dfrac{\delta t}{C2}\dfrac{v_{p_j} - v_{id_j}}{R_{d2}} \\ \text{if}\left[v_{e_j} > s_{w_j}, 0, i_{m_j} + \dfrac{\delta t}{L_m}\left(-v_{p_j} - R_{w_r} i_{r_j}\right)\right] \end{bmatrix} \tag{1.33}$$

Here, node voltages, instead of capacitive states, and inductive currents are designated as state variables and are all identified in Figure 1.1 [A slight discrepancy of symbols exists between (1.33) and Figure 1.1; e.g. $v_{cd} = v_d$,

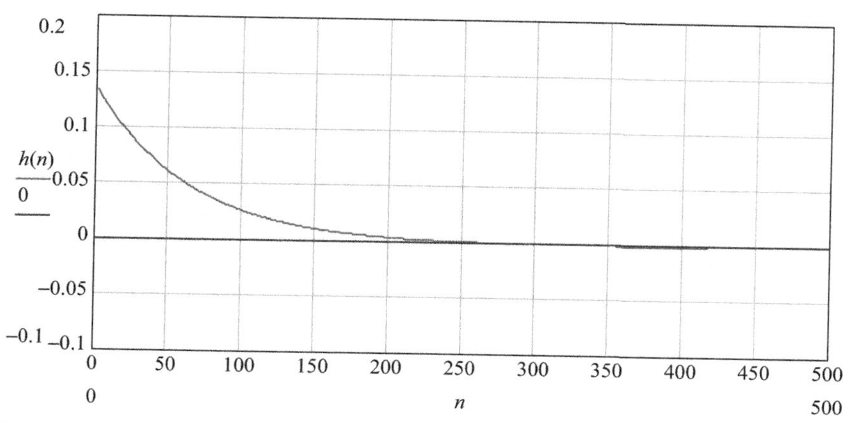

Figure 1.17 ***Impulse Response with Decaying Property.***

$v_{id} = v_{pd}$]. Also, switching decision and action are carried out by the conditional "if (logic, true, false)" statement. In addition, transformer magnetizing current, i_m, and reset, i_r, are also included. The simulation runs a total of 200 switching cycles at a time resolution of 50 points per cycle. At a switching frequency of 125 kHz, the run time comes to 1.6 ms and is considered sufficient for the converter operation to reach the steady state, while the computation step size selected is deemed proper without the risk of undersampling, but not overstretching the computation resource. In addition, for (1.33), part designators corresponding to Figure 1.1 shall be observed: $R_{1a} = R30$, $C_{2a} = C16$, $C_{1a} = C15$, $R_{2a} = R28$, $R_{3a} = R29$, $C_{3a} = C3$.

In the following series of figures, Figure 1.18a–i, both the transient phase and 10 steady-state cycles are given for each variable.

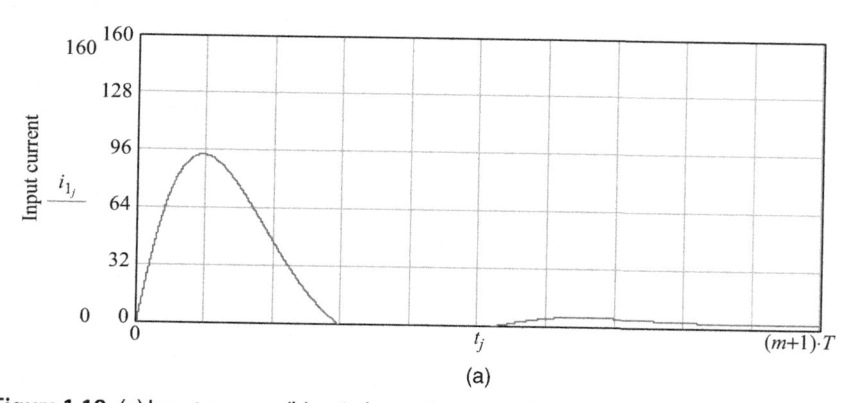

(a)

Figure 1.18 (a) Input current, (b) switch or primary winding current, (c) secondary voltage at D_2 cathode, (d) D_1 current, (e) free-wheel diode D_2 current, (f) output inductor current, (g) output voltage, (h) error voltage, and (i) transformer magnetizing and reset current.

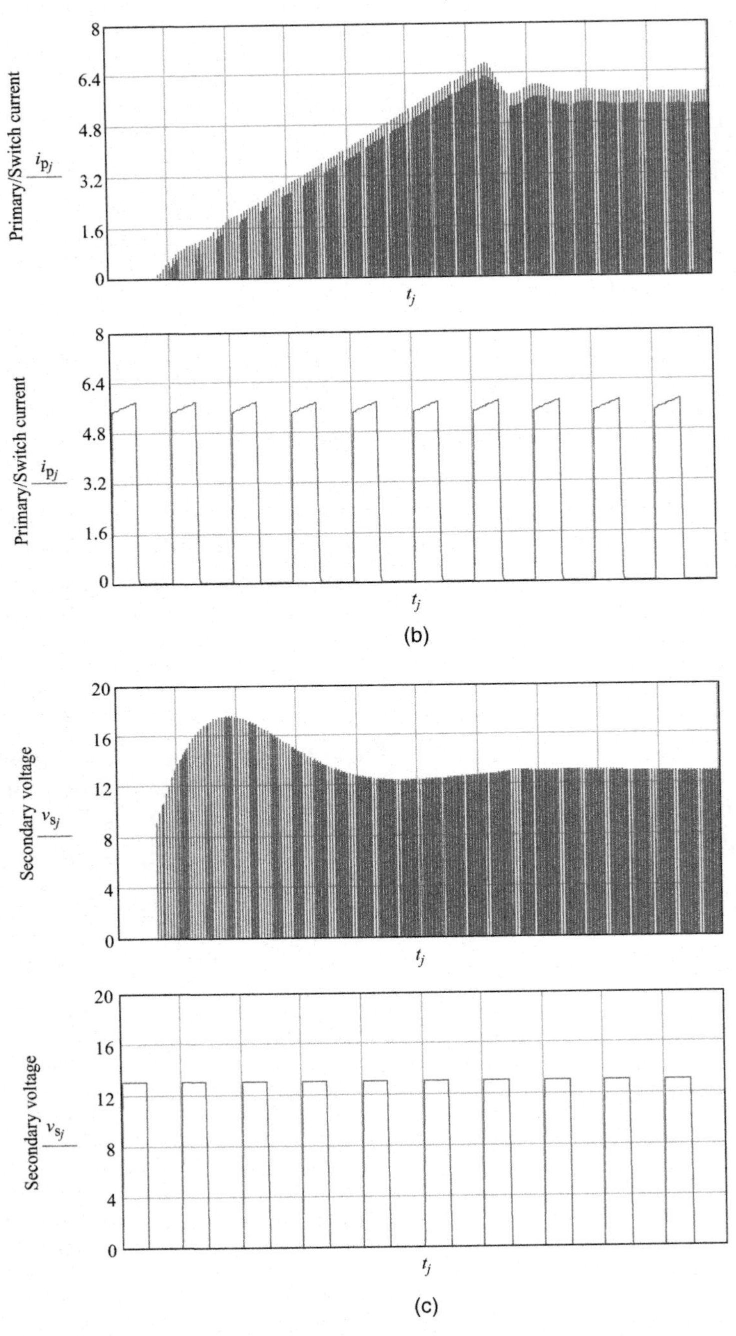

(b)

(c)

Figure 1.18 ***(cont.)***

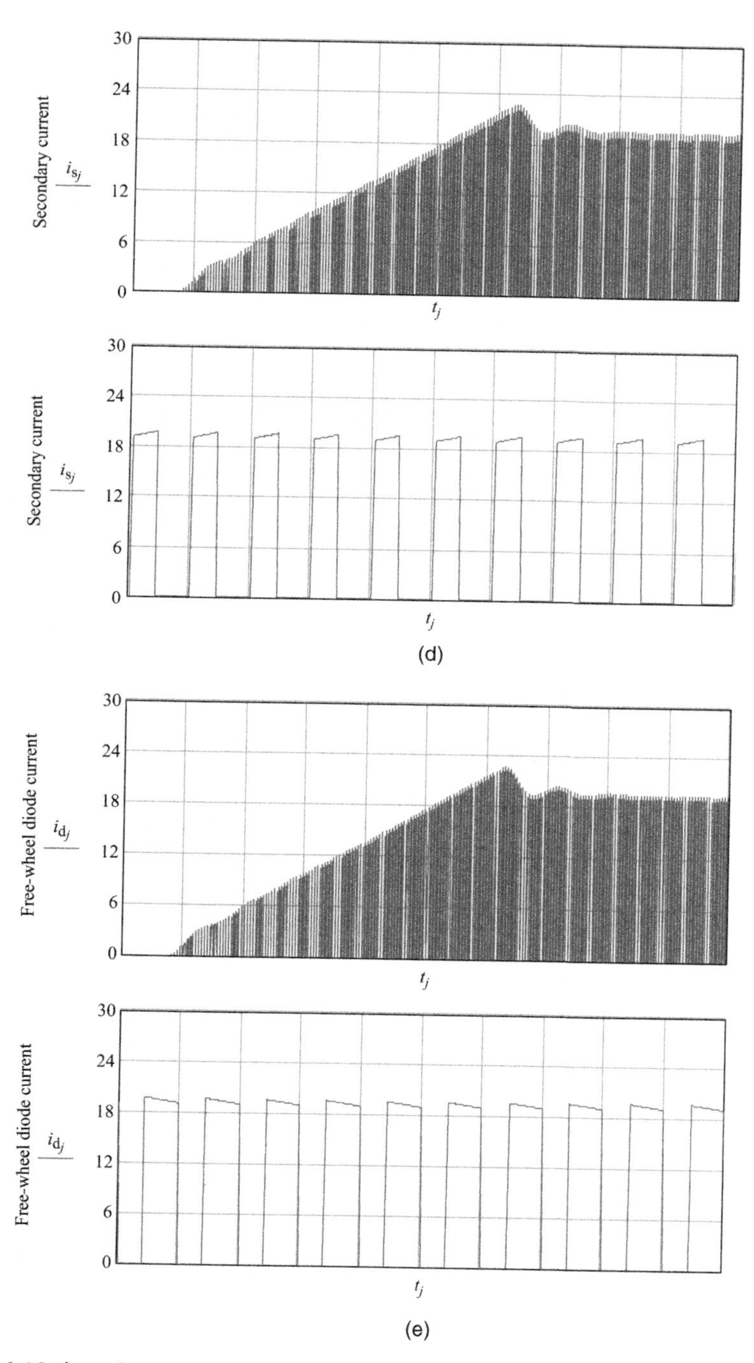

Figure 1.18 ***(cont.)***

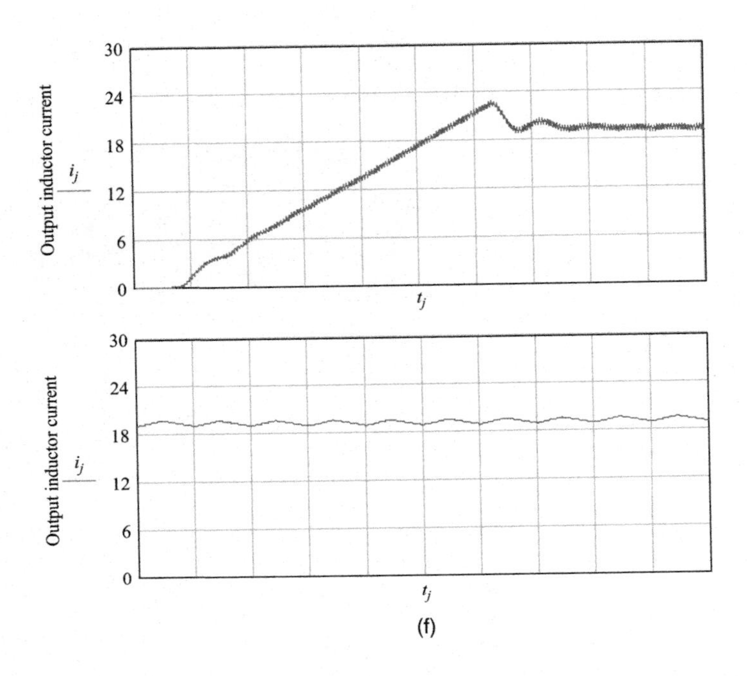

(f)

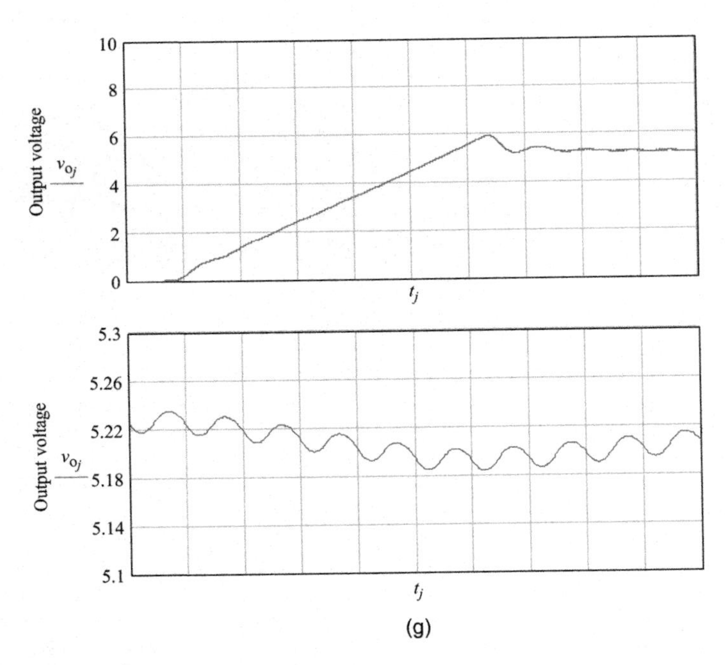

(g)

Figure 1.18 ***(cont.)***

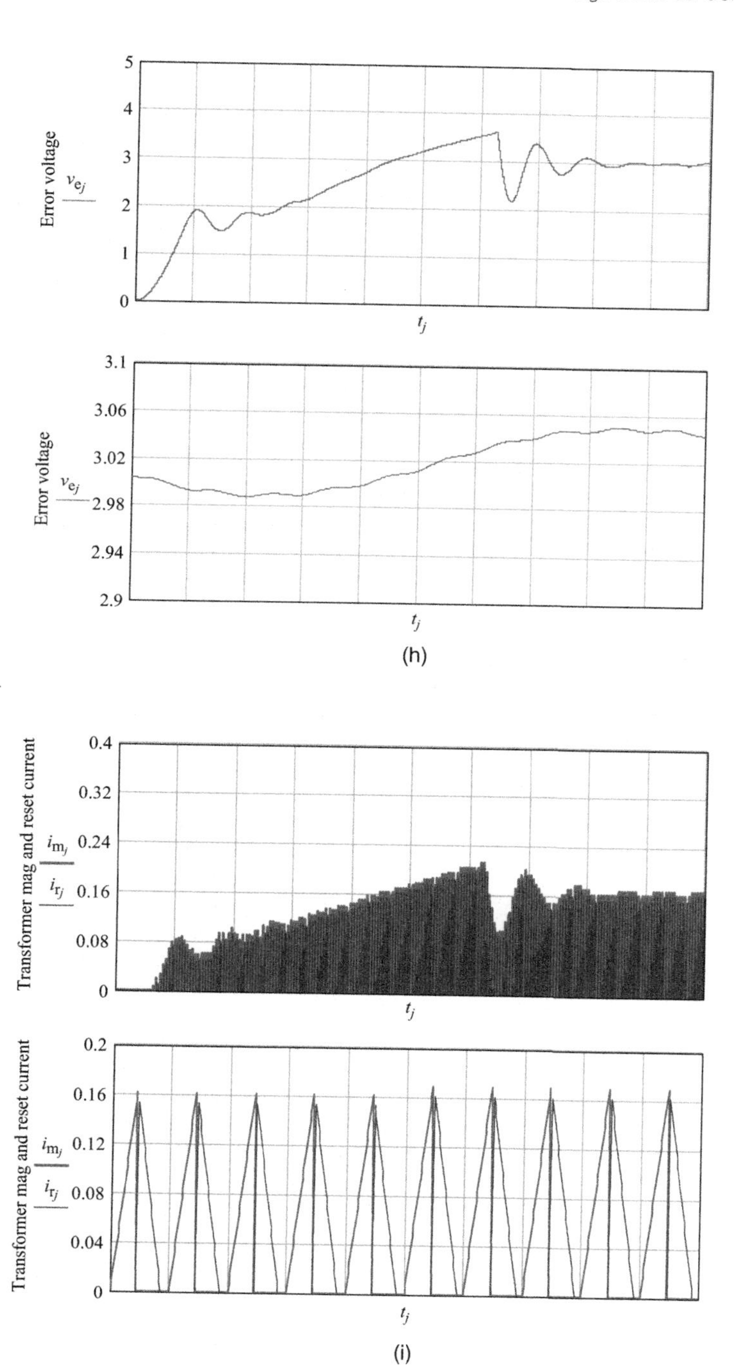

Figure 1.18 ***(cont.)***

The input current shows a 93-A surge peak and about 2.2-A steady state DC. In actual application, a 5-A slow blow inline fuse should be able to sustain the turn-on surge while also providing overload protection.

The simulation seems to show a subharmonic oscillation, or a result yet to reach the steady state. In the case of the former, the error amplifier design may need more refinement or improvement. The latter requires longer simulation, perhaps 300 switching cycles. But we will not attempt either, since design refinement is not the goal of this writing.

Next, we move to the simulation with the digital filter. For this, the governing discrete equations corresponding to Figure 1.8 change slightly. Variables v_a and v_b corresponding to Figure 1.1 no longer exist, while v_e is replaced by y, the digital filter output. Intermediate variable w is also introduced after the transformer reset current. The new equation set is given as

$$\begin{bmatrix} i_{1_{j+1}} \\ v_{p_{j+1}} \\ i_{j+1} \\ v_{o_{j+1}} \\ v_{cd_{j+1}} \\ y_{j+1} \\ i_{m_{j+1}} \\ v_{id_{j+1}} \\ i_{r_{j+1}} \\ w_{j+1} \end{bmatrix} = \begin{bmatrix} i_{1_j} + \left[V_{in} - \left(r_1 i_{1_j} + v_{p_j}\right)\right]\dfrac{\delta t}{L_1} \\ v_{p_j} + \left[i_{1_j} - \left(\dfrac{v_{p_j} - v_{id_j}}{R_{d2}} + \dfrac{N_2}{N_1}\,\text{if}\left(y_j > s_{w_j}, i_j, 0\right) + i_{m_j}\right)\right]\dfrac{\delta t}{C_1} \\ \text{if}\left[i_j < 0, 0, i_j + \dfrac{\delta t}{L}\left[\text{if}\left[y_j > s_{w_j}, \dfrac{N_2}{N_1}\left[v_{p_j} - R_{on}\left(\dfrac{N_2}{N_1} i_j + i_{m_j}\right)\right] - V_d, -V_d\right] - \left(r i_j + v_{o_j}\right)\right]\right] \\ v_{o_j} + \dfrac{\delta t}{C}\left[i_j - \left(\dfrac{v_{o_j} - v_{cd_j}}{R_d} + \dfrac{v_{o_j}}{R_L}\right)\right] \\ v_{cd_j} + \dfrac{\delta t}{C_d}\dfrac{v_{o_j} - v_{cd_j}}{R_d} \\ \text{if}\left[y_j < 0, 0, \text{if}\left[y_j > 12, 12, v_r + a_0\left[\left(v_r - k_f v_{o_j}\right) - b_1 w_j - b_2 w_{j-1} - b_3 w_{j-2} + a_1 w_j + a_2 w_{j-1} + a_3 w_{j-2}\right]\right]\right] \\ \text{if}\left[y_j > s_{w_j}, i_{m_j} + \dfrac{\delta t}{L_m}\left[v_{p_j} - \left(\dfrac{N_2}{N_1} i_j + i_{m_j}\right)R_{on}\right], 0\right] \\ v_{id_j} + \dfrac{\delta t}{C_2}\dfrac{v_{p_j} - v_{id_j}}{R_{d2}} \\ \text{if}\left[y_j > s_{w_j}, 0, i_{m_j} + i_{r_j} + \dfrac{\delta t}{L_m}\left(-v_{p_j} - R_{w_r} i_{r_j}\right)\right] \\ \left(v_r - k_f v_{o_j}\right) - b_1 w_j - b_2 w_{j-1} - b_3 w_{j-2} \end{bmatrix} \tag{1.34}$$

In contrast to the analog version, simulation of the digital version is running at exactly the sampling rate, 2 MHz, which is the base for selecting the bilinear transform scaling constant, C. Therefore, 16 points per switching cycle are computed in this case, and controlled starting of reference voltage 2.5 V is still maintained (Figure 1.19).

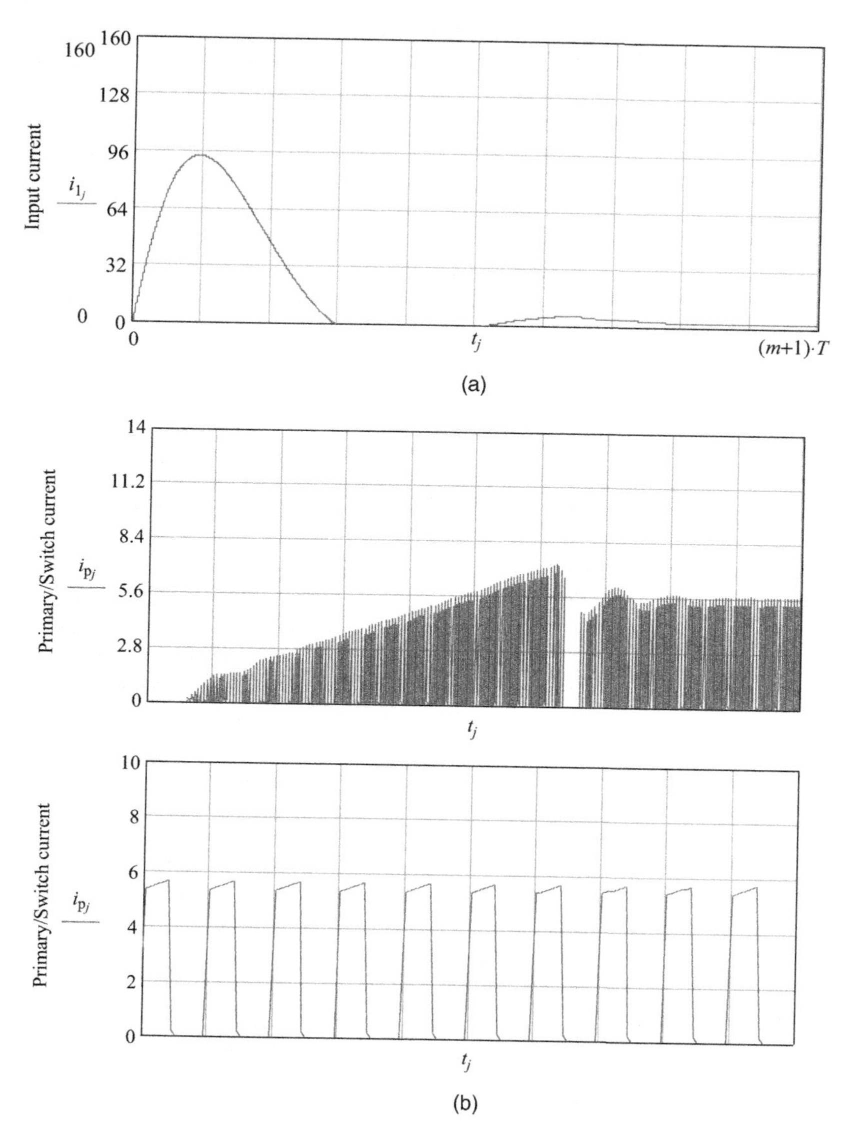

Figure 1.19 (a) Input current, (b) switch, or primary winding, current, (c) D_2 cathode voltage, (d) D_1 current, (e) D_2 current, (f) output inductor current, (g) output voltage, (h) digital filter output, and (i) transformer magnetizing and reset current.

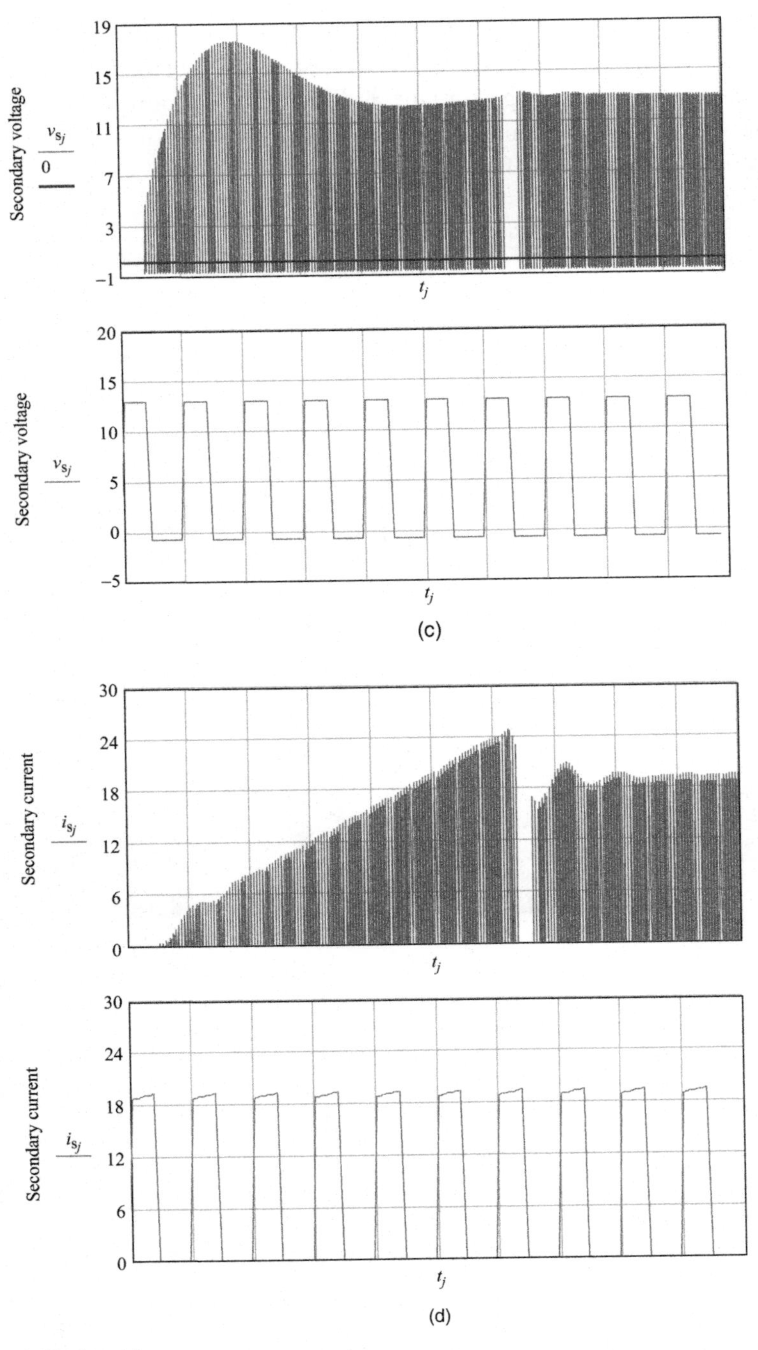

Figure 1.19 ***(cont.)***

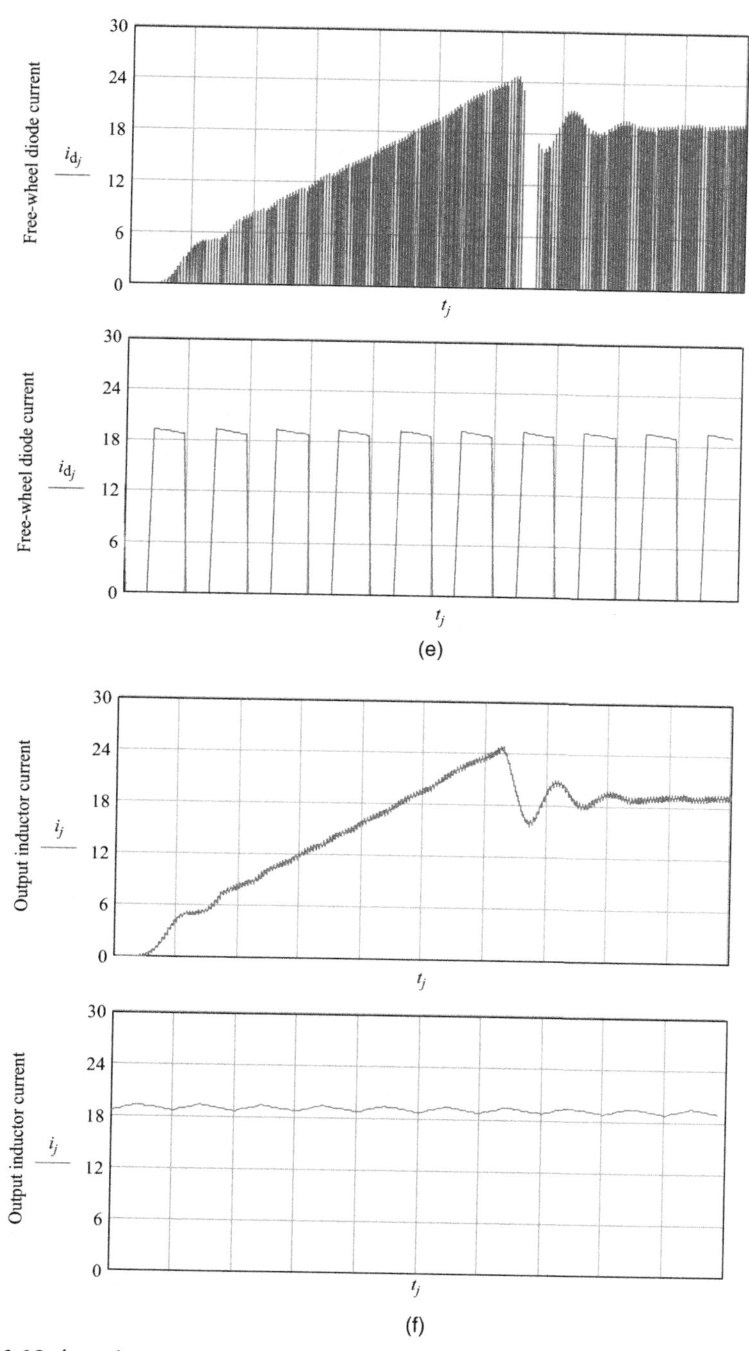

Figure 1.19 ***(cont.)***

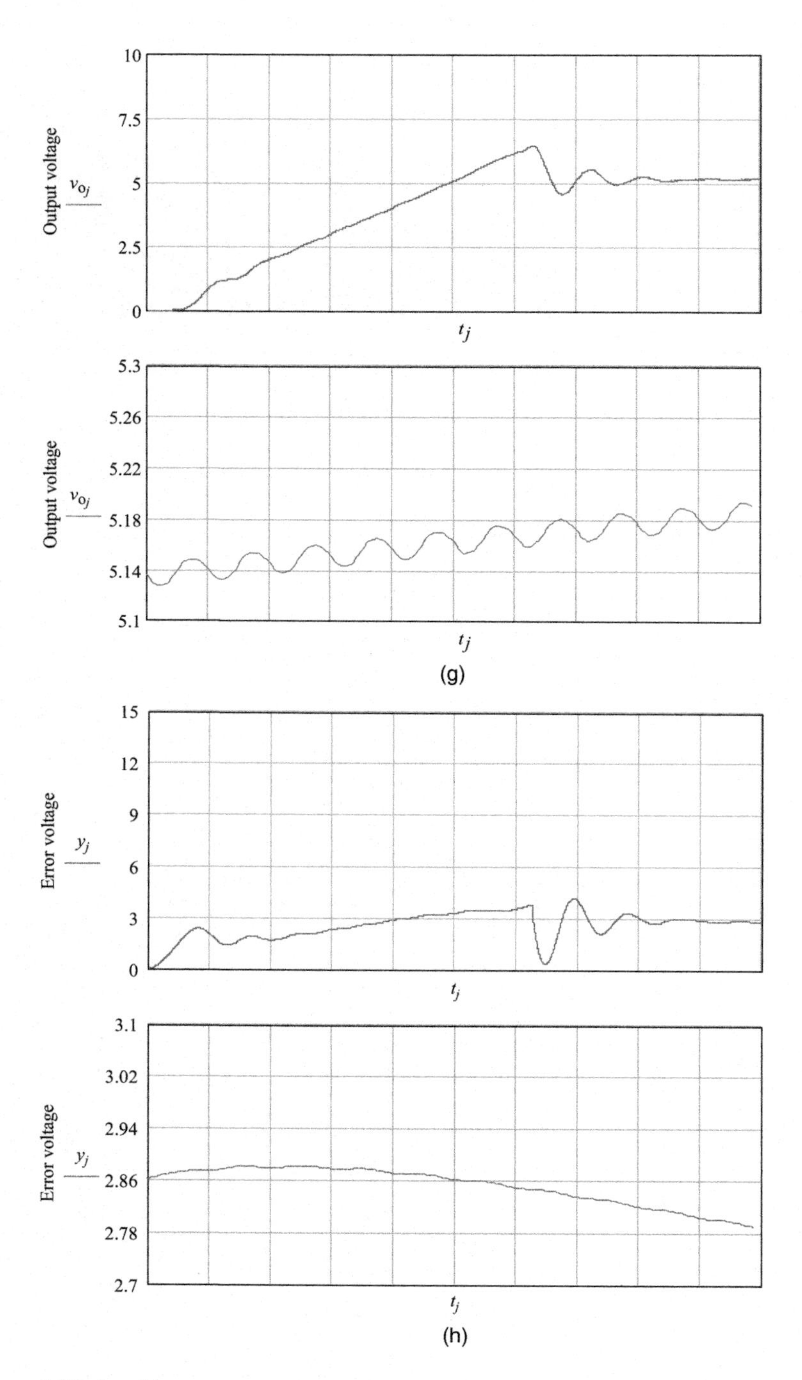

Figure 1.19 ***(cont.)***

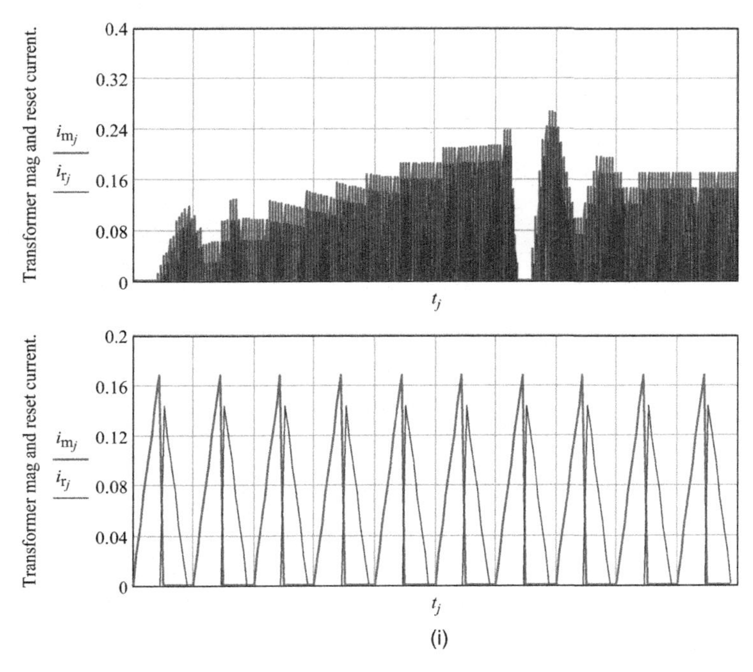

Figure 1.19 *(cont.)*

As observed, the converter performance based on the digital version matches that of the analog.

1.10 SIMULATIONS BASED ON MATLAB® SIMULINK

We now turn attention to another simulation approach: MATLAB SIMULINK.

In the past decades, particularly in designing analog integrated circuits, SPICE simulation has been the dominant tool at the device level. Starting in the early 1980s, it began to spread into the domain of discrete circuit design, for instance, power circuits with components of large geometry. Without exception, power converters also follow suit. Switch-mode power supply (SMPS), conceived in the early 1970s, took longer to join the rank. It turns out that the tool intended early on for linear, low-frequency, integrated circuits (LIC) at the device level did not work well for nonlinear, switching circuits operating at high frequency. The Ebers–Moll transistor

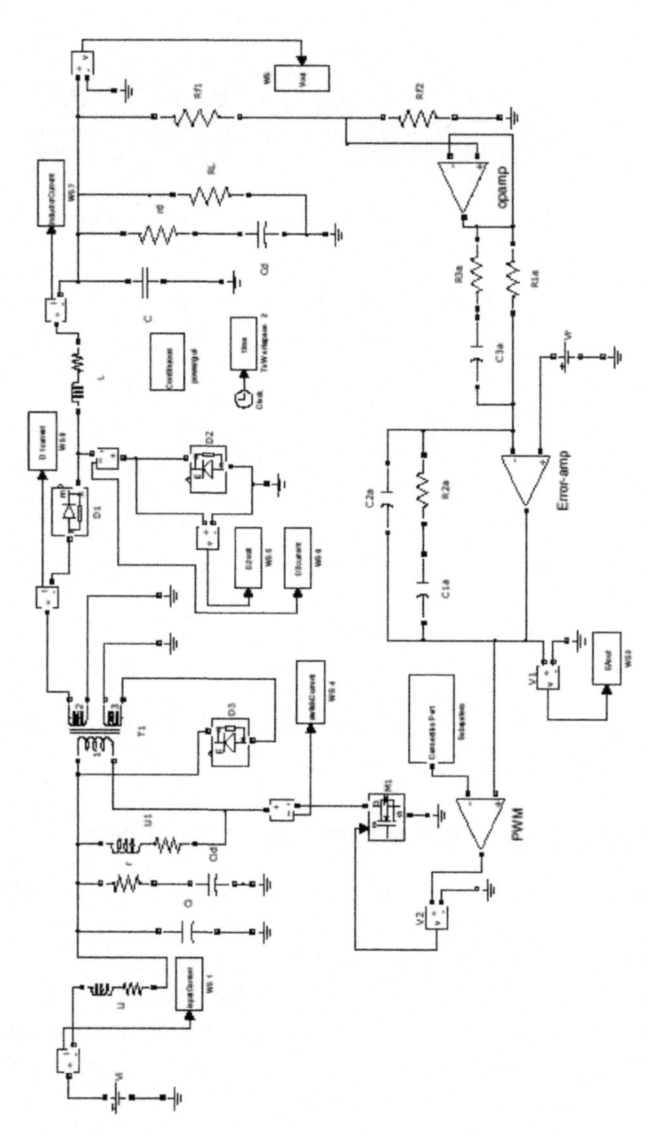

Figure 1.20 ***SIMULINK Model with Error Amp Represented by Physical Device.*** (a) Output voltage, (b) error voltage, (c) D_1 and D_2 cathode, (d) input current, (e) switch current, (f) inductor current, (g) D_1 current, and (h) D_2 current.

model with added parasitic elements works for LIC. It, however, ends up as a waste of resources in simulating a switch operating in an on/off fashion at high speed. Basically, the tool encountered a massive challenge resolving multiple convergences at switching edges in the time domain. Any single divergence holds up the entire simulation and stops everything cold. SMPS with unruly on/off transients make it worse. In essence, SPICE is not the right tool for power converter simulation. Then, here it comes SIMULINK, which uses a mix of behavior and physical models. Figure 1.20 has the error amplifier represented by a macromodel and RC passive devices. Figure 1.21

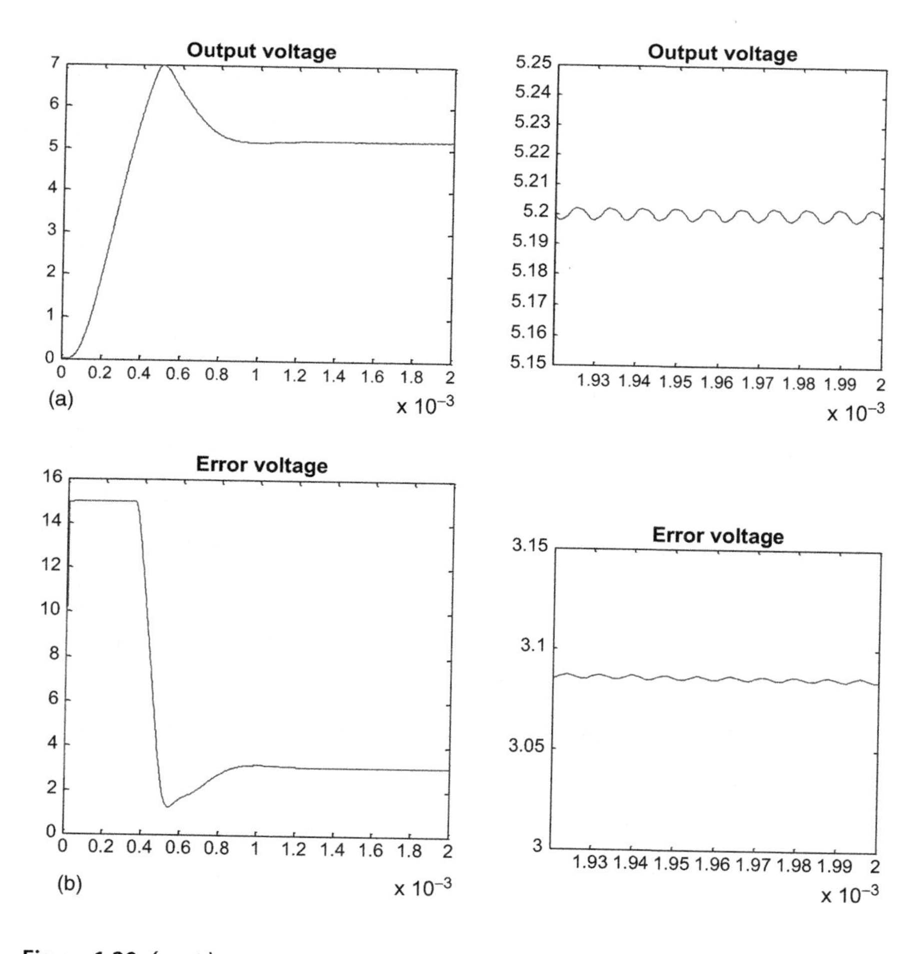

Figure 1.20 ***(cont.)***

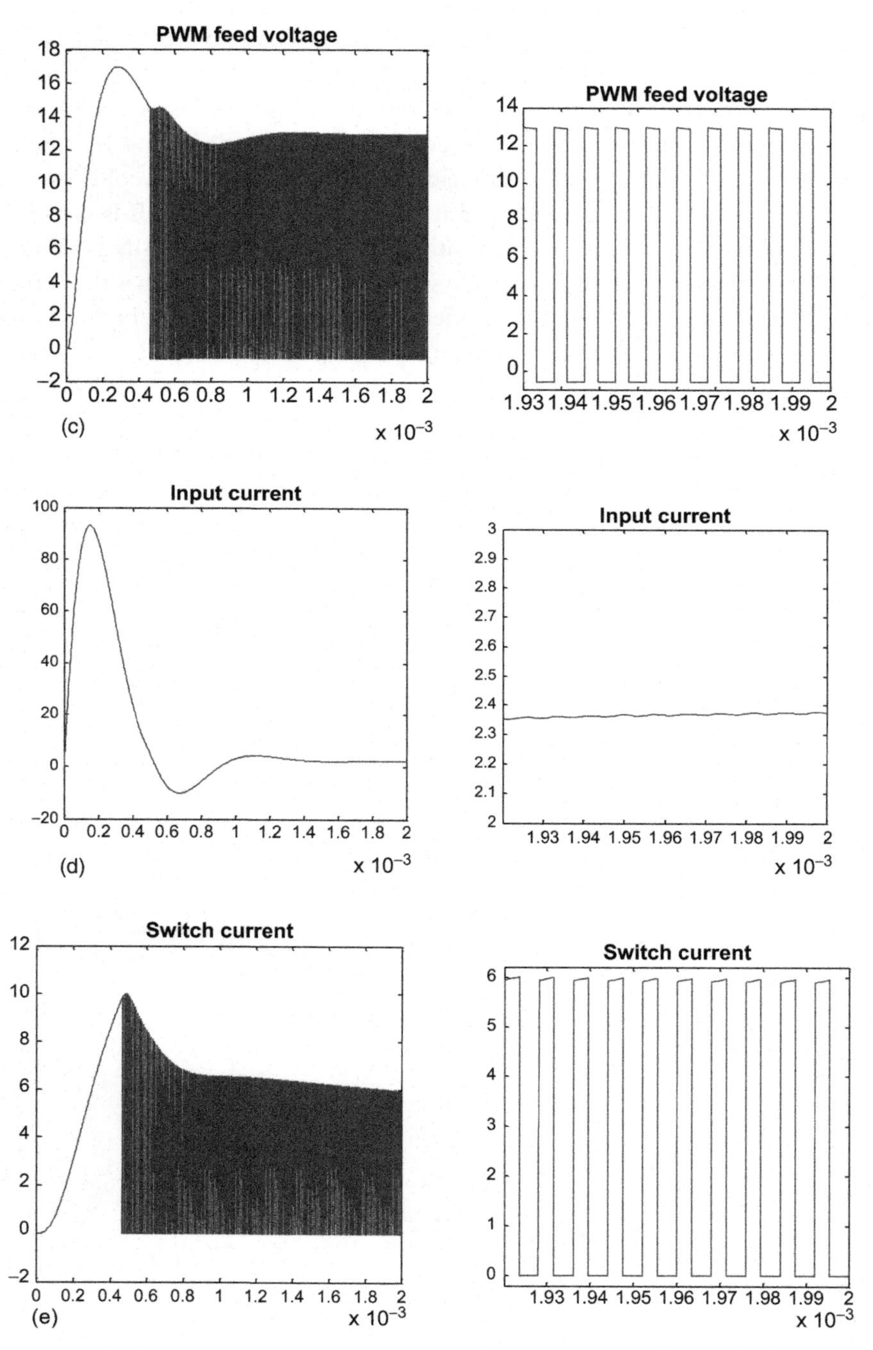

Figure 1.20 *(cont.)*

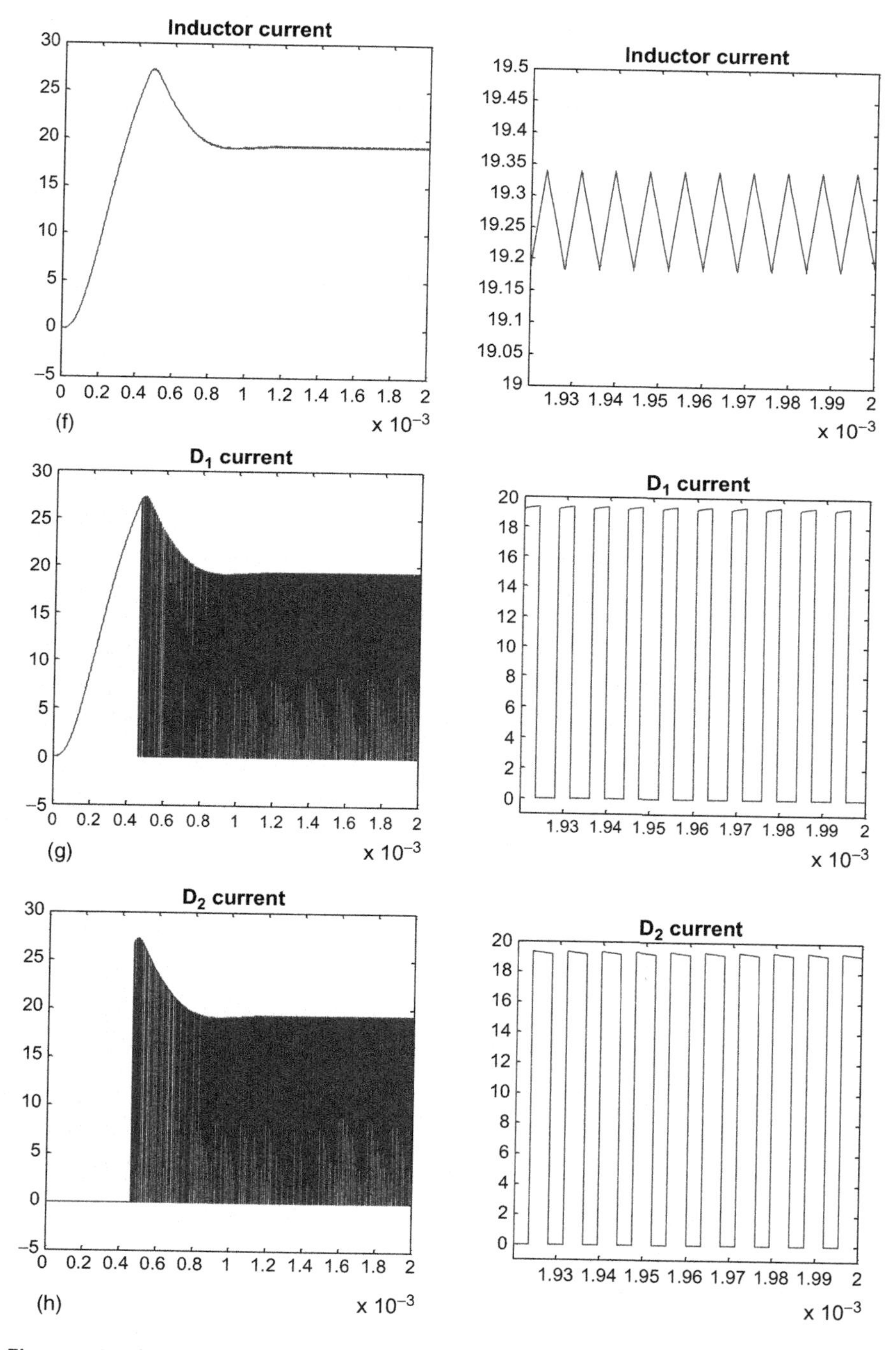

Figure 1.20 *(cont.)*

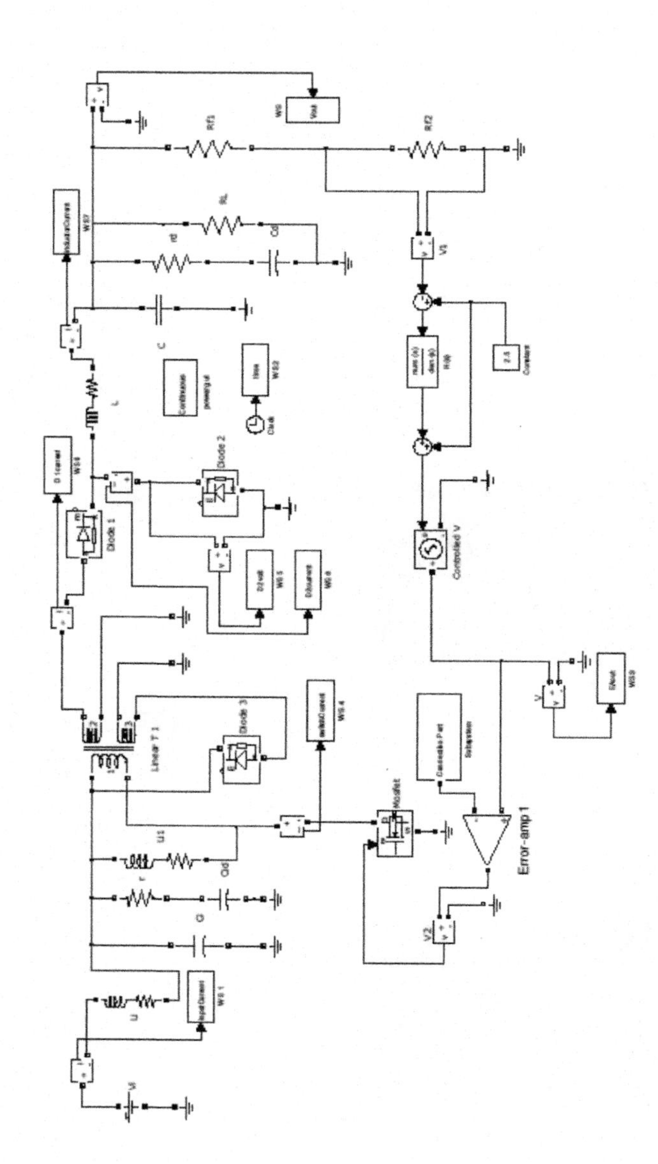

Figure 1.21 ***SIMULINK Model with Error Amp in H(s) Form.*** (a) Output voltage, (b) error voltage, (c) D_1 and D_2 cathode, (d) input current, (e) switch current, (f) inductor current, (g) D_1 current, and (h) D_2 current.

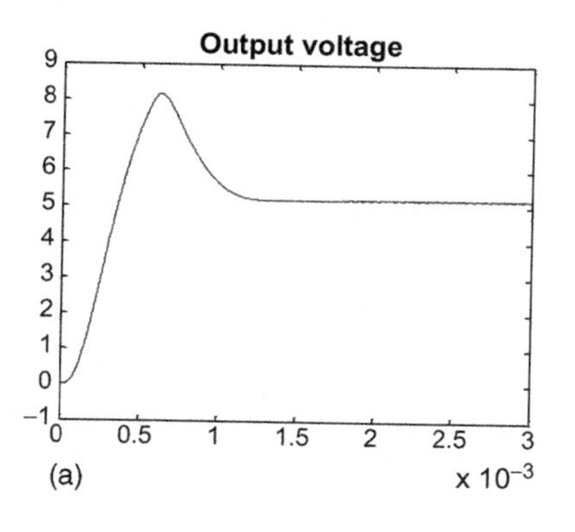

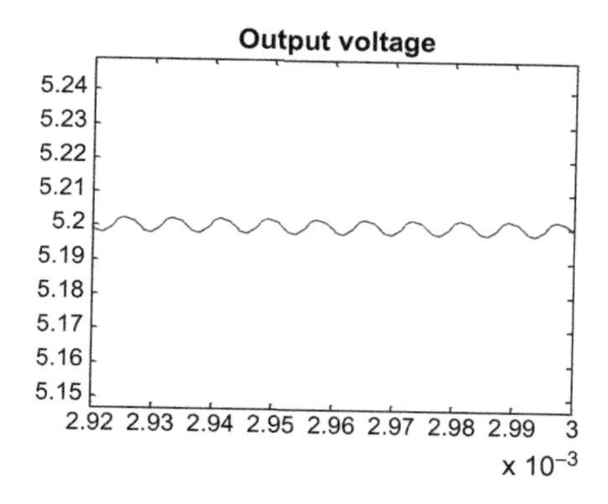

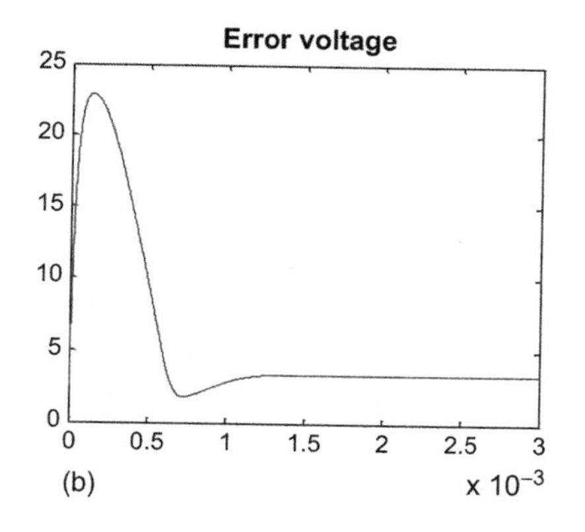

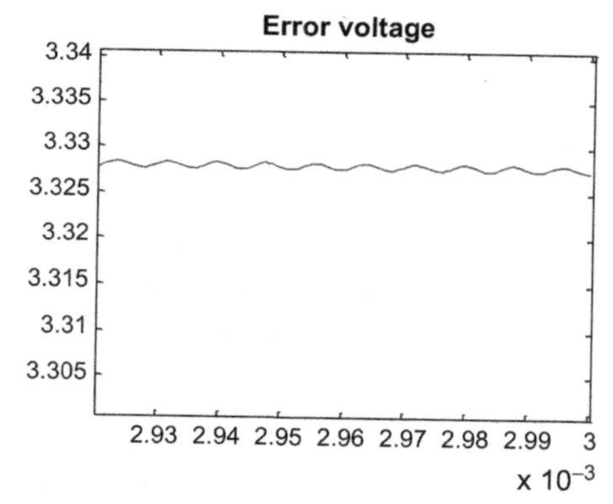

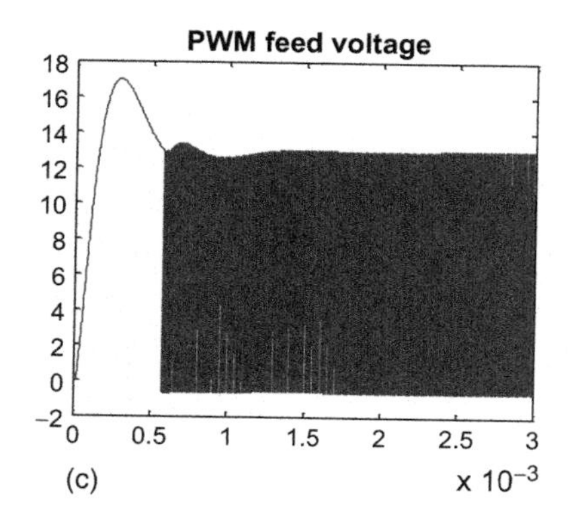

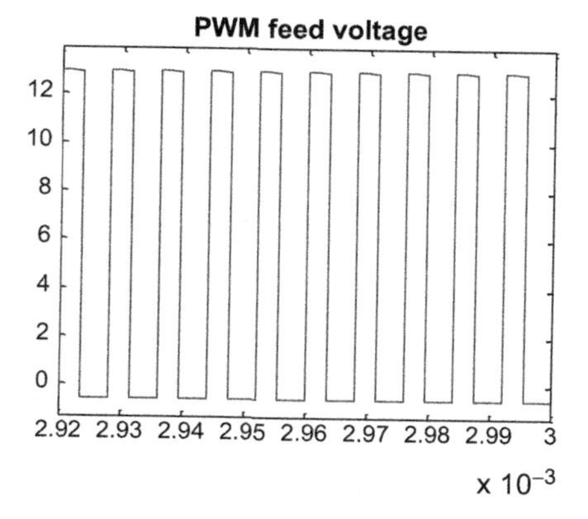

Figure 1.21 ***(cont.)***

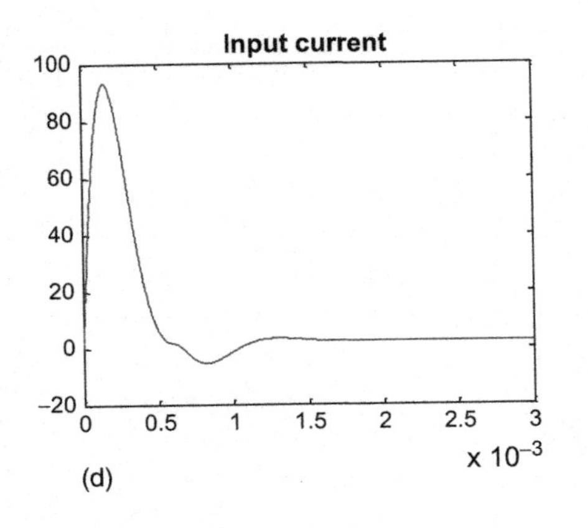

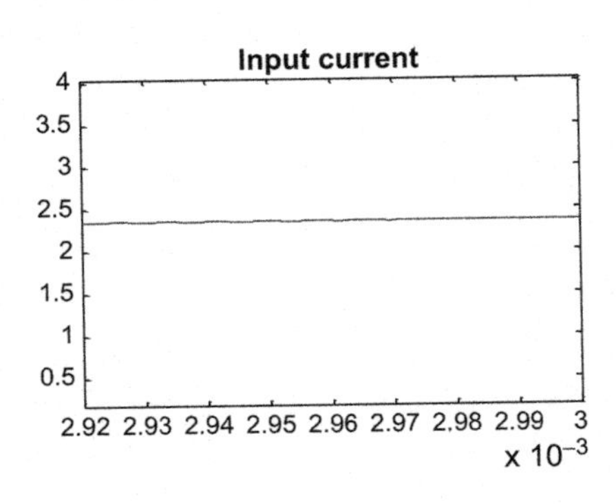

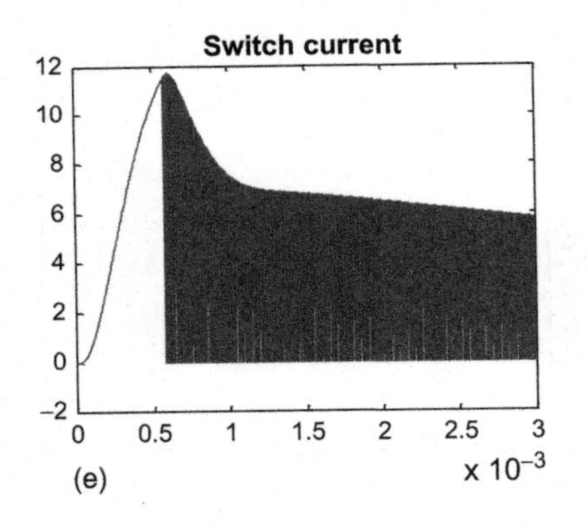

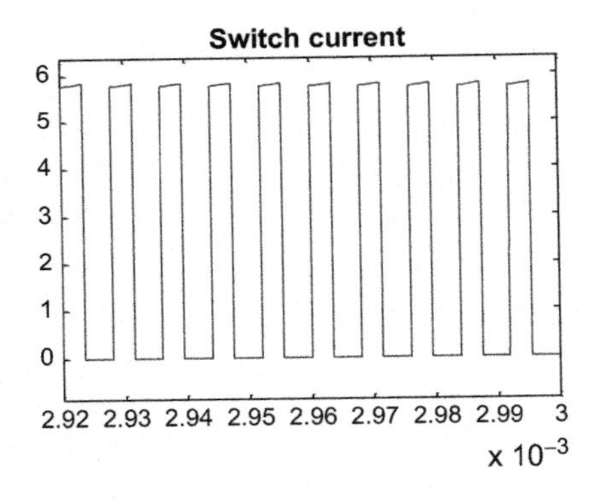

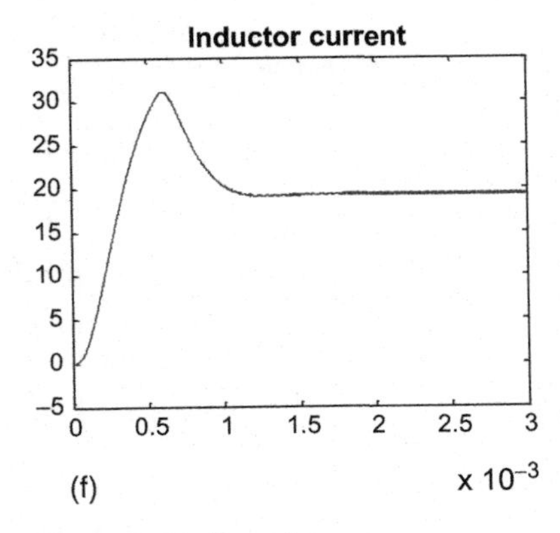

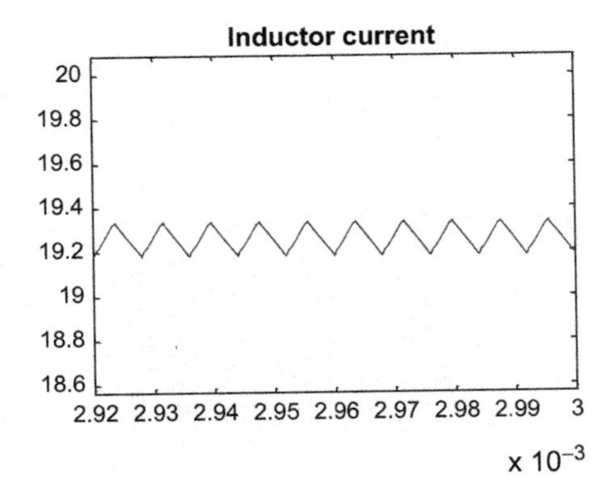

Figure 1.21 ***(cont.)***

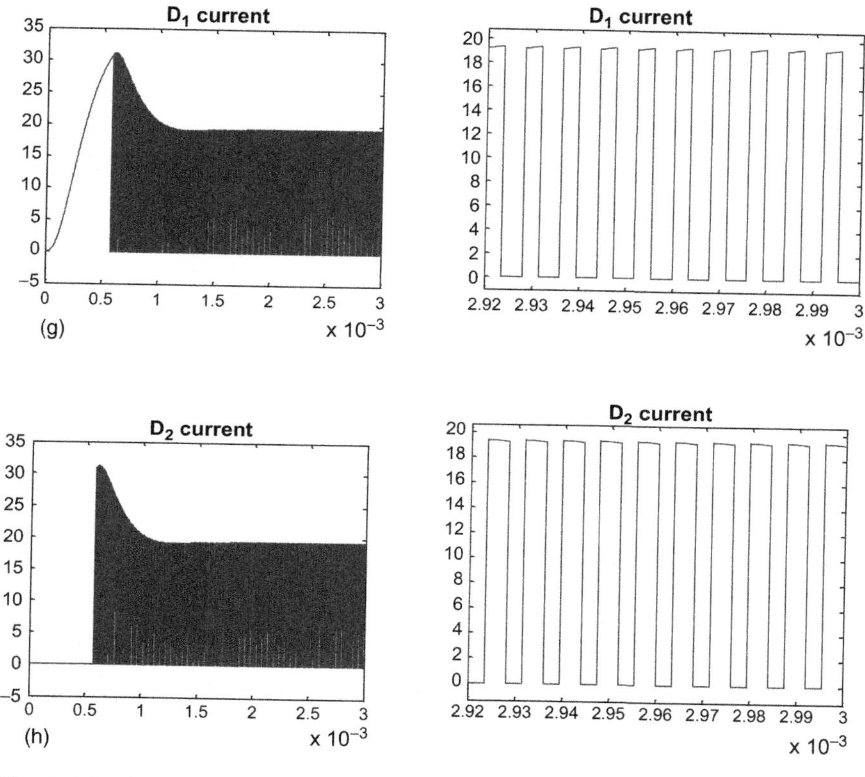

Figure 1.21 *(cont.)*

replaces the error amplifier with analog transfer function. Figure 1.22 takes the final step with the error amplifier in digital filter form.

1.11 DIGITAL PWM

By this time, some readers may already take note that the feedback loop presented in Figure 1.22, the SIMULINK diagram, is not in a totally digital form yet, in particular the block feeding the inverting input of "error amp 1." That is, the generator that creates the triangular clock (often named sawtooth in power electronics literature) is still built with analog approach, Figure 1.23. Basically, a constant current source is charging a timing capacitor linearly, while a pulse clock with low duty cycle and at a selected frequency discharges the same capacitor periodically and rapidly via a low resistance path. The combination of charging current magnitude, the timing

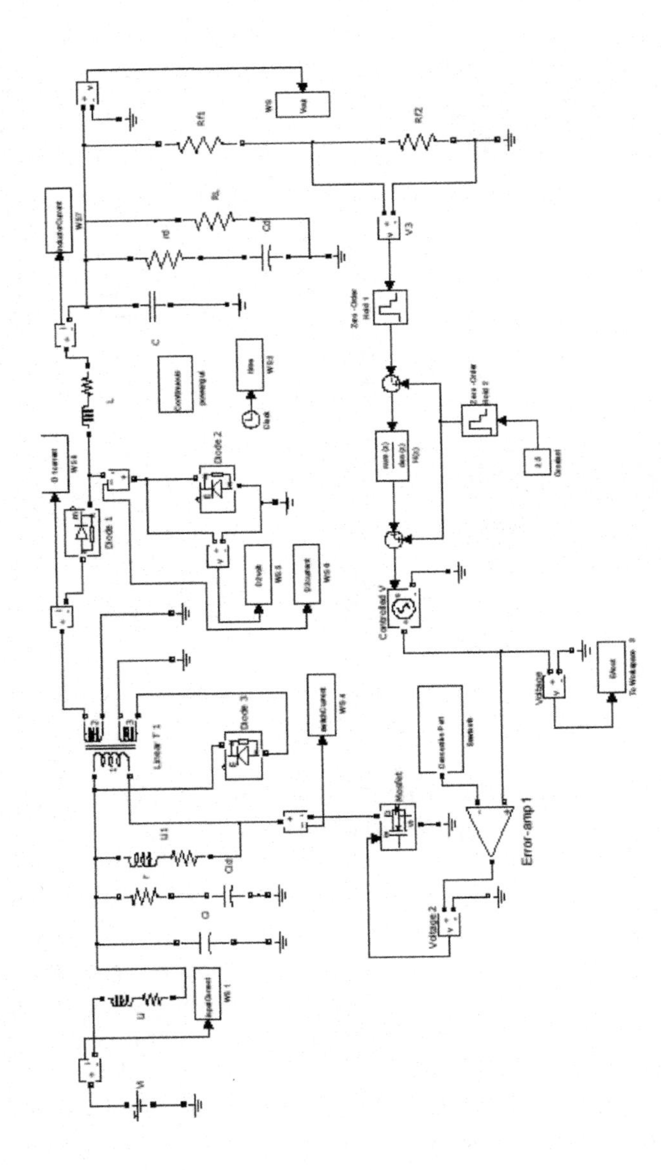

Figure 1.22 ***SIMULINK Model with Error Amp in*** **H(z)** ***Form.*** (a) Output voltage, (b) error voltage, (c) D_1 and D_2 cathode, (d) input current, (e) switch current, (f) inductor current (g) D_1 current, and (h) D_2 current.

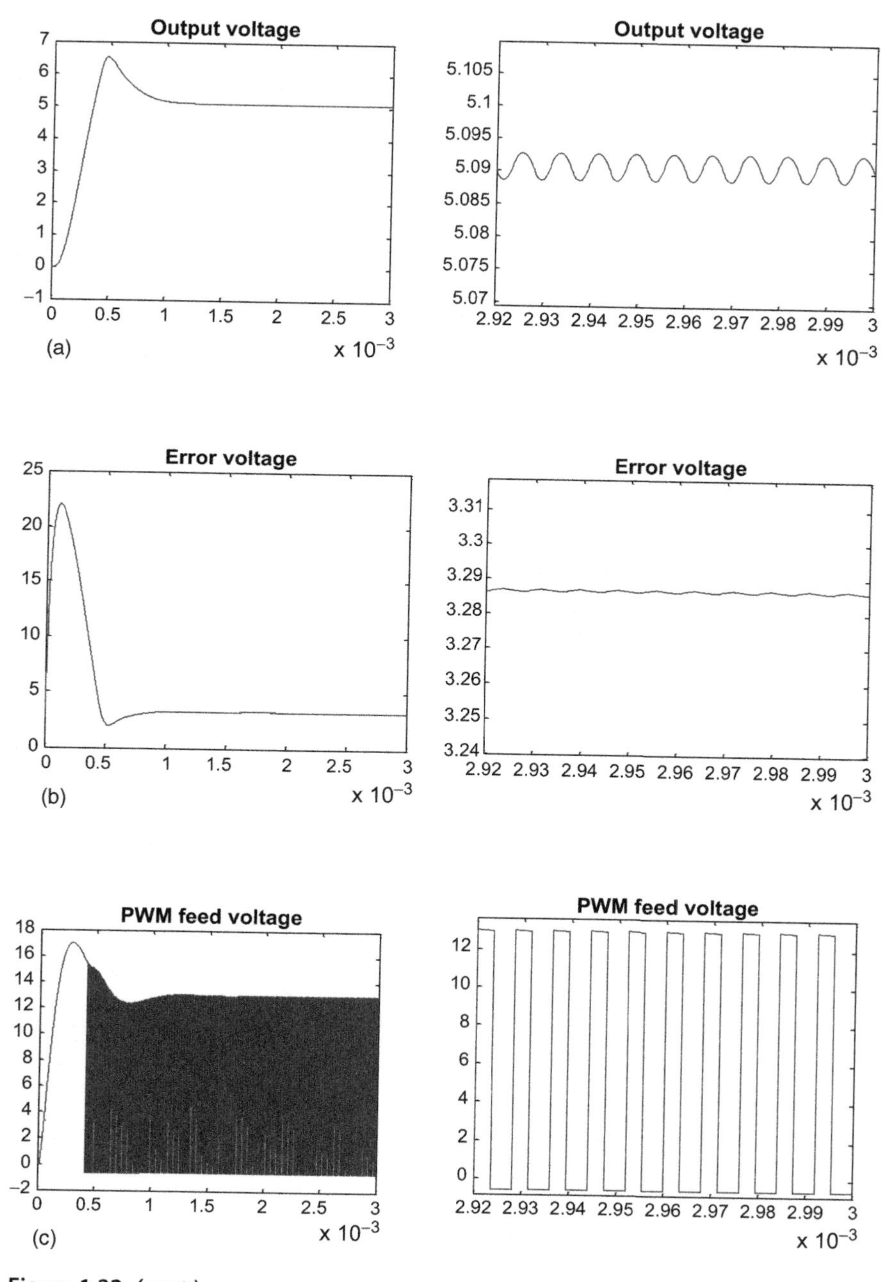

Figure 1.22 *(cont.)*

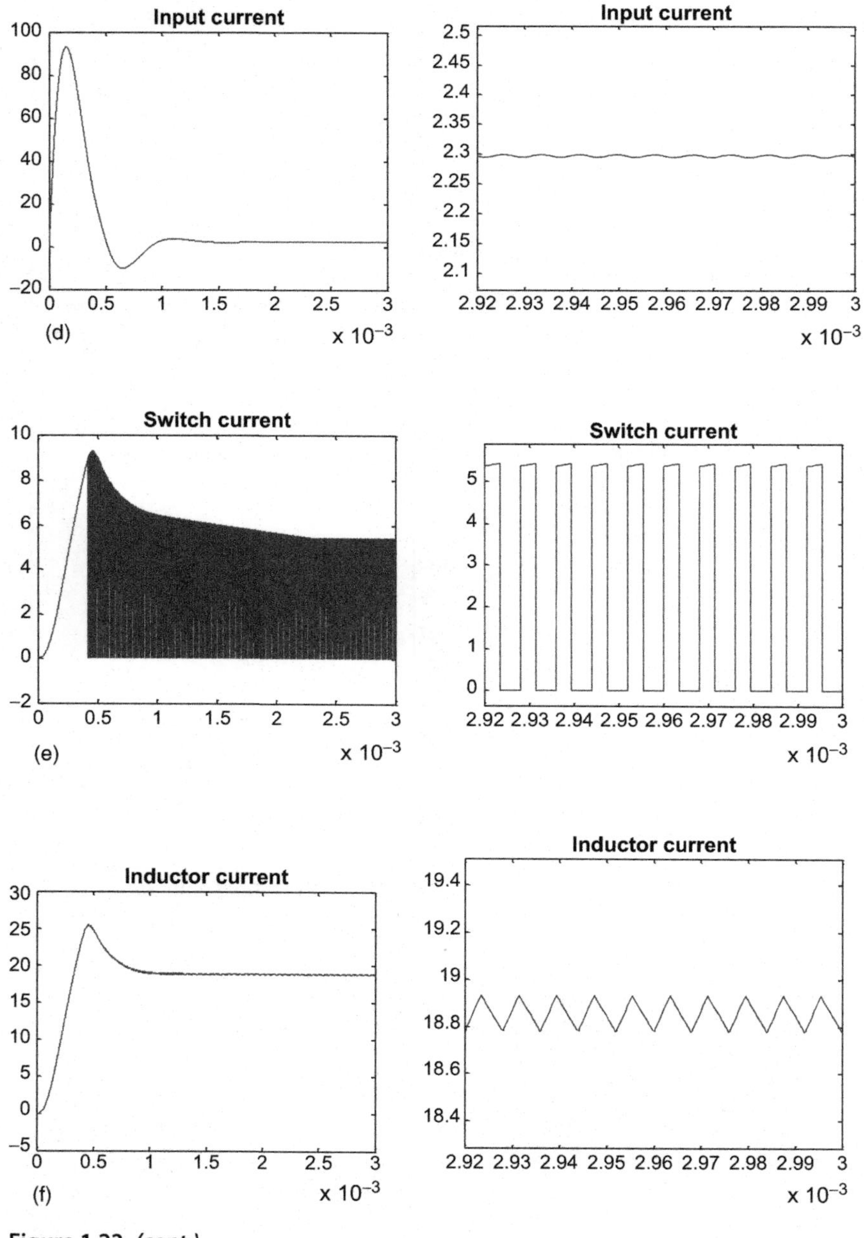

Figure 1.22 *(cont.)*

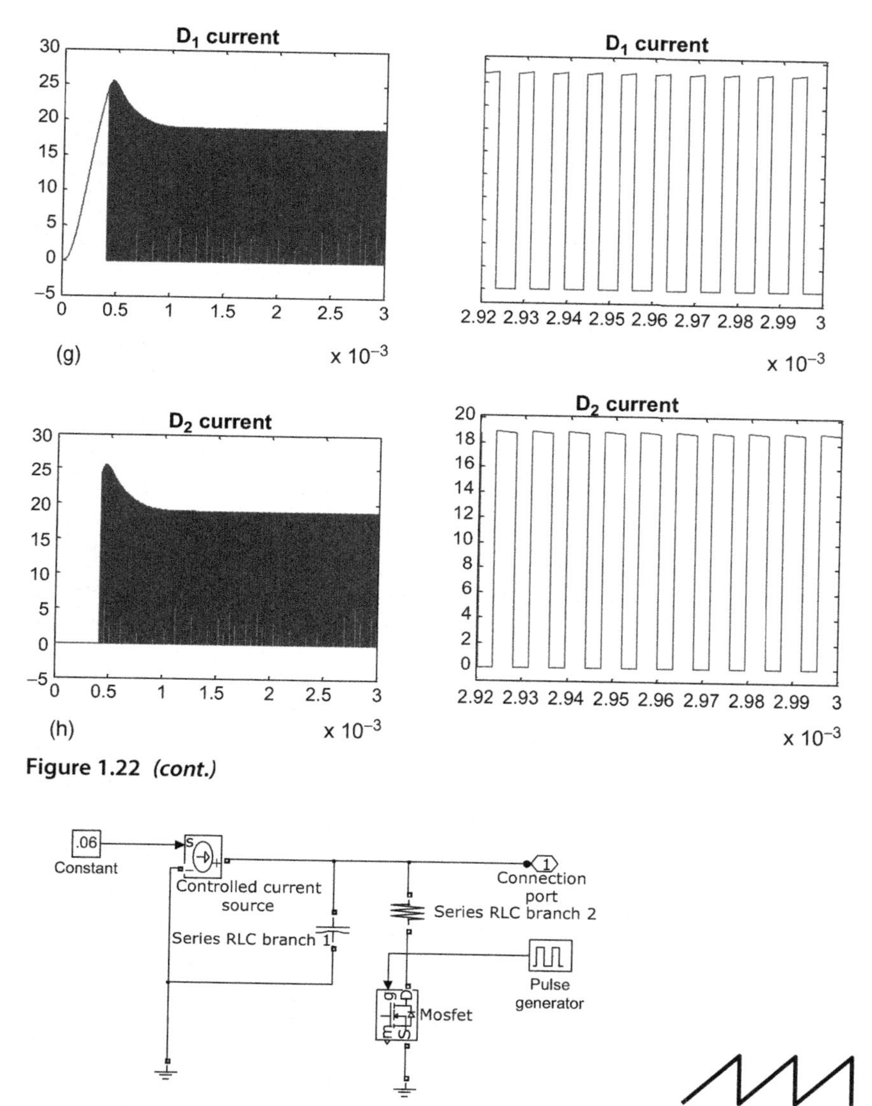

Figure 1.23 ***Analog Sawtooth Generator, Waveform Output at Pin 1.***

capacitor value, and the discharging resistor value at the desired frequency enables one to set the sawtooth waveform feeding the inverting input of the PWM comparator.

The triangular waveform shall now be replaced by one created completely on the basis of discrete, numerical processing alone without any analog function. Figure 1.24 shows the process in diagrammatic form.

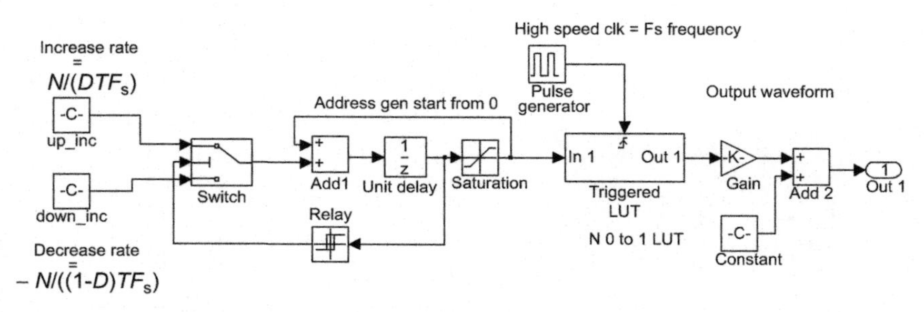

Figure 1.24 ***Digital Sawtooth Generator.***

There are two clocks involved in the digital generator: the power switch ON/OFF frequency $F = 1/T$ and the internal high speed sampling frequency $F_s = 1/T_s$. There are also four wave-forming parameters: N, the sawtooth amplitude full-count resolution; top, the sawtooth peak; bottom, the sawtooth trough; and D, the up-ramp duty cycle. For the example used in this chapter, $F = 125$ kHz, $F_s = 20$ MHz, $N = 2048$, top $= 5.7$, bottom $= 1$, and $D = 0.98$.

The wave generation process begins by defining a two-ramp slope: the up-ramp and the down-ramp. Starting from zero, the up-ramp must reach the full count within the designated duty cycle time. Therefore, the up-ramp slope is

$$\frac{NT_s}{DT} = \frac{N}{DTF_s} \tag{1.35}$$

In reverse and starting from the full count, the down-ramp must return to zero. Therefore, the down-ramp slope is

$$-\frac{NT_s}{(1-D)T} = -\frac{N}{(1-D)TF_s} \tag{1.36}$$

Both slopes are fed to a controlled "Switch," with the positive rate entry designated as "input #1," the negative entry as "input #3," and the control port "input #2."

During the up swing, the "Switch" passes "input #1." A "Unit delay" with $1/F_s$ resolution holds "Adder" output. By summing the positive slope and the previous adder's output, the "Adder" counts up at the F_s clock rate and generates a "staircase up."

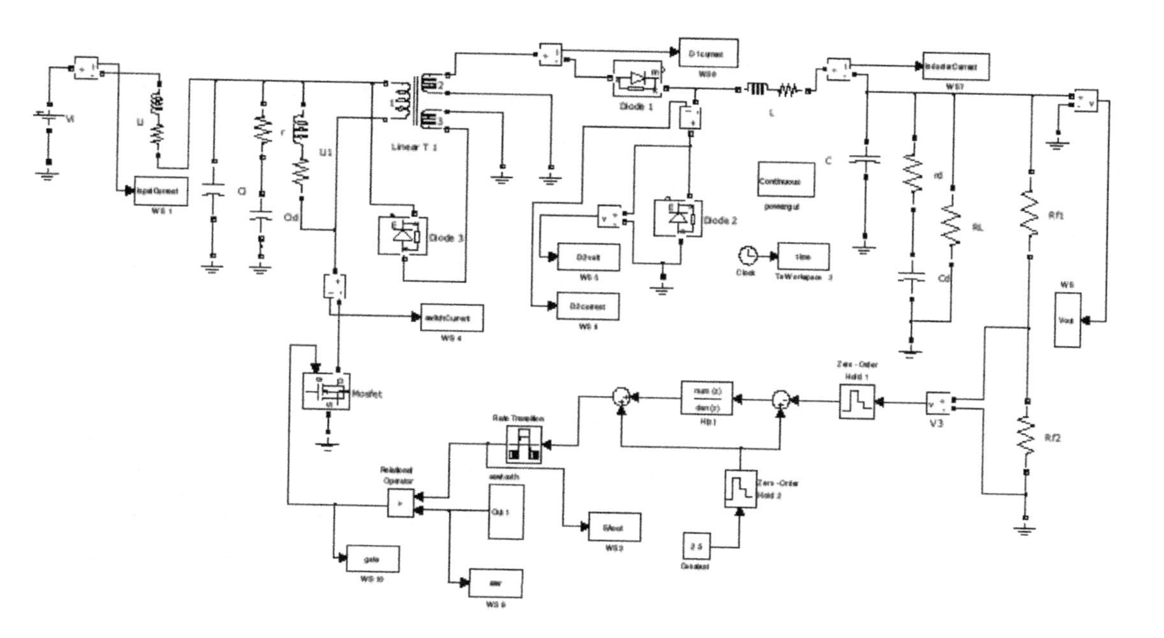

Figure 1.25 ***Forward Converter with All Digital Feedback.*** (a) Output voltage, (b) error voltage, (c) D_1 and D_2 cathode, (d) input current, (e) switch current, (f) inductor current, (g) D_1 current, and (h) D_2 current.

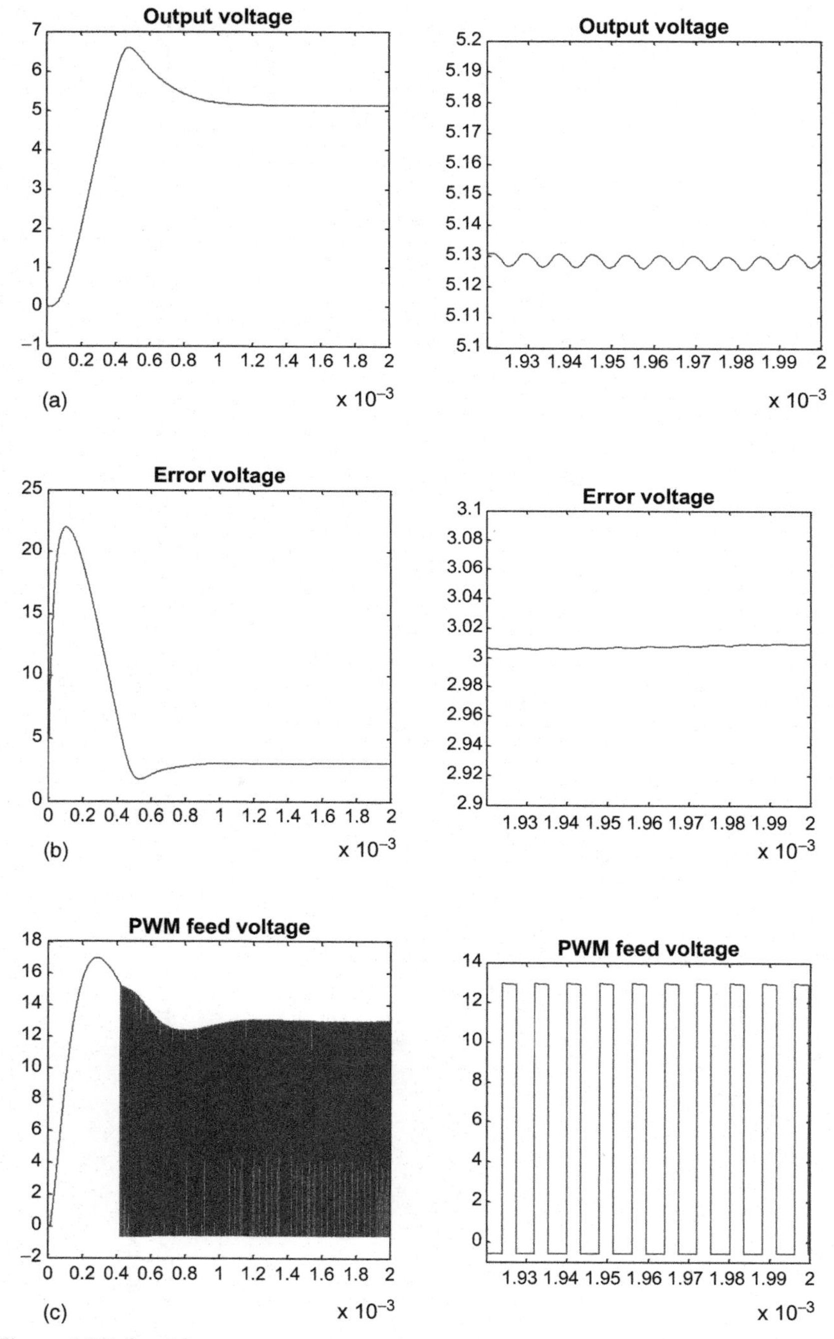

Figure 1.25 ***(cont.)***

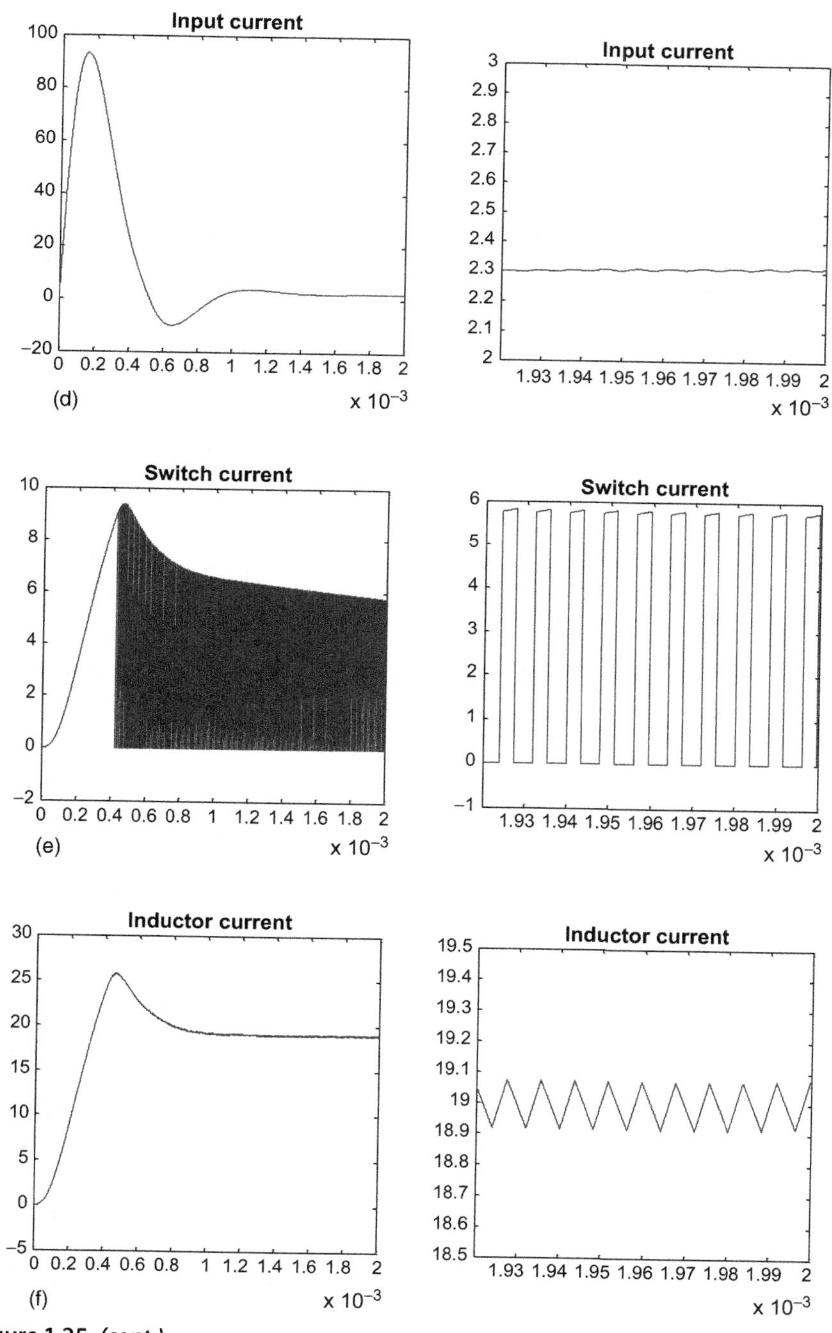

Figure 1.25 ***(cont.)***

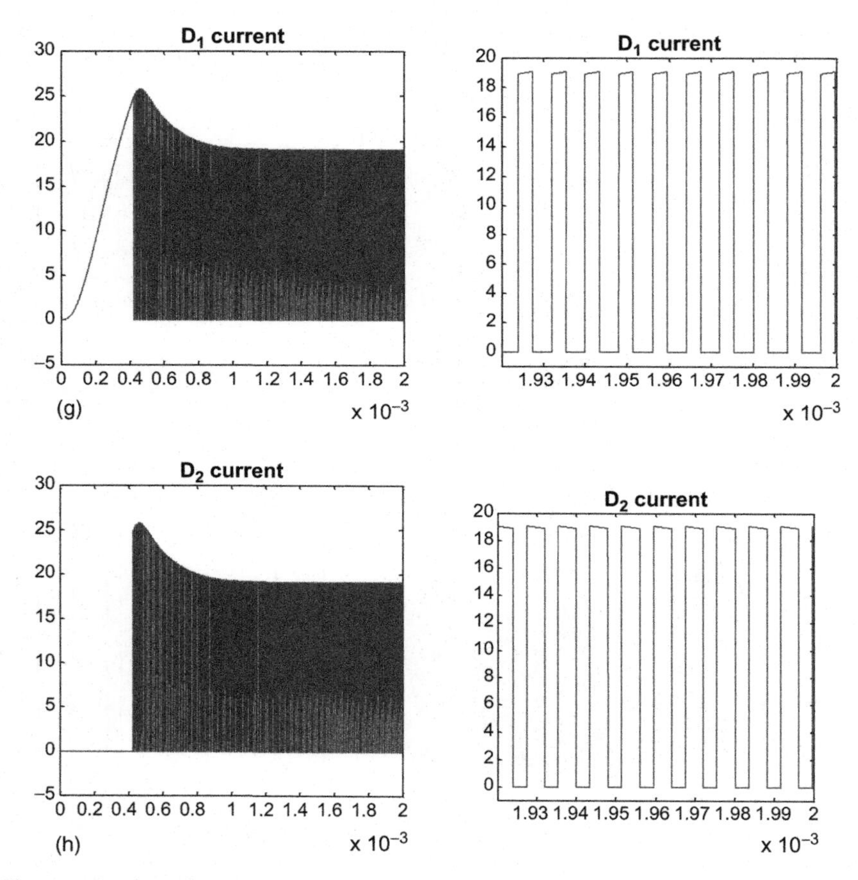

Figure 1.25 ***(cont.)***

When the "Adder" up-count output reaches "$N-1$," the saturation count, so does the "unit delay". At this point, the "Relay" output switches to "0" state. This action passed to the "Switch" input #2 forces the "Switch" to pass its "input #3," and begins down-ramp process also at F_s clock rate.

During the down swing and when the "Adder" and the "Unit delay" reaches "0," the "Relay" output switches to "1" state. This action forces the "Switch" to pass its "input #1" and repeats another up-ramp cycle.

The previous process generates a cyclic staircase, up and down, sawtooth with "0 to normalized full scale 1" swing. The sawtooth stream feeds the "Look Up Table," where the sampling clock, F_s, scans it.

"Gain" block magnifies the sawtooth swing from 0–1 to 0–4.7. A constant bias "$C = 1$" added to the sawtooth swing makes the output swing 1–5.7.

This concludes the digital sawtooth construct and it is ready to replace the analog version. Once the replacement is done, the SIMULINK model given in Figure 1.22 becomes Figure 1.25.

Compared with Figure 1.22, where PWM is performed by an analog comparator, Figure 1.25 performs the PWM with a digital operator " > " that examines the relative magnitude between two inputs: the digital error signal and the digital sawtooth.

Running Figure 1.25 for 2 ms, the real operation time, Figure 1.25a–h give its performance.

CHAPTER 2

Forward Converter with Current-Mode Control

2.1 SCHEMATIC WITH ANALOG CONTROLLER AND CURRENT FEEDBACK

In contrast to Figure 1.1, the voltage-mode control, current-mode control Figure 2.1 has an inner current feedback loop. The current feedback may be implemented in two approaches: a magnetic current sensor as shown, or a low-value resistive element R_s 10–100 mΩ, in series with the primary power switch. The former provides isolation, but it is bulkier, more expensive, requires cumbersome installation, and may be less effective in high frequency. The latter is simpler and cheaper, but it is less efficient.

As shown, there is basically no change in the power train. The power stage gain, $G_{vd}(s)$, designated as G_{vd} in the following, is totally identical to the one derived in Section 1.2. However, a huge difference for the PWM gain does exist between the two modes of control. With voltage-mode control, the PWM gain F_m is quite simple. In the case of peak current-mode control, the simplicity vanishes. Briefly, the outer voltage loop generates a voltage error, V_e. When the switch-peak current signal intercepts the voltage loop error signal, it terminates the switch-on state by resetting the reset–set (RS) flip-flop. A new conduction interval is then initiated by the next clock-pulse, and the cycle repeats.

2.2 DERIVATION OF PWM GAIN

Referring to the output filter of Figure 2.1, the inductor current is understood to have a steady state profile, as given in Figure 2.2.

Two distinctive segments are identified. The ramping up corresponds to the state that both the main switch and rectifier D_1 are conducting. The ramping down is associated with the condition of main switch off and free-wheel diode D_2 on. The ramp up current is reflected to the transformer's primary side, and on top of that, the primary magnetizing current, due to L_p, is also added. The instantaneous ramp-up current that the current sensor sees can be written as:

Power Converters with Digital Filter Feedback Control
http://dx.doi.org/10.1016/B978-0-12-804298-4.00002-2

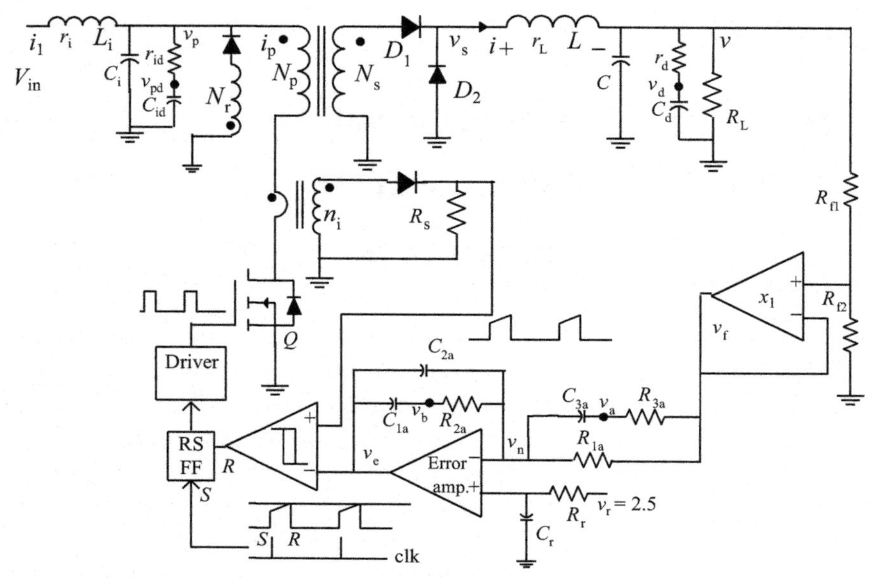

Figure 2.1 ***Forward Converter with Peak Current-Mode Control.***

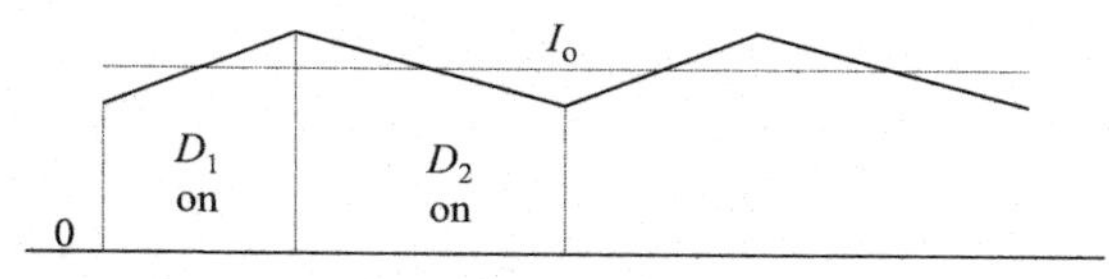

Figure 2.2 ***Output Inductor Current in Steady State.***

$$i_p(t) = \frac{N_s}{N_p}\left(I_o - \frac{((N_s/N_p)V_{in} - V_D - V_o)D}{2Lf_s} + \frac{(N_s/N_p)V_{in} - V_D - V_o}{L}t\right) + \frac{V_{in}}{L_p}t \tag{2.1}$$

where rectifier drop, V_D, and switching frequency, f_s, are accounted for. The ramp-up current passes through a 1–to–n_i current transformer. The steady state open-loop duty cycle is then determined by the action in which the ramp up signal intercepts the error voltage. That is,

$$\frac{i_p(DT_s)}{n_i}R_s = V_e \tag{2.2}$$

Equations (2.1) and (2.2) result in the duty cycle expression under open loop.

$$D(V_e, V_{in}, V_o) = \frac{(n_i V_e / R_s) - (N_s/N_p)(V_o/R_L)}{(N_s/N_p)\left[\left[\left((N_s/N_p)V_{in}\right) - V_D - V_o\right]/2Lf_s\right] + (V_{in}/L_p f_s)} \tag{2.3}$$

The expression indicates that the open loop duty cycle is a dynamic function of three variables; error voltage V_e, input V_{in}, and output V_o. It therefore has a total derivative:

$$\begin{aligned} dD &= \frac{\partial D}{\partial V_e}\delta V_e + \frac{\partial D}{\partial V_{in}}\delta V_{in} + \frac{\partial D}{\partial V_o}\delta V_o \\ &= F_m \cdot \delta V_o + F_g \cdot \delta V_{in} + F_v \cdot \delta V_o \end{aligned} \tag{2.4}$$

where, for instance:

$$F_m = \frac{\partial D(V_e, V_{in}, V_o)}{\partial V_e} = \frac{1}{(N_s/N_p)\left[\left((N_s/N_p)V_{in} - V_D - V_o\right)/2Lf_s\right] + (V_{in}/L_p f_s)}\frac{n_i}{R_s} \tag{2.5}$$

The other two gain coefficients, F_g and F_v, in symbolic forms, are extremely burdensome to write and are omitted in print form. Both are easily available by way of numerical computation. Expressing (2.4) and power stage gain in block diagram form, we have Figure 2.3.

For loop gain evaluation, Figure 2.3 may be simplified by assuming a constant input. This simplification leads to Figure 2.4.

By absorbing the inner current loop, the block diagram is further reduced, Figure 2.5.

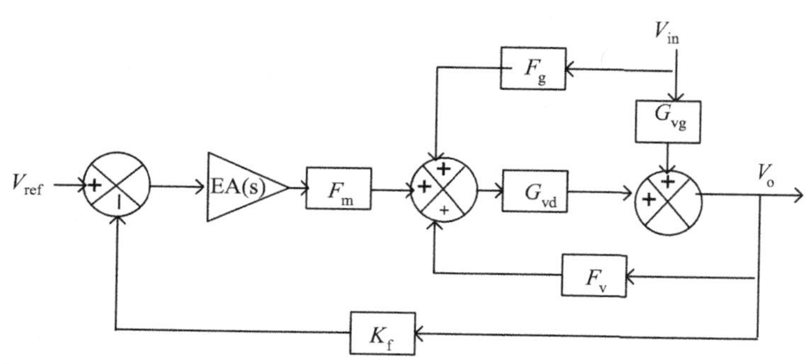

Figure 2.3 ***AC Block Diagram for Figure 2.1.***

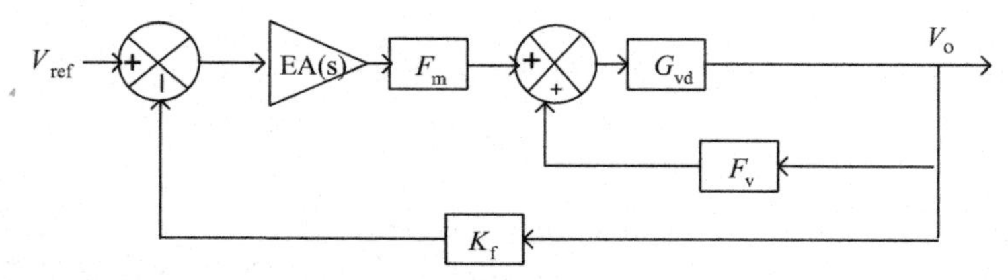

Figure 2.4 ***AC Block Diagram with Constant Input.***

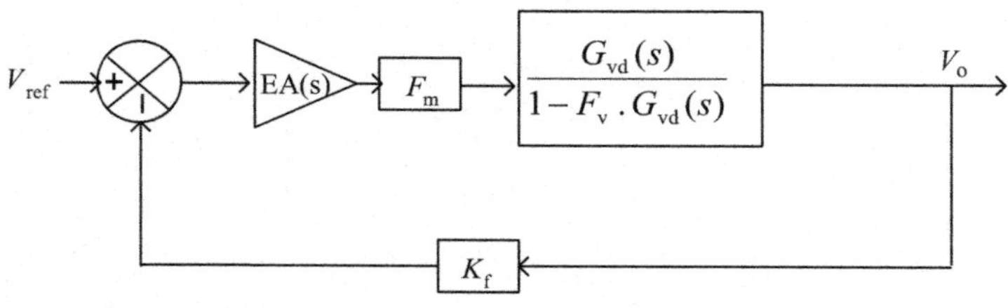

Figure 2.5 ***Close-Loop Diagram with Current Loop Absorbed.***

Therefore, the modulator gain for current-mode control is:

$$M(s) = K_f F_m \frac{G_{vd}(s)}{1 - F_v G_{vd}(s)} \tag{2.6}$$

Of course, gain block F_v needs to be figured out numerically too.

Once (2.6) is done, the rest of the procedure is identical to what had been presented in Section 1.3. Again, the example discussed further tells the story better.

2.3 EXAMPLE

For this example, the magnetic current sensor has a 1:50 turn ratio, 50 Ω R_s, and 1 mH primary magnetizing inductance. All other parts in the power train and operating conditions stay unchanged. PWM gain computation yields F_m = 5.304 and F_v = −2.242, both in the unit of fraction/volt. The resulting modulator gain has been shown here in Figure 2.6.

If a desired close-loop crossover frequency of 10 kHz is picked, the plot here shows a modulator gain of −2.7 db and −82.6° phase. Again, specify the desired phase margin of 60°, and repeat the same process as outlined in Section 1.3; a type-III error amplifier with the following components is identified. R_{1a} = 2K, C_{2a} = 5.8 × 10^{-9}, C_{1a} = 9.3 × 10^{-9}, R_{2a} = 2.76K, R_{3a} = 1.26K, C_{3a} = 7.8 × 10^{-9}. Placing the analog error amplifier and its

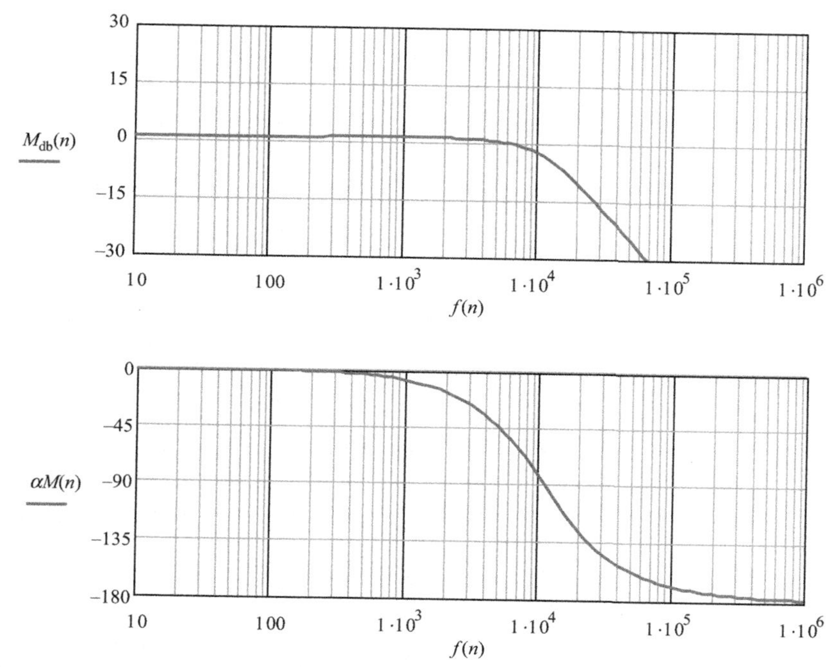

Figure 2.6 ***Modulator Gain and Phase in Frequency Domain.***

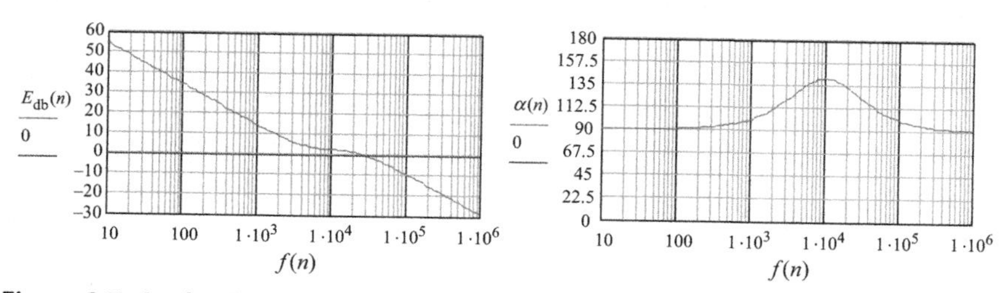

Figure 2.7 ***Analog Type-III Error Amplifier's Frequency Response and the Overall Loop Gain.***

components in H(s) equation (1.16) form, it shows a frequency response of Figures 2.7 and loop gain Figure 2.8.

Compared with Figure 1.12, current-mode control seems to offer a wider bandwidth. It is because the output inductor's *di/dt* also plays a direct role in setting the duty cycle. This fact results in a much better transient response caused by either input or load change.

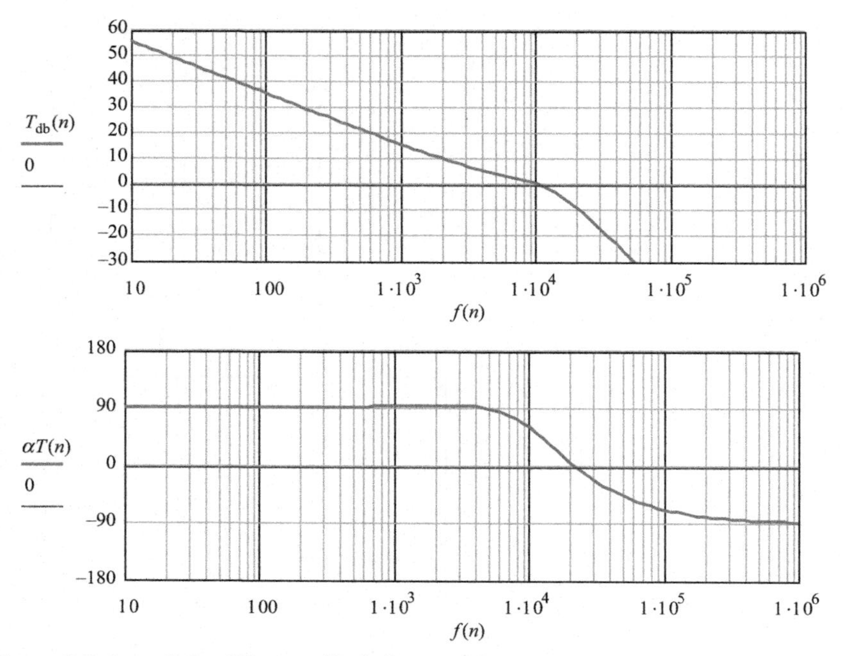

Figure 2.8 ***Loop Gain of Current-Mode Forward Converter.***

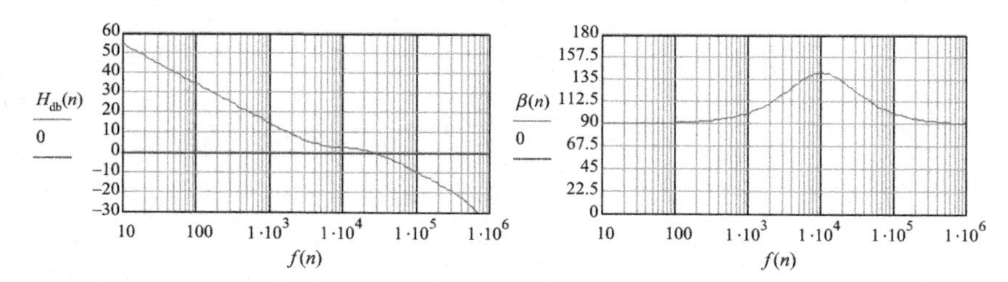

Figure 2.9 ***Digital Filter's Frequency Response.***

Next, s-to-z bilinear mapping with the same scaling constant C ($= 4 \times 10^6$) is performed. The following $H_{III}(z)$, from equation (1.23), coefficients are obtained.

$$a_0 = 53.870821 \times 10^{-3} \qquad a_1 = -51.788383 \times 10^{-3}$$
$$a_2 = -53.850696 \times 10^{-3} \qquad a_3 = 51.808508 \times 10^{-3}$$
$$b_1 = -2.901354 \times 10^0 \qquad b_2 = 2.805141 \times 10^0$$
$$b_3 = -903.786802 \times 10^{-3}$$

Its frequency response shows more attenuation at high frequency, from 400 kHz to 1 MHz (Figure 2.9).

2.4 SIMULATION AND PERFORMANCE VERIFICATION

Prior to performing simulation, it shall be noted that the peak current-mode PWM output requires a RS flip-flop. A pulse clock at the switching frequency sets the RSFF, while a switch current up-ramp intercepting the error signal resets the RSFF. This RSFF action is implemented (MathCAD logic operation) in the first statement of the iterative difference equation set, (2.7). In this case, q-output of the RSFF serves as a switch control, turning ON/OFF the main switch. By contrast, voltage-mode control, Figure 1.1 and Chapter 1, directly enlist PWM comparator output as switch control. It shall also be note that slow start is implemented. In addition, for equation (2.7), part designators corresponding to Figure 2.1 shall be observed: $R_{1a} = R30$, $C_{2a} = C16$, $C_{1a} = C15$, $R_{2a} = R28$, $R_{3a} = R29$, and $C_{3a} = C3$.

$$\begin{pmatrix} q_{j+1} \\ i_{1_{j+1}} \\ v_{P_{j+1}} \\ i_{j+1} \\ v_{o_{j+1}} \\ v_{cd_{j+1}} \\ v_{a_{j+1}} \\ v_{b_{j+1}} \\ v_{e_{j+1}} \\ i_{m_{j+1}} \\ v_{id_{j+1}} \\ i_{r_{j+1}} \end{pmatrix} = \begin{bmatrix} \neg\left[\left[\text{if}\left[(R_s/n_i)\left(i_{m_j} + (N_2/N_1)i_j q_j\right) \geq v_{e_j}, 1, 0\right]\right] \vee \neg\left[q_j \vee \left(\text{if}(c_{lk_j} = 1, 1, 0)\right)\right]\right] \\ i_{1_j} + \left[V_{in} - \left(r_1 i_{1_j} + v_{P_j}\right)\right](\delta t/L_1) \\ v_{P_j} + \left[i_{1_j} - \left(\left(v_{P_j} - v_{id_j}\right)/R_{d2} + (N_2/N_1)i_j q_j + i_{m_j}\right)\right](\delta t/C_1) \\ \text{if}\left[i_j < 0, 0, i_j + (\delta t/L_1)\left[\text{if}\left[\begin{matrix} q_j = 0, -V_d\cdot(N_2/N_1) \\ \left[v_{P_j} - R_{on}\left((N_2/N_1)i_j q_j + i_{m_j}\right)\right] - V_d \end{matrix}\right] - \left(r i_j + v_{o_j}\right)\right]\right] \\ v_{o_j} + (\delta t/C)\left[i_j - \left(\left(v_{o_j} - v_{cd_j}\right)/R_d + \left(v_{o_j}/R_L\right)\right)\right] \\ v_{cd_j} + (\delta t/C_d)\left(\left(v_{o_j} - v_{cd_j}\right)/R_d\right) \\ v_{a_j} + (\delta t/C_3)\left[\left(k_f v_{o_j} - v_{a_j}\right)/R29\right] \\ v_{bj} + \delta t\left[\begin{matrix} \left(\left(v_{r_j} - v_{b_j}\right)/R28C16\right) - \left(\left(k_f v_{o_j} - v_{r_j}\right)/R30C16\right) - \\ \left(\left(k_f v_{o_j} - v_{a_j}\right)/R29C16\right) + \left(\left(v_{r_j} - v_{b_j}\right)/R28C15\right) \end{matrix}\right] \\ \text{if}\left[v_{e_j} < 0, 0, \text{if}\left[v_{e_j} > 15, 15, v_{e_j} + (\delta t/C16)\left(\begin{matrix} \left(v_{r_j} - v_{b_j}\right)/R28 - \left(k_f v_{o_j} - v_{r_j}\right)/ \\ R30 - \left(k_f v_{o_j} - v_{a_j}\right)/R29 \end{matrix}\right)\right]\right] \\ \text{if}\left[q_j = 0, 0, i_{m_j} + (\delta t/L_m)\left[v_{P_j} - R_{on}\left((N_2/N_1)i_j q_j + i_{m_j}\right)\right]\right] \\ v_{id_j} + (\delta t / C_2)\left(v_{P_j} - v_{id_j}\right)/R_{d2} \\ \text{if}\left[q_j = 0, i_{m_j} + i_{r_j} + (\delta t/L_m)\left(-v_{P_j} - R_{w_r} i_{r_j}\right), 0\right] \end{bmatrix} \tag{2.7}$$

Compared with the performance of voltage-mode control, Figure 1.17a–i, current mode control performance, Figure 2.10a–h show better quality in terms of faster settling and less sinusoidal ringing tail.

Next, let's see how current-mode control works with digital filter. The difference equation set is given in (2.8), in which direct form implementation of the digital filter, represented by variables y and w, replaces analog counterparts. At 2 MHz sampling rate, the same time coverage, 200 cycles

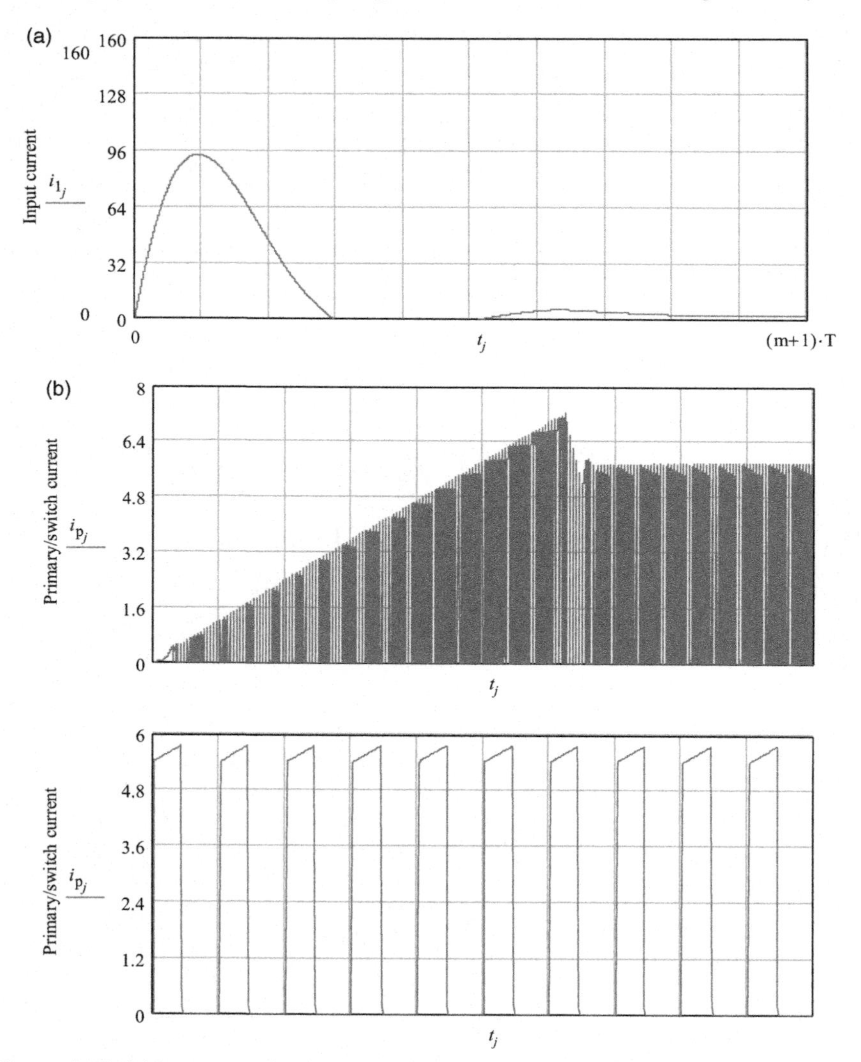

Figure 2.10 (a) Input current, (b) primary winding/switch current. (c) D_2 cathode voltage, (d) D_1 current, (e) D_2 current, (f) output inductor current, (g) output voltage, (h) error voltage.

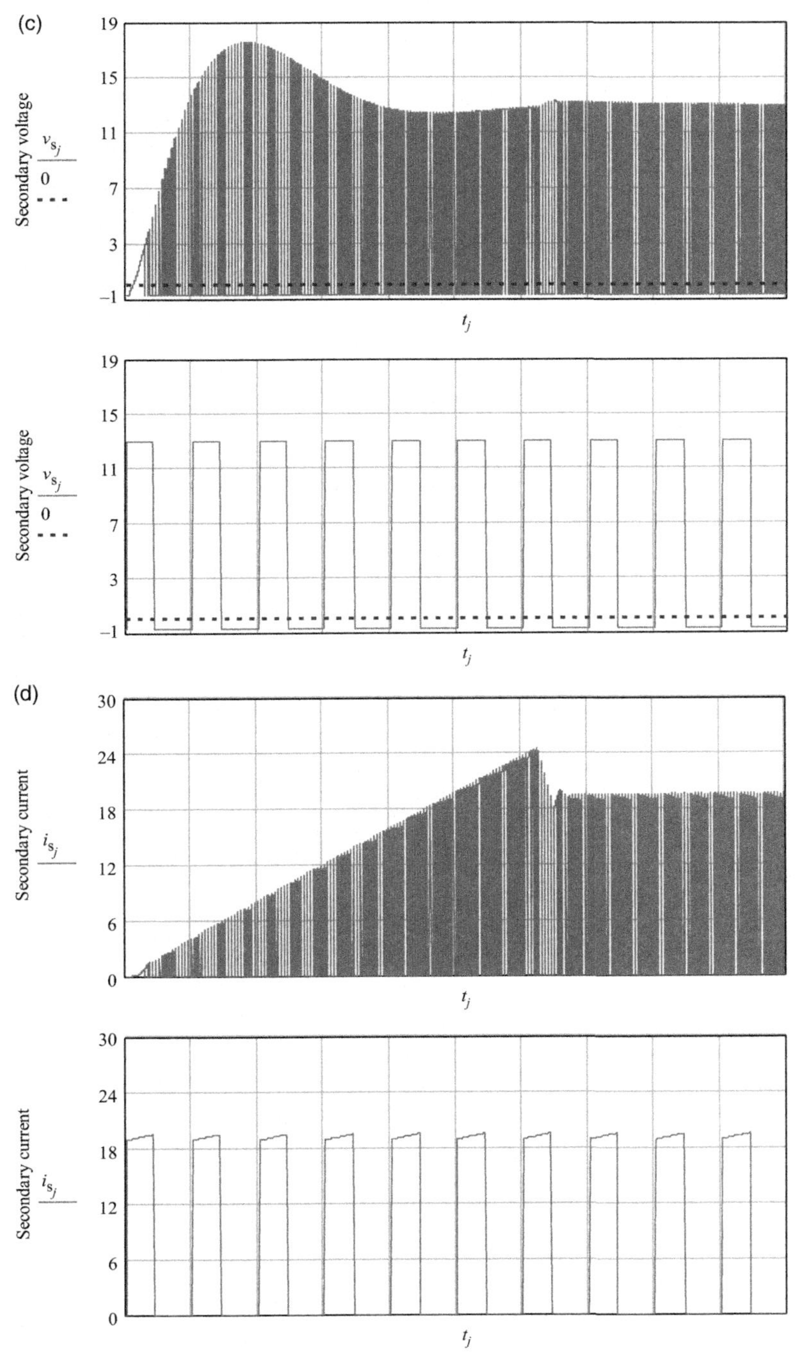

Figure 2.10 ***(cont.)***

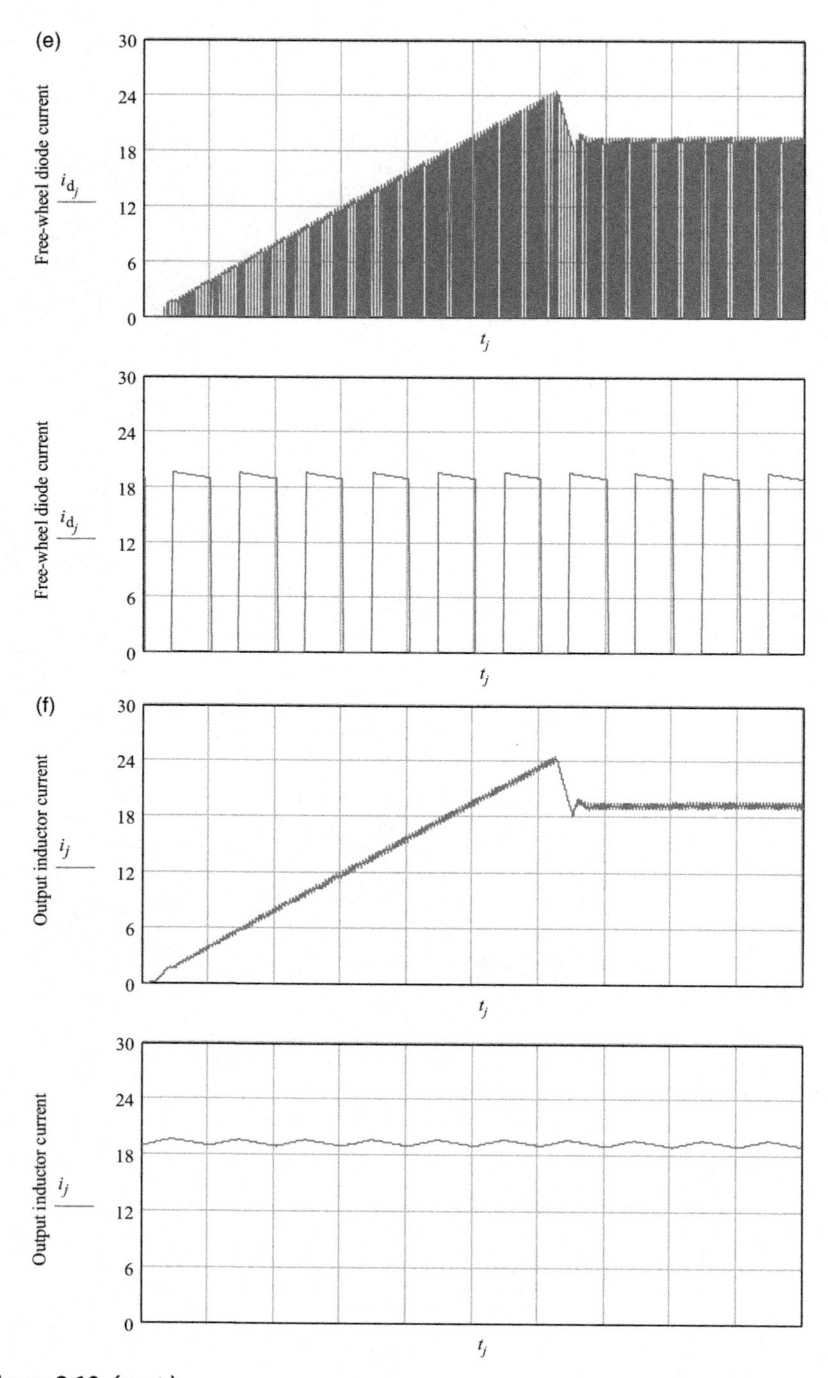

Figure 2.10 ***(cont.)***

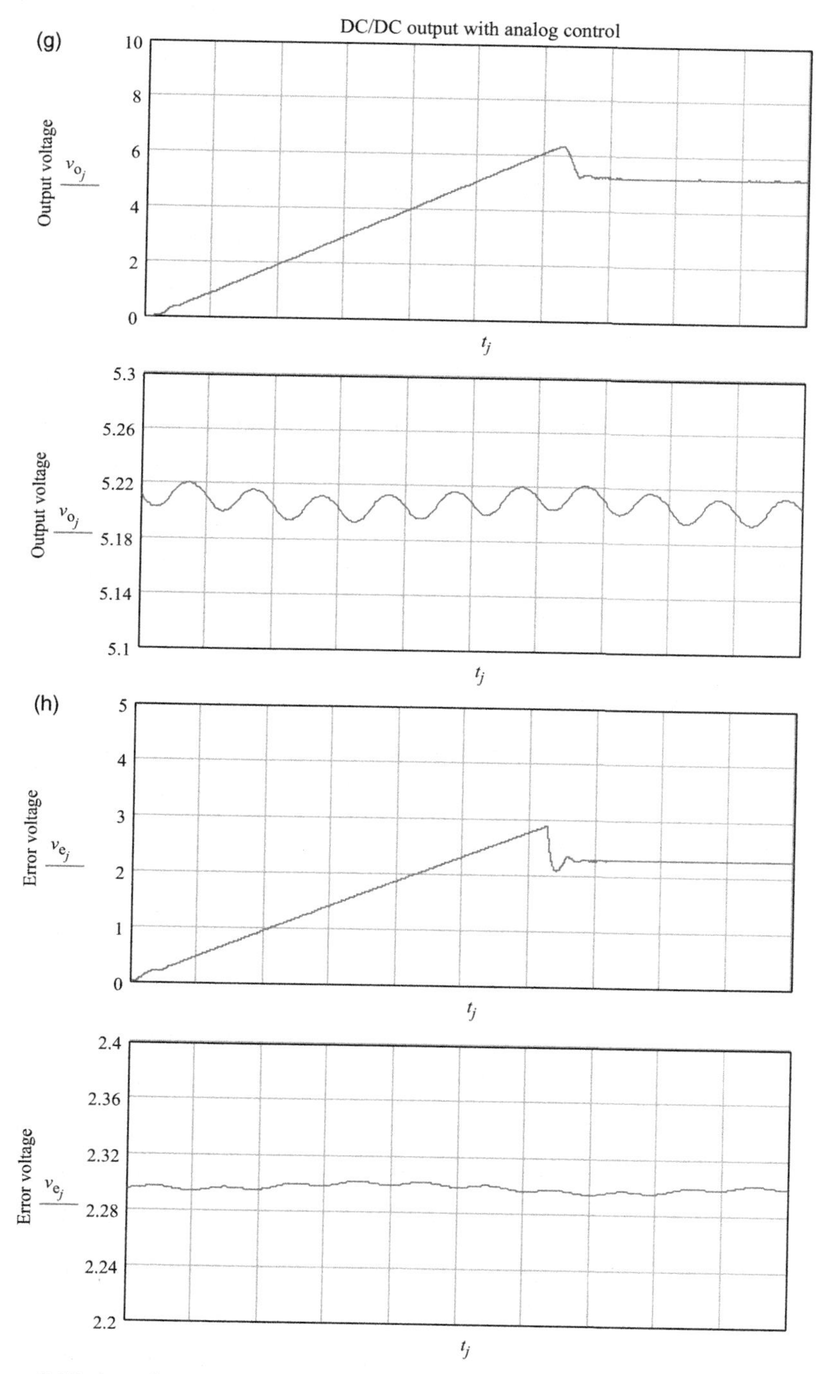

Figure 2.10 ***(cont.)***

at the switching frequency, as all previous simulations, is conducted. Time domain plots, Figure 2.11a–i, follow the difference equation set.

$$
\begin{pmatrix} q_{j+1} \\ i_{1_{j+1}} \\ v_{\mathrm{p}_{j+1}} \\ i_{j+1} \\ v_{\mathrm{o}_{j+1}} \\ v_{\mathrm{cd}_{j+1}} \\ \gamma_{j+1} \\ i_{\mathrm{m}_{j+1}} \\ v_{\mathrm{id}_{j+1}} \\ i_{\mathrm{r}_{j+1}} \\ w_{j+1} \end{pmatrix}
=
\begin{bmatrix}
\neg\begin{bmatrix} \left[\mathrm{if}\left[\left(R_{\mathrm{s}}/n_i\right)\left(i_{\mathrm{m}_j}+\left(N_2/N_1\right)i_j q_j\right)\geq \gamma_j,1,0\right]\right] \\ \vee\neg\left[q_j \vee\left(\mathrm{if}\left(c_{\mathrm{lk}_j}=1,1,0\right)\right)\right] \end{bmatrix} \\
i_{1_j}+\left[V_{\mathrm{in}}-\left(r_1\cdot i_{1_j}+v_{\mathrm{p}_j}\right)\right]\left(\delta t/L_1\right) \\
v_{\mathrm{p}_j}+\left[i_{1_j}-\left(\left(v_{\mathrm{p}_j}-v_{\mathrm{id}_j}\right)/R_{\mathrm{d2}}+\left(N_2/N_1\right)i_j q_j+i_{\mathrm{m}_j}\right)\right]\left(\delta t/C_1\right) \\
\mathrm{if}\left[i_j<0,0,i_j+\left(\delta t/L_1\right)\left[\mathrm{if}\begin{bmatrix} q_j=0,-Vd,\left(N_2/N_1\right) \\ \left[v_{\mathrm{p}_j}-R_{on}\left(\left(N_2/N_1\right)i_j q_j+i_{\mathrm{m}_j}\right)\right]-V_{\mathrm{d}}\end{bmatrix}-\left(ri_j+v_{\mathrm{o}_j}\right)\right]\right] \\
v_{\mathrm{o}_j}+(\delta t/C)\left[i_j-\left(\left(v_{\mathrm{o}_j}-v_{\mathrm{cd}_j}\right)/R_{\mathrm{d}}+\left(v_{\mathrm{o}_j}/R_{\mathrm{L}}\right)\right)\right] \\
v_{\mathrm{cd}_j}+\left(\delta t/C_{\mathrm{d}}\right)\left(v_{\mathrm{o}_j}-v_{\mathrm{cd}_j}\right)/R_{\mathrm{d}} \\
\mathrm{if}\left[\gamma_j<0,0,\ \mathrm{if}\left[\gamma_j>12,\ 12,\ v_{\mathrm{r}}+a_0\left[\left(v_{\mathrm{r}}-k_{\mathrm{f}}v_{\mathrm{o}_j}\right)-b_1w_{\mathrm{j}}-b_2\cdot w_{j-1}-b_3w_{j-2}\right]+ a_1w_j+a_2w_{j-1}+a_3w_{j-2}\right]\right] \\
\mathrm{if}\left[q_j=0,0,i_{\mathrm{m}_j}+\left(\delta_{\mathrm{t}}/L_{\mathrm{m}}\right)\left[v_{\mathrm{p}_j}-\left(\left(N_2/N_1\right)i_j q_j+i_{\mathrm{m}_j}\right)R_{on}\right]\right] \\
v_{\mathrm{id}_j}+\left(\delta t/C_2\right)\left[\left(v_{\mathrm{p}_j}-v_{\mathrm{id}_j}\right)/R_{\mathrm{d2}}\right] \\
\mathrm{if}\left[q_j=0,i_{\mathrm{m}_j}+i_{\mathrm{r}_j}+\frac{\delta t}{L_{\mathrm{m}}}\left(-v_{\mathrm{p}_j}-R_{\mathrm{wr}}i_{\mathrm{r}_j}\right),0\right] \\
\left(v_{\mathrm{r}}-k_{\mathrm{f}}v_{\mathrm{o}_j}\right)-b_1w_j-b_2w_{j-1}-b_3w_{j-2}
\end{bmatrix}
\tag{2.8}
$$

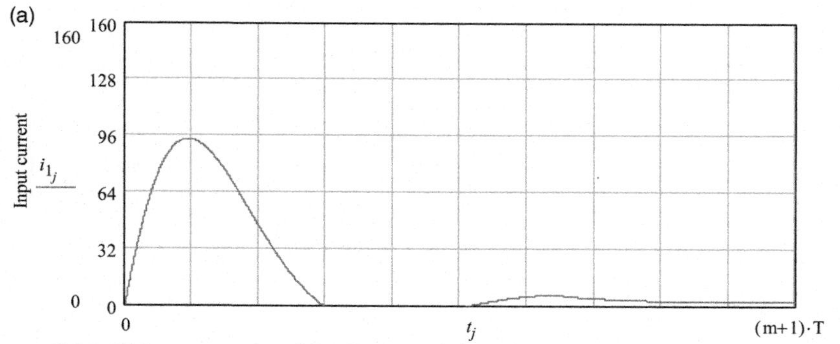

Figure 2.11 (a) Input current, (b) primary winding, or switch, current, (c) D_2 cathode voltage, (d) D_1 current (e) D_2 current, (f) output inductor current, (g) output voltage, (h) error voltage, (i) magnetizing and reset current.

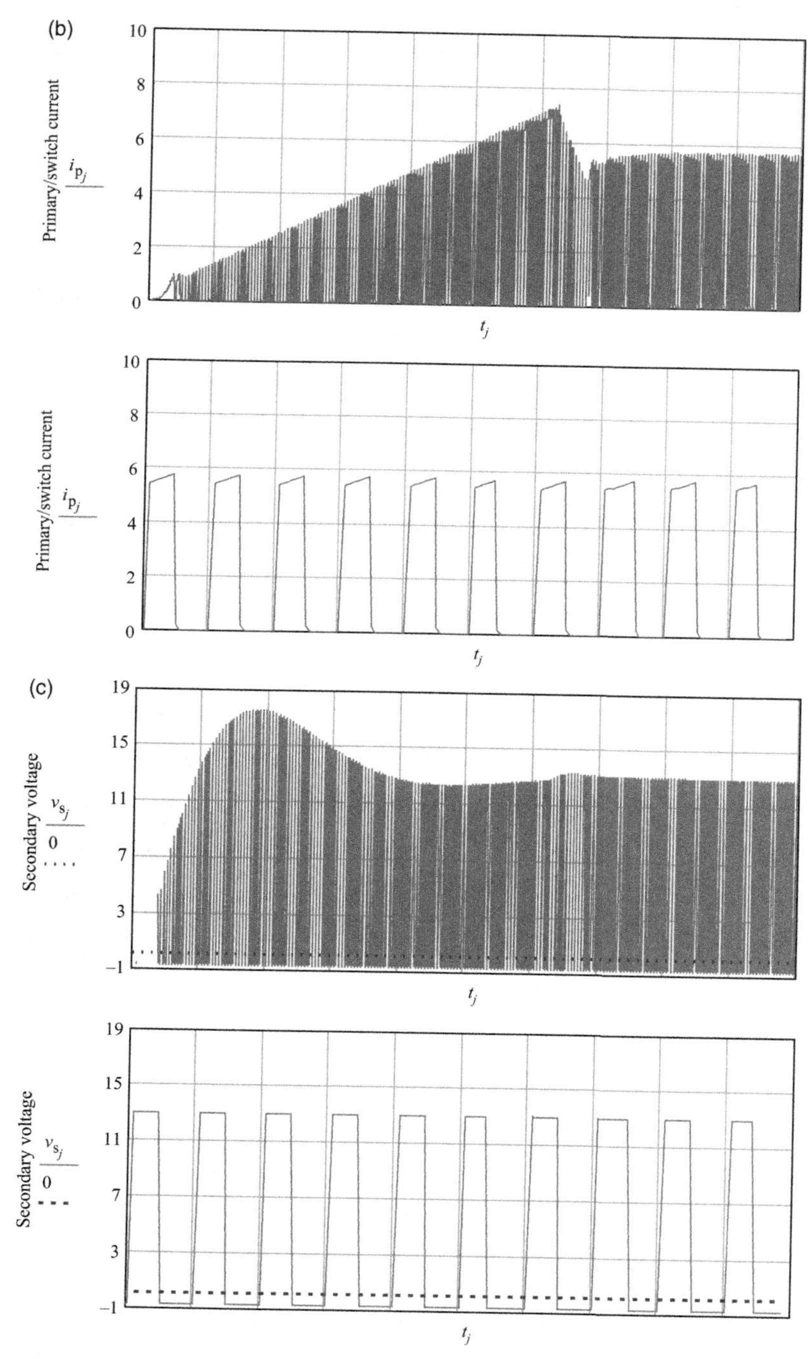

Figure 2.11 ***(cont.)***

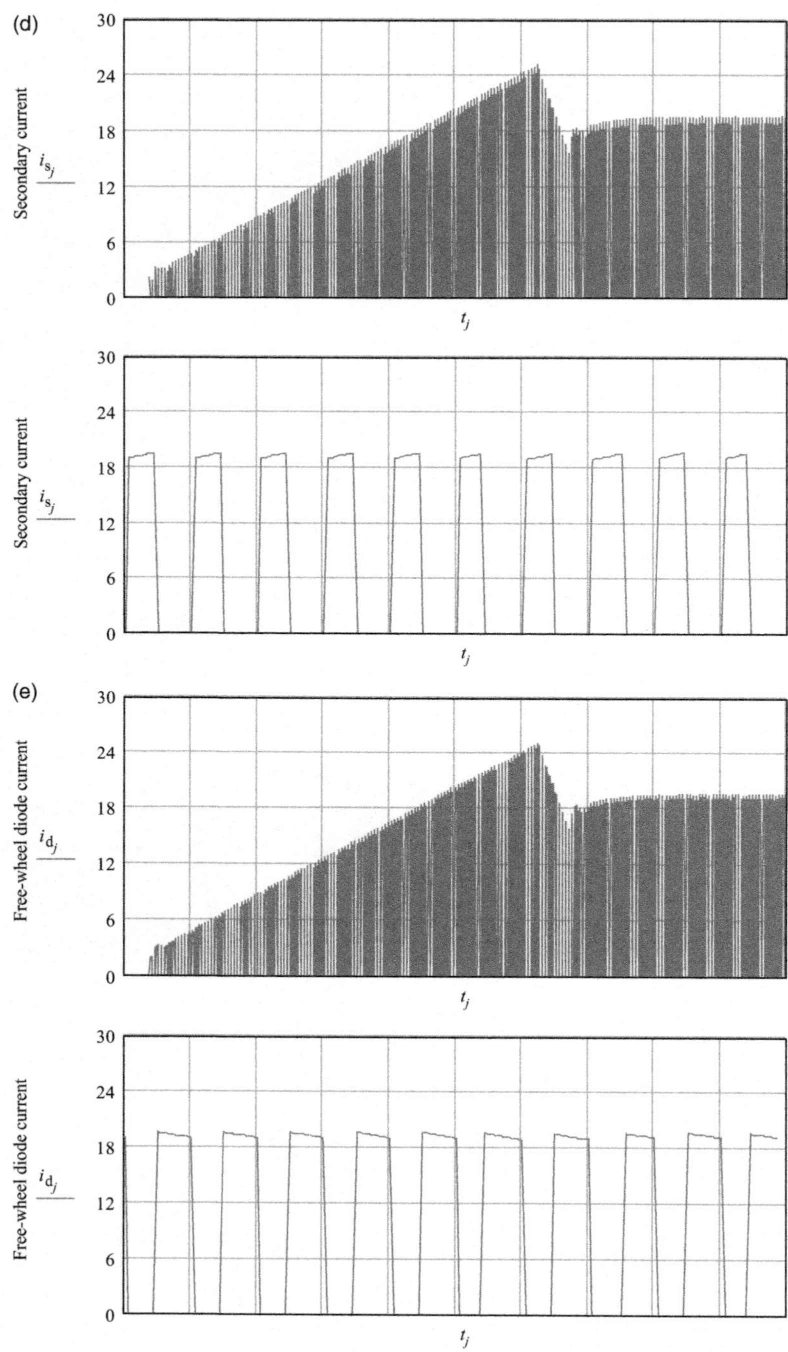

Figure 2.11 ***(cont.)***

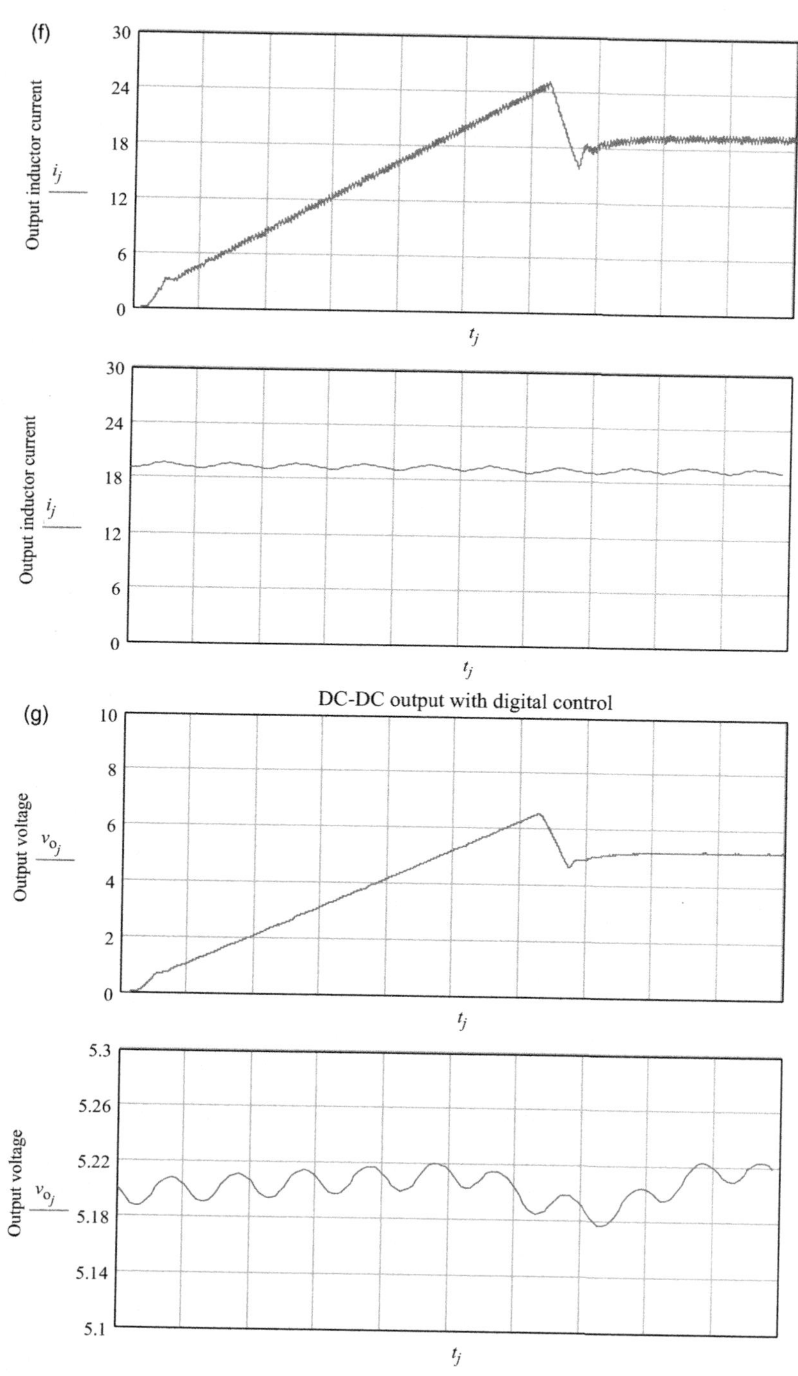

Figure 2.11 ***(cont.)***

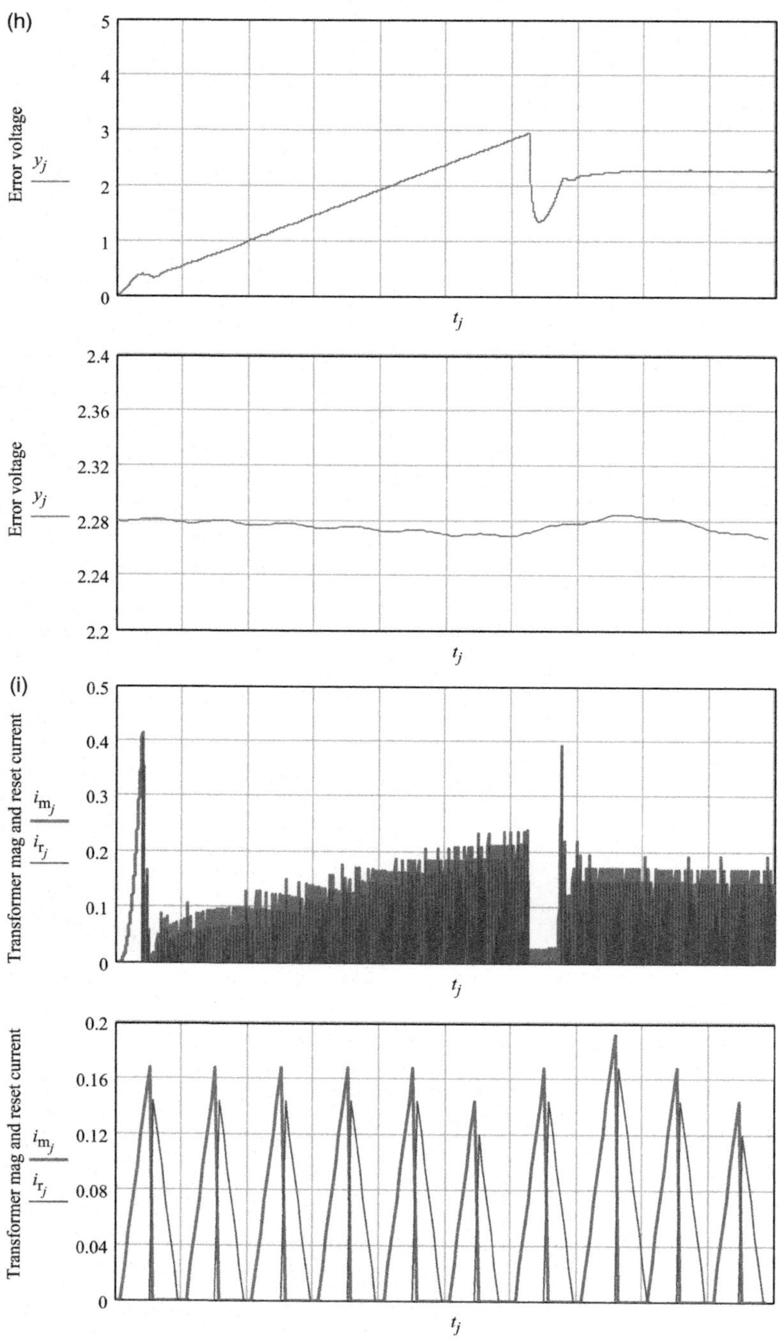

Figure 2.11 ***(cont.)***

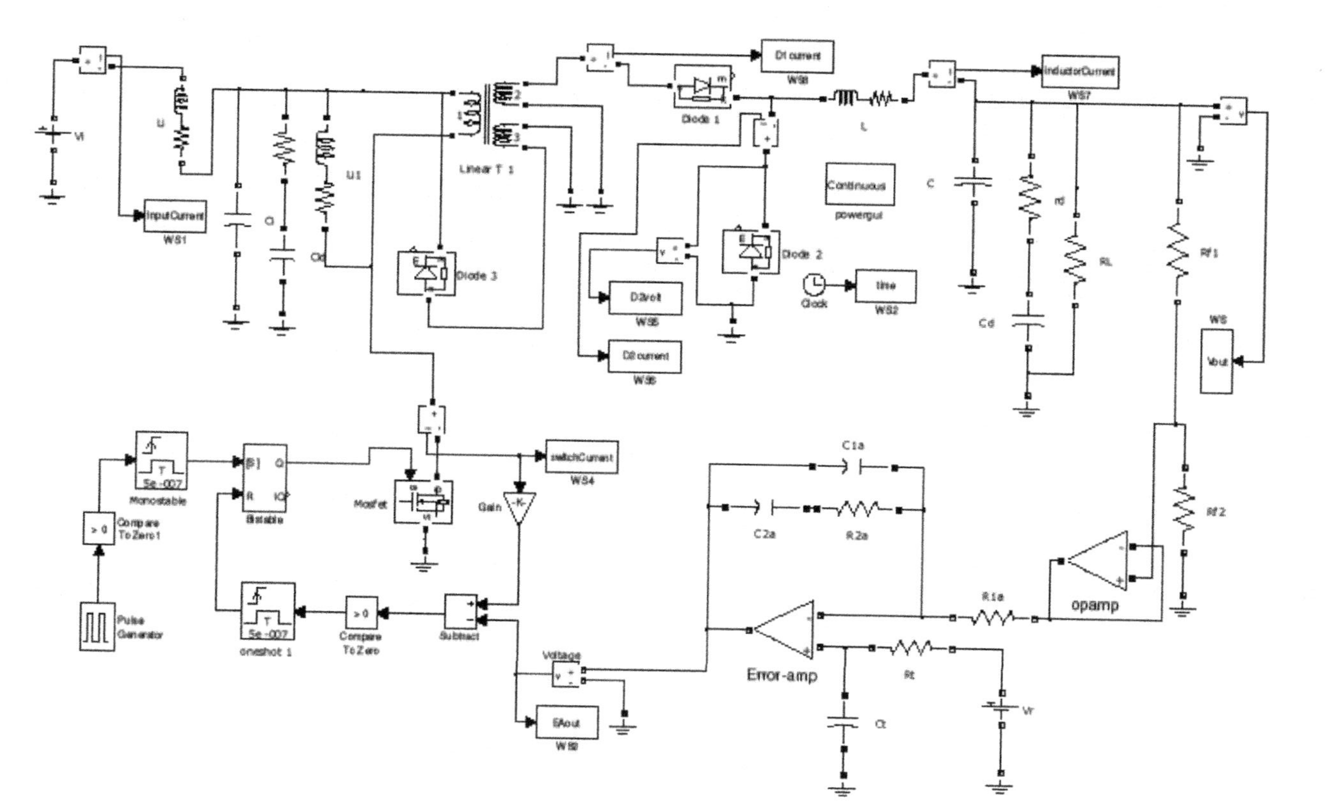

Figure 2.12 ***SIMULINK Model with Error Amp Represented by Physical Device.*** (a) Output voltage, (b) error voltage, (c) D_1 and D_2 cathode, (d) input current, (e) switch current, (f) inductor current, (g) D_1 current, (h) D_2 current.

2.5 MATLAB SIMULINK SIMULATION

In both Figures 2.1 and 2.2, a type-III error amplifier was employed for demonstration purpose as well as for a 60°-phase margin. It turns out that if the desired phase margin is reduced to 45° at the same crossover frequency of 10 kHz, a type-II error amplifier with two less components can perform just as well for current-mode control. The reader is invited to confirm, with $R_{1a} = 2K$ preselected, $C_{1a} = 0.0029$ μF, $C_{2a} = 0.009$ μF, and $R_{2a} = 3.6K$. In the depictedSIMULINK models, type-II error amplifiers are shown. Figure 2.12 gives a physical device model and Figure 2.12a–h present simulation plots. Figure 2.13 replaces the physical model with continuous transfer function H(s), while Figure 2.14 plugs in the corresponding digital filter.

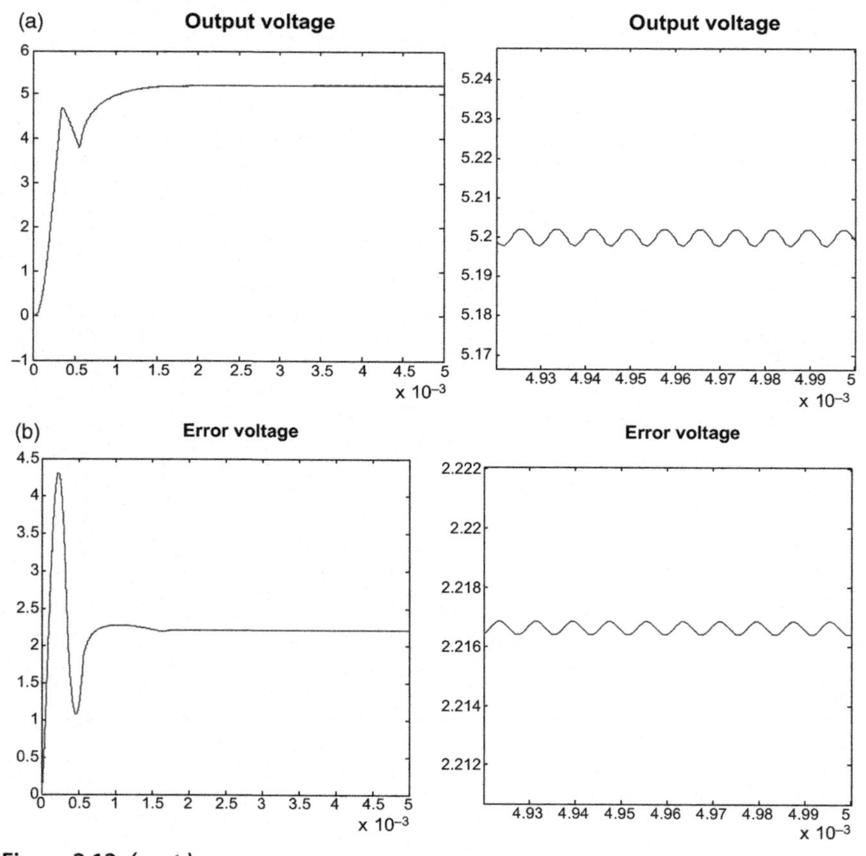

Figure 2.12 *(cont.)*

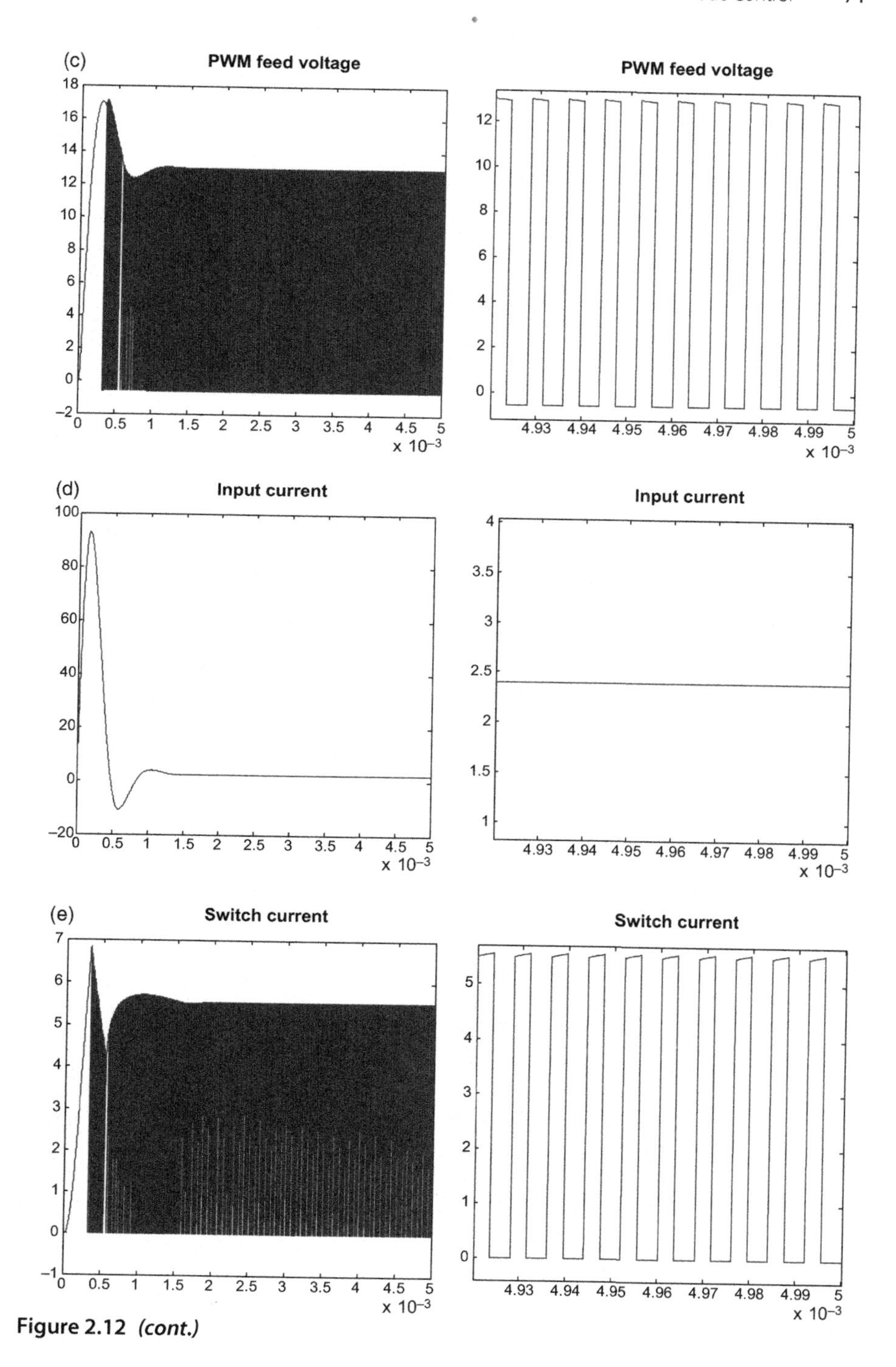

Figure 2.12 *(cont.)*

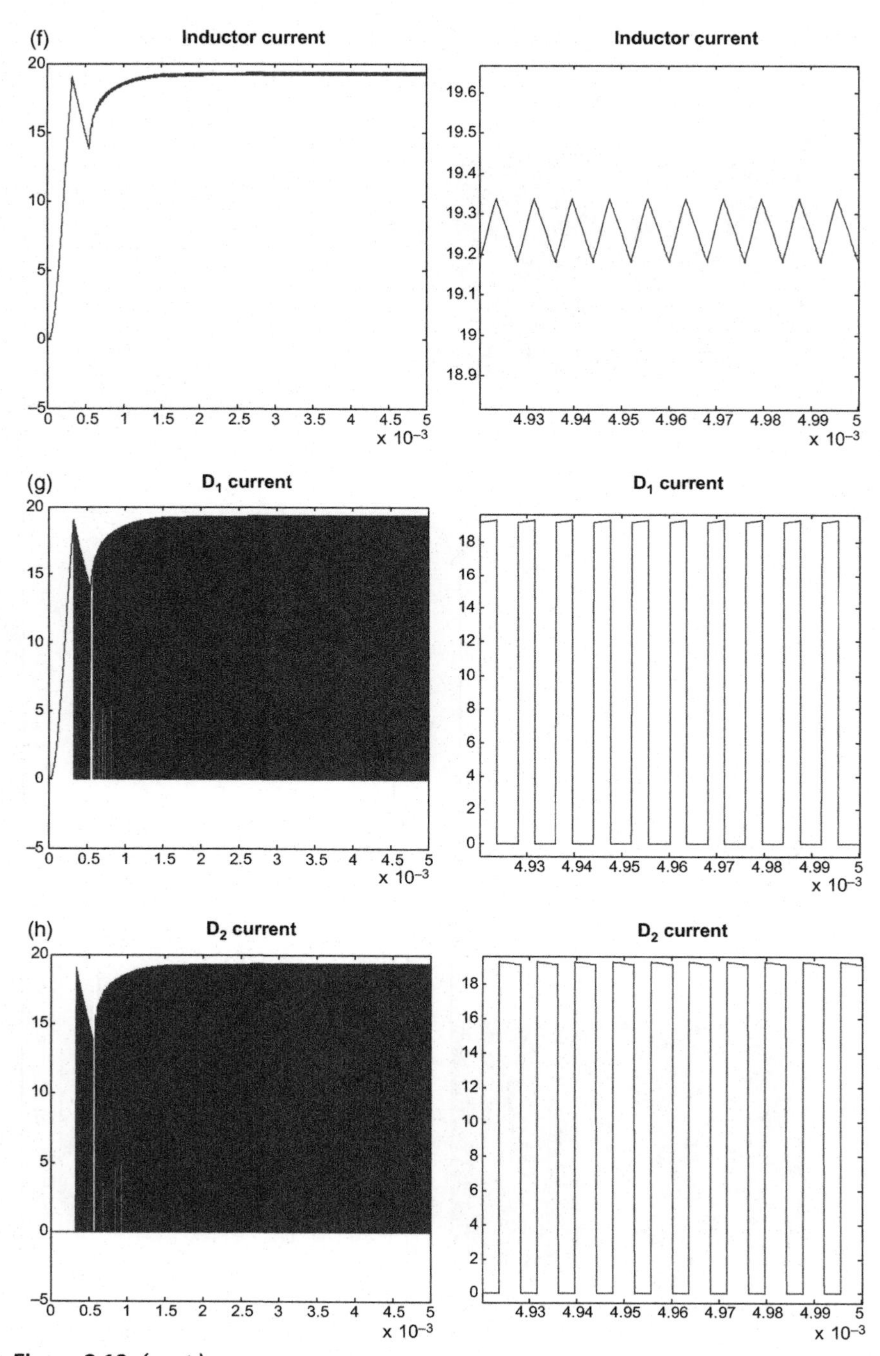

Figure 2.12 *(cont.)*

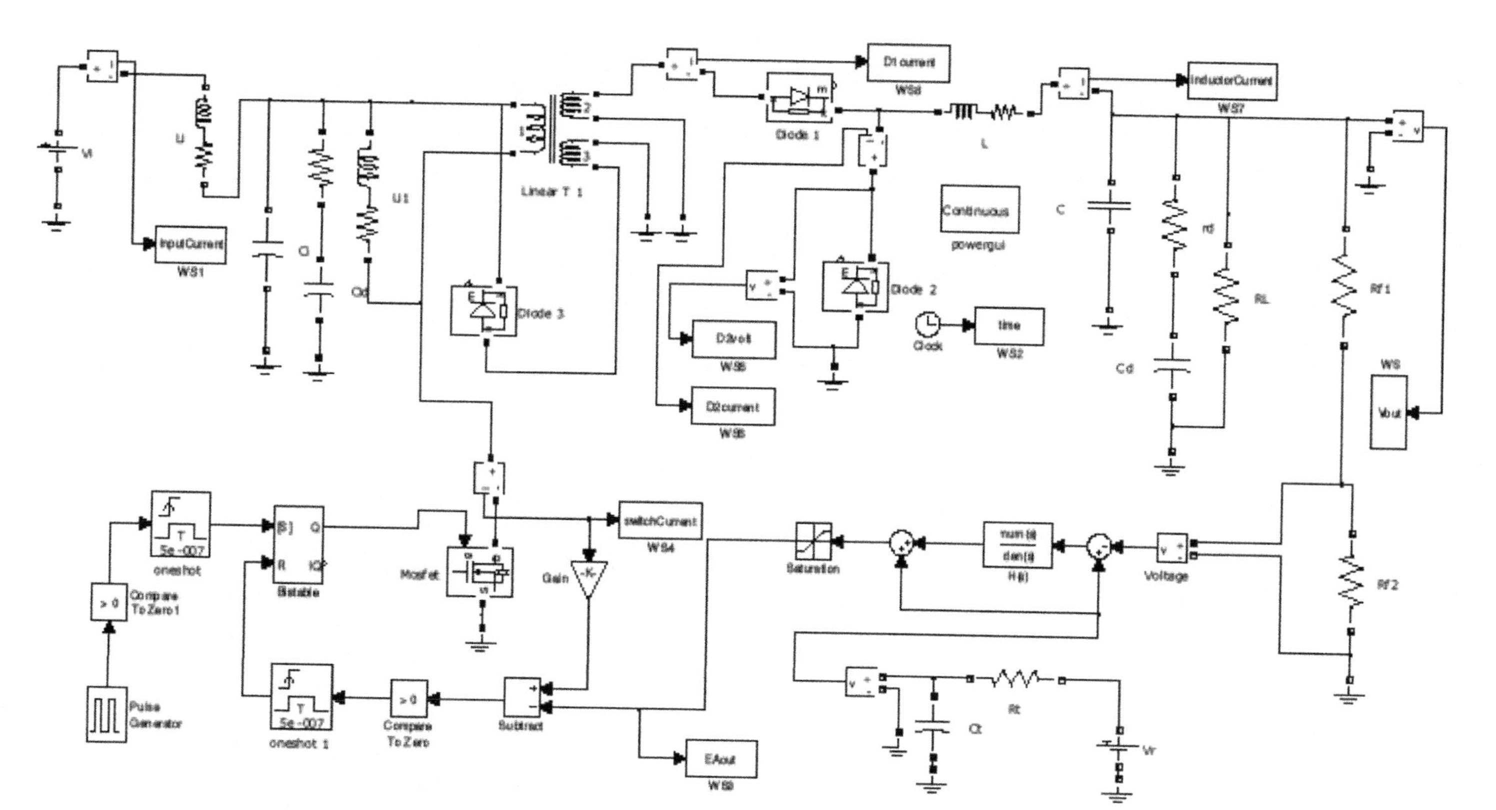

Figure 2.13 ***SIMULINK Model with Error Amp in H(s) Form.*** (a) Output voltage. (b) error voltage. (c) D_1 and D_2 cathode. (d) input current. (e) switch current. (f) inductor current. (g) D_1 current. (h) D_2 current.

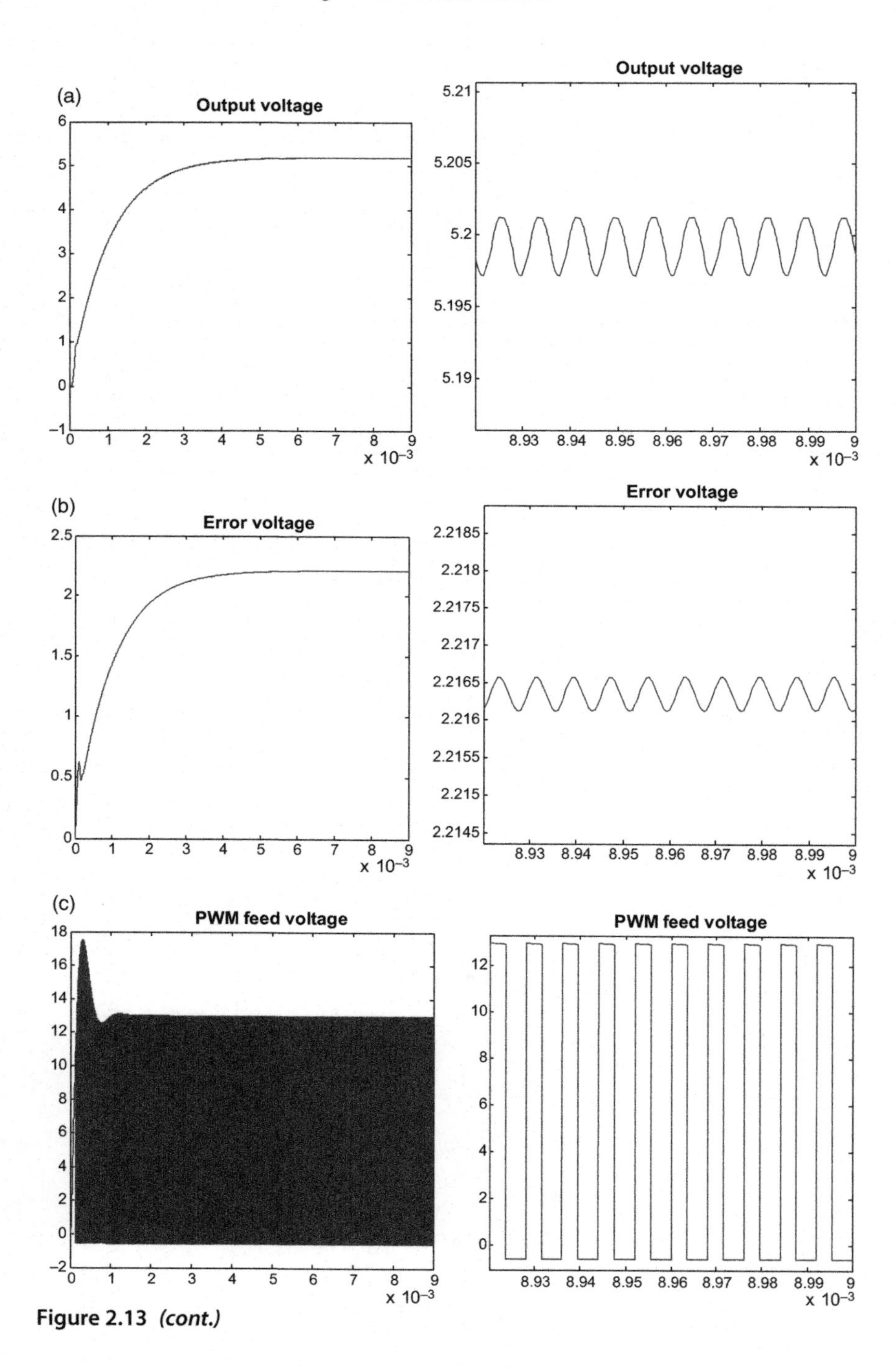

Figure 2.13 ***(cont.)***

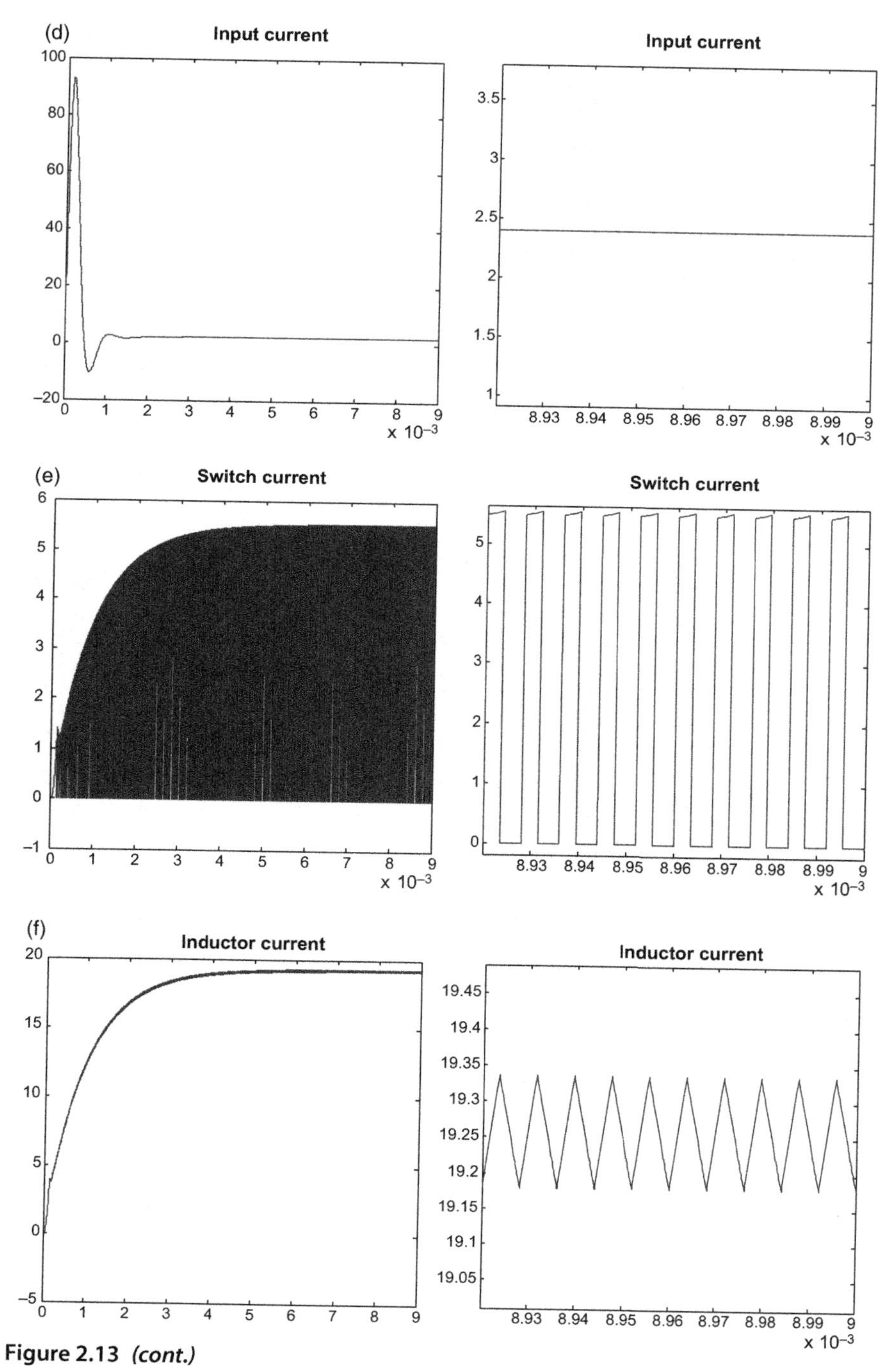

Figure 2.13 ***(cont.)***

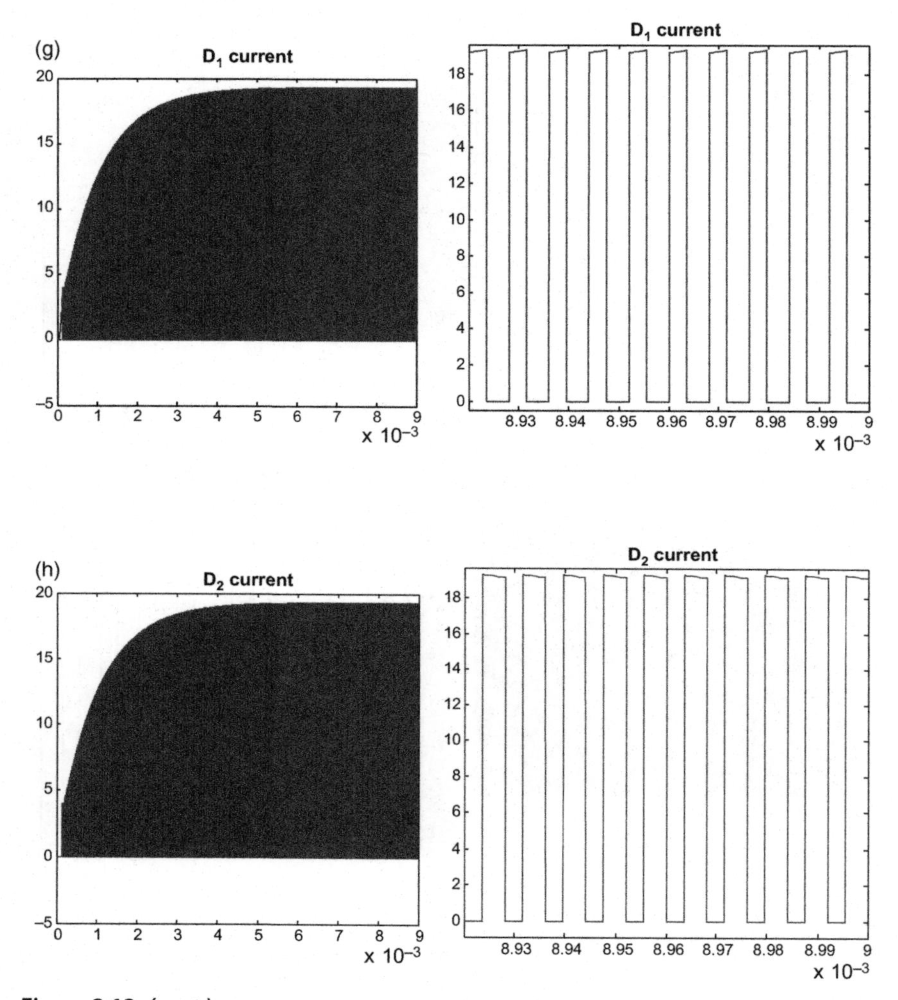

Figure 2.13 ***(cont.)***

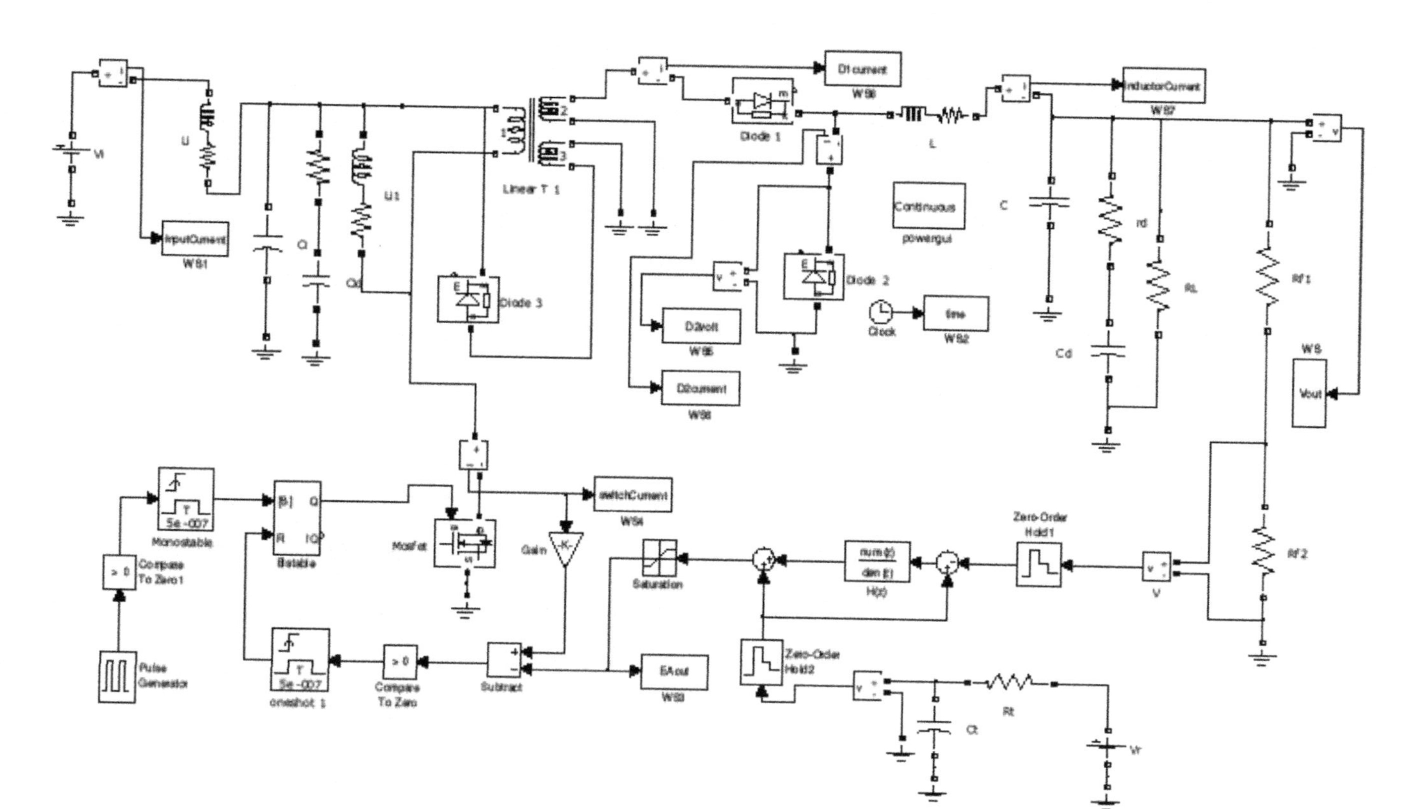

Figure 2.14 ***SIMULINK Model with Error Amp in Hz Form.*** (a) Output voltage, (b) error voltage, (c) D_1 and D_2 cathode, (d) input current. (e) switch current, (f) inductor current, (g) D_1 current, (h) D_2 current.

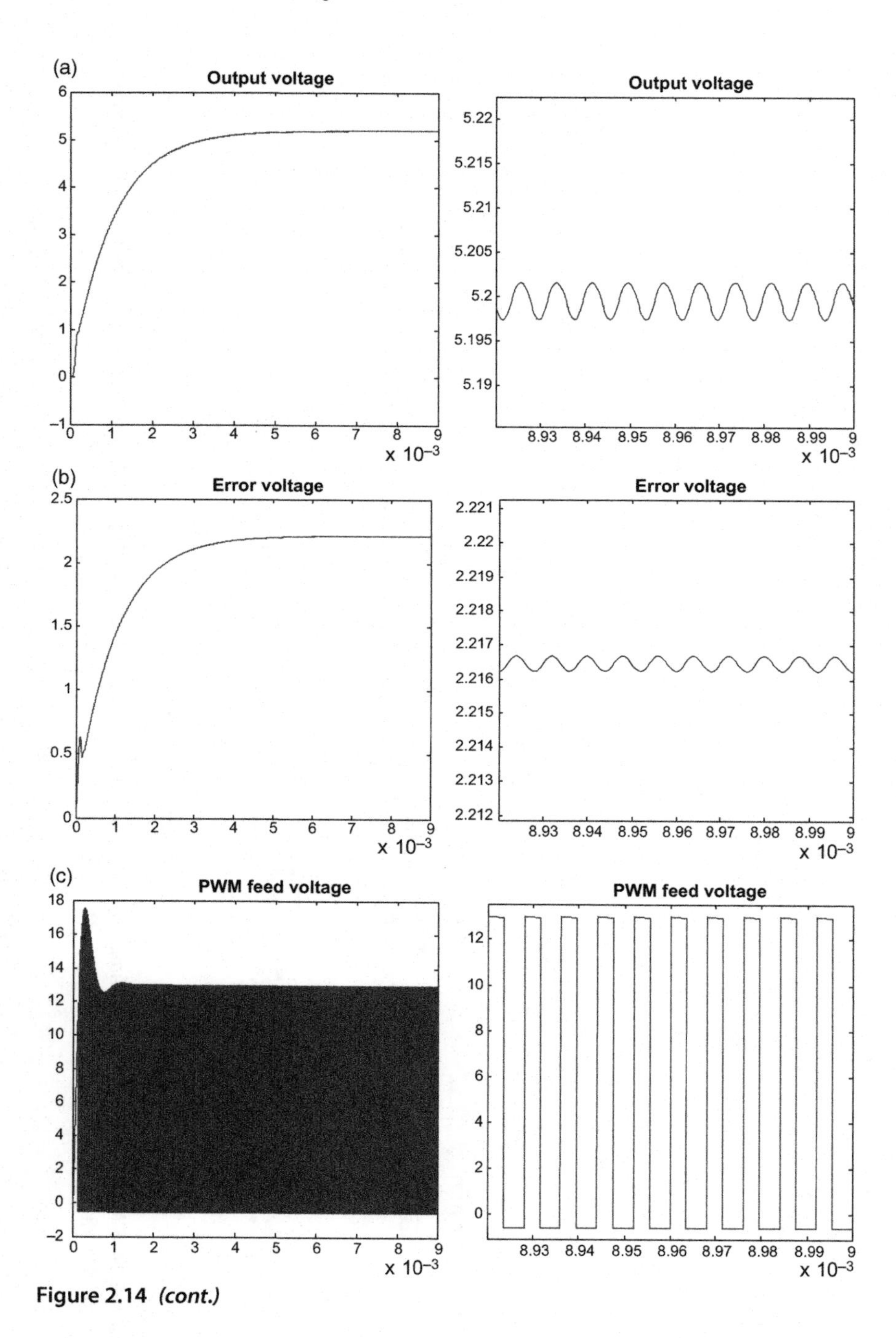

Figure 2.14 *(cont.)*

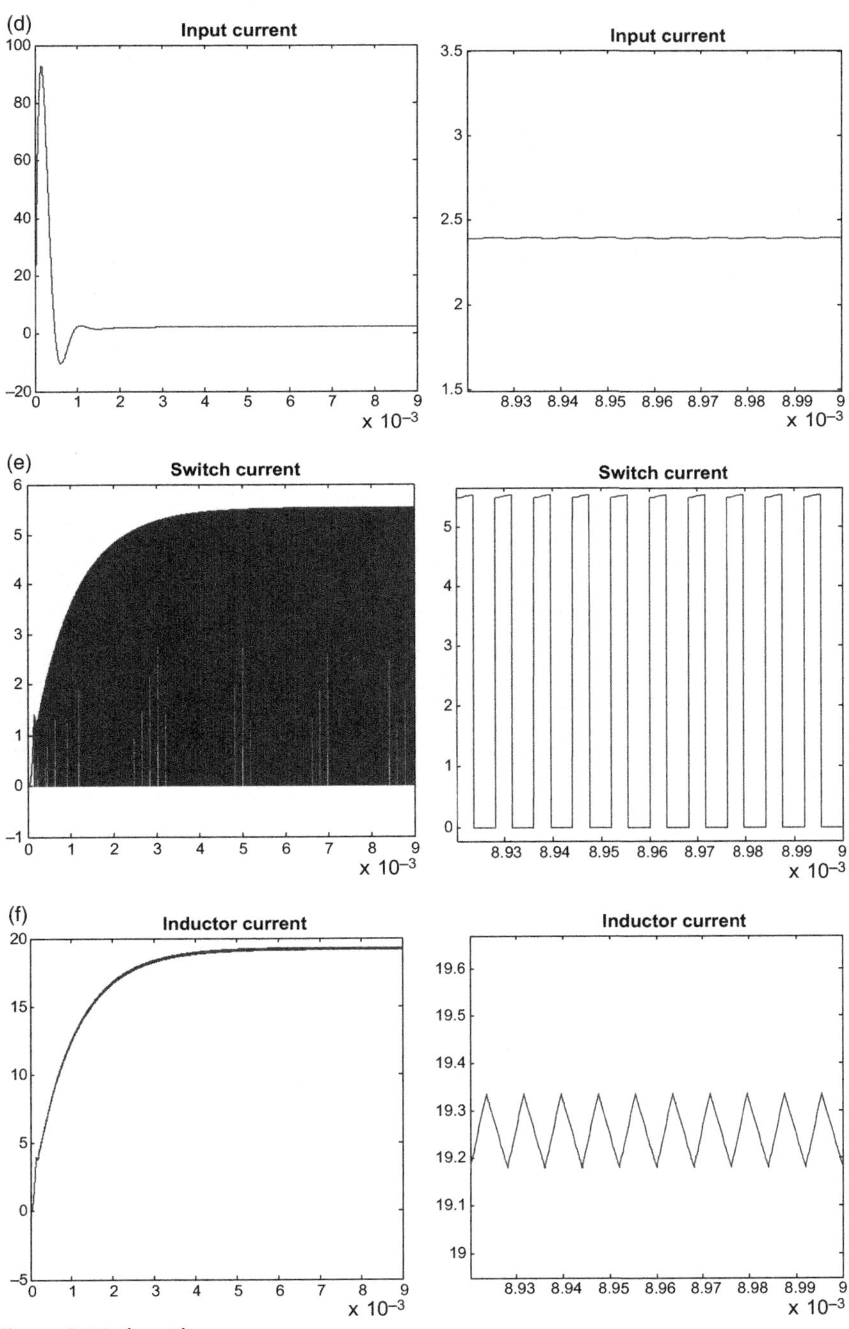

Figure 2.14 ***(cont.)***

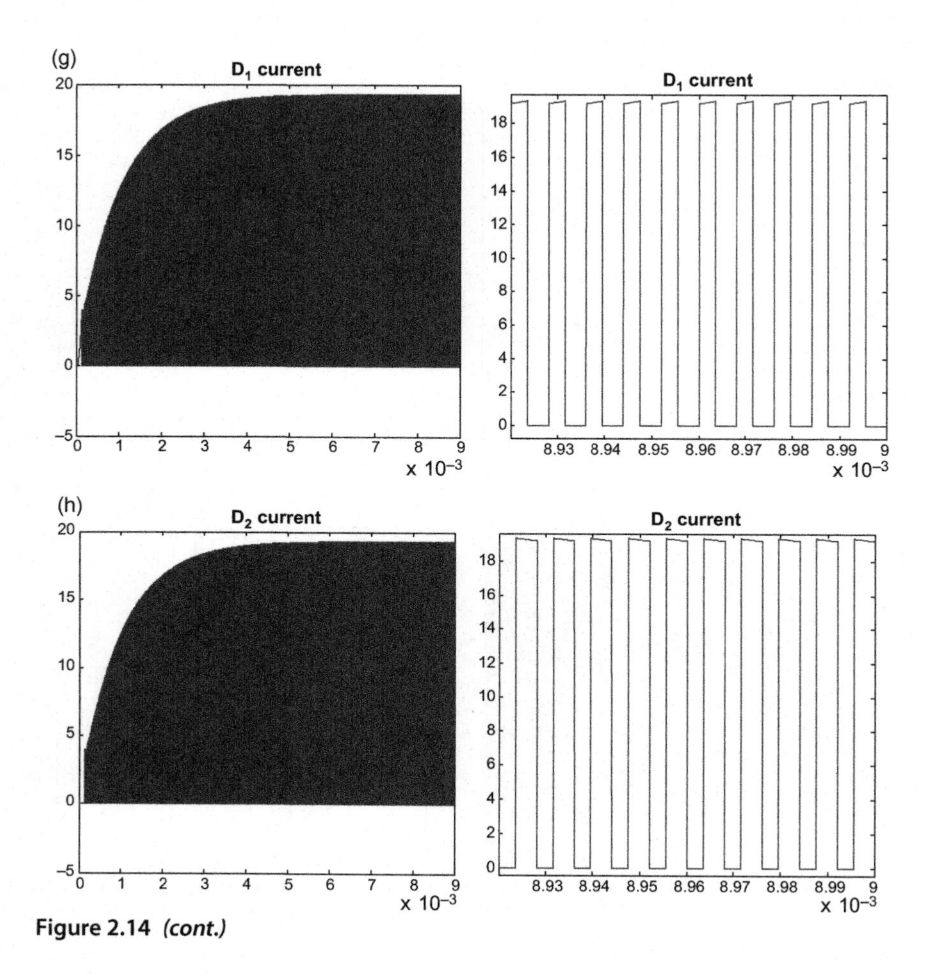

Figure 2.14 ***(cont.)***

PART II

Flyback Converter

The mechanisms of energy transfer employed by the forward converter presented in Part I and by the flyback converter in this part are quite different.

Forward converter transfers energy directly from the input to the output load with minimum storage in between. In Figure 1.1 and Figure 2.1, neither the transformer nor the output filter serves as storage device.

By contrast, the flyback converter transfers energy sort of indirectly by storing a bucket in one-time step then dumping it (to the load) in the other. And the cycle repeats. The storing act is instituted via an inductor.

Along the same coverage sequence, Part II treats both voltage-mode and current-mode control. In theory, flyback converters can work in two conduction modes, continuous and discontinuous, attributed to the initial current of inductor storage. When it initiates from zero, reaches a peak, and ends in zero, discontinuous conduction mode (DCM) is in operation. This is often the prevailing design for most flyback power converters. In the continuous conduction mode (CCM), the storage inductor's current never runs dry. However, in CCM, the issue of right-half-plane zero exists and makes control loop stability an issue that requires special treatment. For that reason, flyback converters operating in CCM are not covered in this part.

CHAPTER 3

Flyback Converter with Voltage-Mode Control

3.1 DESIGN OF DCM POWER STAGE

A typical flyback converter's power stage is shown in Figure 3.1. If multiple outputs are required, the secondary side circuits are duplicated as needed.

In DCM operation, the primary and secondary windings exhibit the following current waveforms, Figure 3.2, in which the primary current I_p starts from zero, ends at a peak; then the secondary starts at a peak and ends at zero. In effect, a complete energy transfer cycle is performed and repeated. This input to output coupling mechanism offers one very important property. The output receives power injection when the input is decoupled. An output overload is isolated from the input line.

Given an input range, V_{in_min}, V_{in}, V_{in_max}, and a specified output requirement, V_o at I_o, the design procedure begins by selecting first a desired switching frequency, f_s, and $T = 1/f_s$. The secondary inductance required to support the specified output is approximately equal to

$$L_s = \frac{\left(V_o + V_{CR}\right) D_2^2 T}{2I_o}, \quad I_s = \frac{2I_o}{D_2} \tag{3.1}$$

in which, at the discretion of the designer, duty cycle parameters D_1 and D_2 are preselected and partitioned such that $0 < D_1 + D_2 < 1$. The rectifier drop is also accounted for. What (3.1) gives is a general form. It is applicable for other output winding, if more than one is needed. However, the approximation contains a very high percentage error, about 15–20%. A more accurate, highly analytical approach is given in Appendix D.

The determination of the primary inductance is less straightforward. The primary current profile is described by

$$i_p(t) = \frac{V_{in}}{r_p}\left(1 - e^{-\left[t/(L_p/r_p)\right]}\right) \tag{3.2}$$

in which the primary winding resistance and switch on resistance are included in r_p. At the transition boundary $t = D_1 T$, ampere-turn conservation,

http://dx.doi.org/10.1016/B978-0-12-804298-4.00003-4

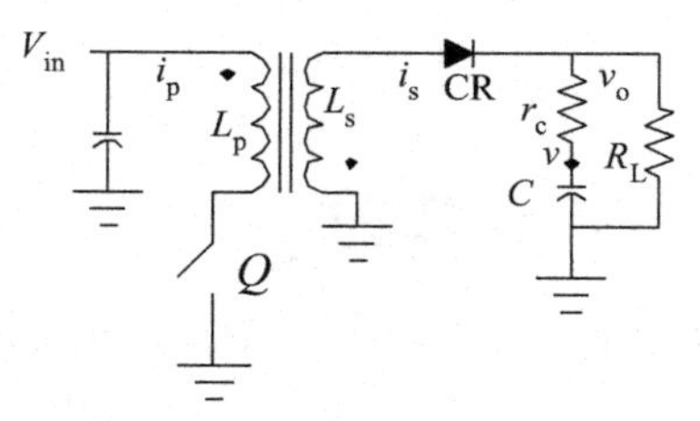

Figure 3.1 ***Power Stage of a Flyback Converter.***

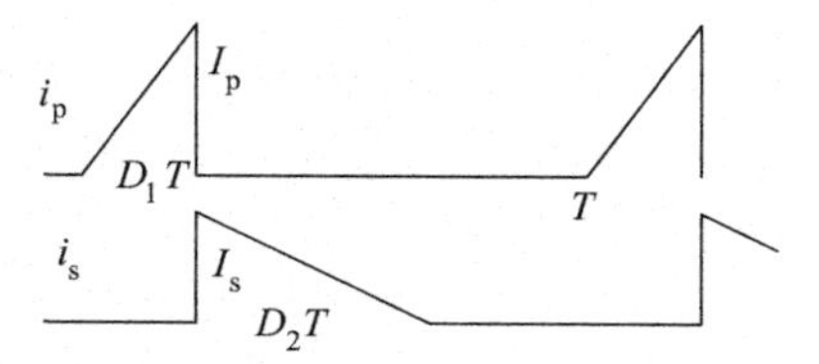

Figure 3.2 ***DCM Current Waveform.***

or continuity, must hold. In order to make the point more clearly, the case of two outputs is presented here. Ampere-turn continuity demands

$$\sqrt{\frac{L_{s1}}{L_p}}\frac{2I_{o1}}{D_2}+\sqrt{\frac{L_{s2}}{L_p}}\frac{2I_{o2}}{D_2}=\frac{V_{in}}{r_p}\left(1-e^{-\left[D_1T/(L_p/r_p)\right]}\right) \tag{3.3}$$

Both L_{s1} and L_{s2} are selected according to (3.1) and a chosen D_2. Using a good mathematical software, (3.3) enables a designer to find and size the primary inductance L_p. An important note shall be attached here. The turn-ratio of windings is expressed in squared root of inductance ratio, since inductance is proportional to the square of turn number. Designers shall also take note that the size of all inductance is a strong function of the switching frequency. It shall be selected on the basis of reasonable inductance size. It is also feasible to have different D_2 for both outputs. In other words, D_{2a} for output one and D_{2b} for output two. Because of the approximation nature of (3.1), (3.3) also contains the same error percentage-wise.

The output circuit traverses three phases. During D_1T, the output capacitor discharges and supports the load alone. During D_2T, the stored energy in L_s replenishes the output capacitor and supports the load. Then, during the rest of a switch period, the output capacitor again discharges to the load. In phase one, the output capacitor voltage and the output are governed by:

$$\begin{aligned} v_{C1}(t) &= V_{o1}\,e^{-(t/\tau_c)}, \quad v_{o1} = k_R v_{C1}(t) \\ \tau_C &= (r_C + R_L)C, \quad k_R = \frac{R_L}{r_C + R_L} \end{aligned} \tag{3.4}$$

in phase two by:

$$
\begin{aligned}
R_{\mathrm{p}} &= \frac{r_{\mathrm{C}} R_{\mathrm{L}}}{r_{\mathrm{C}} + R_{\mathrm{L}}}, \tau_{\mathrm{S}} = \frac{L_{\mathrm{s}}}{r_{\mathrm{CR}} + R_{\mathrm{p}}}, D(s) = s^2 + \left(\tfrac{1}{\tau_{\mathrm{C}}} + \tfrac{1}{\tau_{\mathrm{S}}}\right)s + \tfrac{k_{\mathrm{R}} R_{\mathrm{L}}}{L_{\mathrm{S}} \tau_{\mathrm{C}}} + \tfrac{1}{\tau_{\mathrm{S}} \tau_{\mathrm{C}}} \\
f_1(t) &= \ell^{-1}\left[\tfrac{s+(1/\tau_{\mathrm{c}})}{D(s)}\right], \quad f_2(t) = \ell^{-1}\left[\tfrac{k_{\mathrm{R}}}{L_{\mathrm{s}} D(s)}\right] \\
i_{\mathrm{s}}(t) &= I_{\mathrm{s}} f_1(t - D_1 T) - V_{\mathrm{o}2} f_2(t - D_1 T) \\
f_3(t) &= \ell^{-1}\left[\tfrac{R_{\mathrm{L}}}{\tau_{\mathrm{C}} D(s)}\right], \quad f_4(t) = \ell^{-1}\left[\tfrac{s+(1/\tau_{\mathrm{s}})}{D(s)}\right] \\
v_{\mathrm{C}2}(t) &= I_{\mathrm{s}} f_3(t - D_1 T) + V_{\mathrm{o}2} f_4(t - D_1 T) \\
v_{\mathrm{o}2}(t) &= R_{\mathrm{p}} i_{\mathrm{s}}(t) + k_{\mathrm{R}} v_{\mathrm{C}2}(t)
\end{aligned}
\tag{3.5}
$$

Noted, inverse Laplace transform, l^{-1}, is invoked.

And in phase three by:

$$
v_{\mathrm{C}3}(t) = V_{\mathrm{o}3}\, \mathrm{e}^{-\left[(t-(D_1+D_2)T)/\tau_{\mathrm{c}}\right]}, \quad v_{\mathrm{o}3} = k_{\mathrm{R}} v_{\mathrm{C}3}(t) \tag{3.6}
$$

Three cyclic starting states, $V_{\mathrm{o}1}$, $V_{\mathrm{o}2}$, and $V_{\mathrm{o}3}$, are linked and determined by the physical law, continuity of state. It dictates that $v_{\mathrm{c}1}(D_1 T) = V_{\mathrm{o}2}$, $v_{\mathrm{c}2}(D_2 T) = V_{\mathrm{o}3}$, and $v_{\mathrm{c}3}(T) = V_{\mathrm{o}1}$. These three wrap-around equalities lead eventually to an important fact: the starting state is actually a function of I_{s}, in (3.1), and four time functions $f_1(t)$ to $f_4(t)$. Appendix D gives details.

3.2 MODULATOR GAIN

With the power stage designed and waveforms formulated, it can now be incorporated with the feedback loop, input filter, PWM block, and switch driver. It results in Figure 3.3, a single output converter, solely for the purpose of showing all basic blocks constituting a flyback topology.

In order to evaluate the control loop gain, Figure 3.3 shall be placed in its corresponding small-signal form, block by block. The feedback factor, K_{f}, is a simple resistive voltage division. The PWM gain remains the same as that explained in the paragraph following (1.10). The error amplifier is yet to be determined and represented only in symbol EA(s). To represent the power train with isolation and line filter, the DCM state-space average model, [1] and [2], is invoked and modified. Figure 3.4 stands for the equivalent model.

There are four dependent current generators: two modulated by duty cycle, the other two by input and output variation. *R* and *C* are not equal to any single or multiple physical elements. Both represent the effective resistive and capacitive loading the power stage experiences. One more step is carried out to absorb the isolation. When it is done, Figure 3.4 is reduced to Figure 3.5.

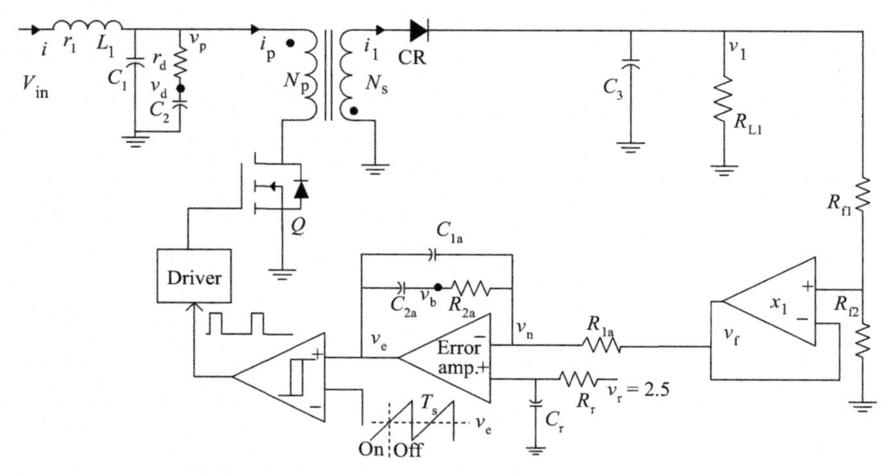

Figure 3.3 ***Flyback Converter with Voltage-Mode Control.***

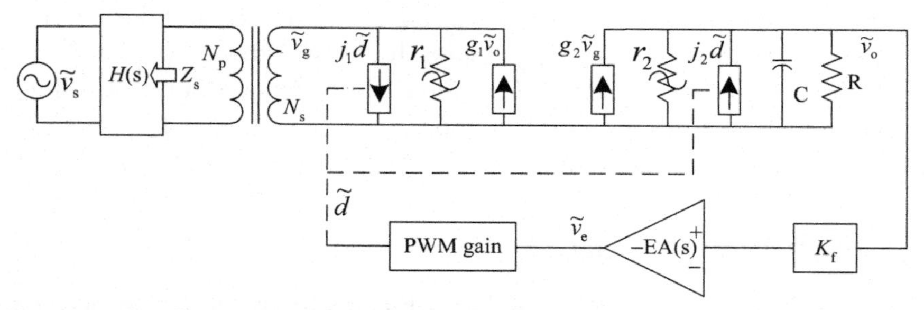

Figure 3.4 ***Small-Signal Model for Flyback Converter in DCM.***

We shall derive the power stage gains, duty cycle-to-output $G_{vd}(s)$ and source-to-output $G_{vs}(s)$. There are two nodes, v_g and v_o. Let $n = N_s/N_p$, two nodes give

$$\begin{aligned}\left(\frac{1}{r_1}+\frac{1}{n^2 Z_s}\right)\tilde{v}_g - g_1\tilde{v}_o &= \frac{H(s)}{nZ_s}\tilde{v}_s - j_1\tilde{d}\\ g_2\tilde{v}_g - \left(\frac{1}{r_2}+\frac{1}{R}+Cs\right)\tilde{v}_o &= -j_2\tilde{d}\end{aligned} \tag{3.7}$$

Two equations in two unknowns give the output

$$\tilde{v}_o = \frac{\left[j_1 g_2 - j_2\left((1/r_1)+(1/n^2 Z_s)\right)\right]\tilde{d} - g_2(H(s)/nZ_s)\tilde{v}_s}{g_1 g_2 - \left((1/r_1)+(1/n^2 Z_s)\right)\left((1/r_2)+(1/R)+Cs\right)} \tag{3.8}$$

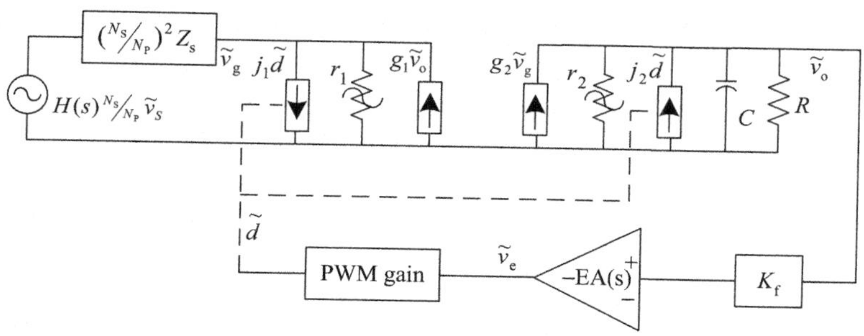

Figure 3.5 ***Flyback Converter in DCM with Isolation and Line Filter Reflected to Secondary Side.***

(3.8) yields duty cycle-to-output and source-to-output gains

$$
\begin{aligned}
G_{vd}(s) &= \frac{j_1 g_2 - j_2\left((1/r_1) + (1/n^2 Z_s)\right)}{g_1 g_2 - \left((1/r_1) + (1/n^2 Z_s)\right)\left((1/r_2) + (1/R) + Cs\right)} \\
G_{vs}(s) &= \frac{-g_2 (H(s)/nZ_s)}{g_1 g_2 - \left((1/r_1) + (1/n^2 Z_s)\right)\left((1/r_2) + (1/R) + Cs\right)}
\end{aligned}
\tag{3.9}
$$

The modulator gain is therefore given by

$$M(s) = K_f F_m G_{vd}(s) \tag{3.10}$$

In form, it looks like (1.11). Yes, K_f and F_m remain almost unchanged. Yet $G_{vd}(s)$ is utterly different. In addition, dependent current generator g_1 was determined to be zero for flyback converter in DCM. Readers shall check out the fact in Ref. [1, p. 152, 155].

Anyway, once the modulator gain is known, the determination process leading to the proper identification of error amplifier and its transformation to digital are the same as presented in the previous chapters. Working out an example shall serve us better.

3.3 EXAMPLE – ONE OUTPUT

A converter is fed by an input of 24V. One output, 5V at 10 A is expected. Based on the definition of Figure 3.1 and by choice, $D_1 = 0.4$ and $D_2 = 0.5$ are assigned. Approach given in Appendix D yields $L_{s1} = 3.67$ μH and $L_p = 33.5$ μH at a switching frequency of $f_s = 20$ kHz selected to allow reasonable inductance value. State-space averaged canonical model parameters depicted in Figure 3.4 are evaluated in sequence according to [1,2]. Extra attention shall be paid to the total equivalent R and C elements. In the original average model, often also named Middlebrook model, these two elements represent only a single output. In this example, $R = 5/10$ and $C = 2{,}000$ μF. For multiple outputs, for instance, two outputs in the second example, R and C elements must account for loading attributed to both outputs. In the process of doing that, turn ratio or square root of inductance ratio shall be included.

The input line filter design takes a little bit of explanation. In the last paragraph, the switching frequency was set at 20 kHz. It is desirable to make the line filter bandwidth less than 1 kHz. 750 Hz is selected. Referring to Figure 3.3 input port, we first preselect $C_1 = 200$ μF. This yields $L_1 = 225$ μH, and, in order to avoid filter peaking in frequency domain, damping resistor $r_d = 0.75$ results and $C_2 = 1{,}000$ μF [4]. Once the line filter is done, (3.9) is ready. By design, the PWM gain equals 0.98/4; that is, 1–5 V sawtooth swing corresponding to 98% duty ratio, and the feedback factor is simply equal to 0.5, given 2.5 V reference voltage and 5 V output. (3.10) yields a plot for modulator gain and phase, Figure 3.6.

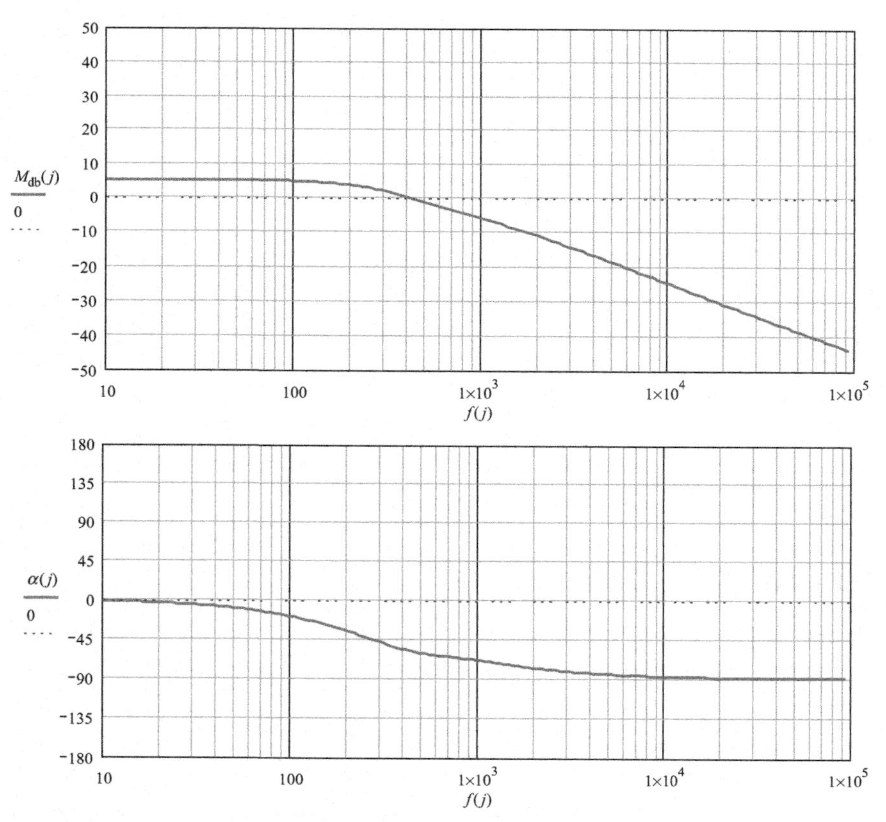

Figure 3.6 ***Modulator of a Voltage-Mode Flyback Converter in DCM.***

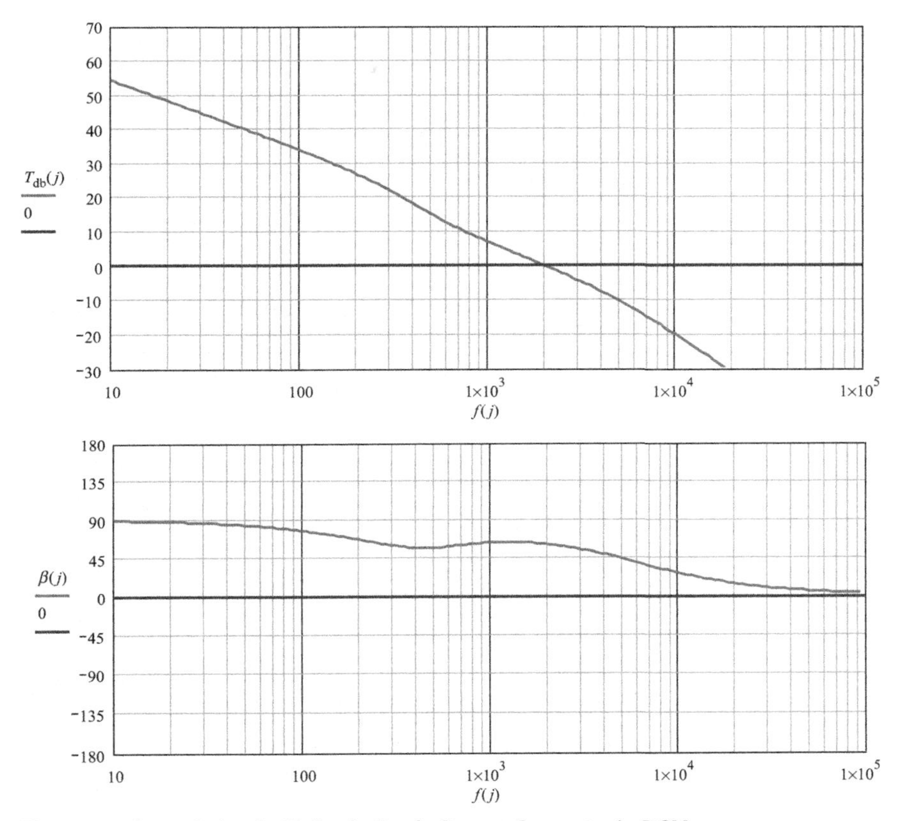

Figure 3.7 ***Loop Gain of a Flyback, Single Output Converter in DCM.***

A 2 kHz crossover frequency is selected. At the frequency, $M_{db} = -11$ and $\alpha_M = -77°$. For 60° phase margin, (1.12) indicates that a phase boost of 47° is required. It means a type-III amplifier is not needed. Instead, a type-II is sufficient to do the job. (1.13) and (1.14) then figure out the pole-zero separation factor, $k = 2.539$, and, subsequently, the amplifier's components, referring to Figure 3.3, $R_{1a} = 2K$, $C_{1a} = 0.0044\ \mu F$, $C_{2a} = 0.024\ \mu F$, and $R_{2a} = 8.4K$. Incorporating the amplifier leads to the loop gain plot Figure 3.7. As expected, the loop crosses at 2 kHz.

3.4 SIMULATION AND PERFORMANCE VERIFICATION – ONE OUTPUT

Once again, the design finished in the previous section shall be verified. Similar to what was done in Chapters 1 and 2, Figure 3.3 circuit gives us the following set of difference equation (3.11), in the order of input line current through L_1, primary winding input voltage v_p, input damping capacitor voltage v_d, primary winding/main switch current i_p, output winding current i_1, output capacitor voltage v_3, error amplifier output v_e, and error amplifier local feedback v_b.

$$\begin{pmatrix} i_{j+1} \\ v_{p_{j+1}} \\ v_{d_{j+1}} \\ i_{p_{j+1}} \\ i_{1_{j+1}} \\ v_{3_{j+1}} \\ v_{e_{j+1}} \\ v_{b_{j+1}} \end{pmatrix} = \left[\begin{array}{c} \left(1-\frac{\delta t}{L_1}\cdot r_1\right) i_j + \frac{\delta t}{L_1}\left(V_{in} - v_{p_j}\right) \\ \left[1-\frac{\delta t}{\left(C_1 r_d\right)}\right] v_{p_j} + \frac{\delta t}{C_1}\left(i_j - i_{p_j} + \frac{1}{r_d} v_{d_j}\right) \\ \left[1-\frac{\delta t}{\left(r_d C_2\right)}\right] v_{d_j} + \frac{\delta t}{\left(r_d C_2\right)} v_{p_j} \\ \text{if}\left[i_{p_j} < 0, 0, \text{if}\left[v_{e_j} > s_{w_j}, \left[1-\frac{\delta t}{L_p}\left(r_p + R_{on}\right)\right] i_{p_j} + \frac{\delta t}{L_p} v_{p_j}, 0\right]\right] \\ \text{if}\left[i_{1_j} < 0, 0, \text{if}\left[\begin{array}{r} v_{e_j} > s_{w_j}, 0, \sqrt{\frac{L_p}{L_{s1}}} i_{p_j} + \left[1-\frac{\delta t}{L_{s1}}\left(r_{s1} + R_{p1}\right)\right] i_{1_j} \\ -\frac{\delta t}{L_{s1}}\left(V_{CR} + k_1 v_{3_j}\right) \end{array} \right]\right] \\ \left[1+\frac{\delta t}{\left(r_3 C_3\right)}\left(k_1 - 1\right)\right] v_{3_j} + \frac{\delta t}{\left(r_3 C_3\right)} R_{p1} i_{1_j} \\ \text{if}\left[v_{e_j} < 0, 0, \text{if}\left[v_{e_j} > 12, 12, v_{e_j} + \left[\begin{array}{r} \frac{\delta t}{C_{1a}}\left(\frac{1}{R_{1a}} + \frac{1}{R_{2a}}\right) VR_j - \frac{\delta t}{\left(R_{2a} C_{1a}\right)} v_{b_j} \\ -\frac{\delta t}{\left(R_{1a} C_{1a}\right)} k_f \left(R_{p1} i_{1_j} + k_1 v_{3_j}\right) \end{array} \right]\right]\right] \\ \begin{array}{r} \left[1-\left[\frac{1}{\left(R_{2a} C_{2a}\right)} + \frac{1}{\left(R_{2a} C_{1a}\right)}\right]\delta t\right] v_{b_j} + \left[\frac{1}{\left(R_{2a} C_{2a}\right)} + \frac{1}{C_{1a}}\left(\frac{1}{R_{1a}} + \frac{1}{R_{2a}}\right)\right]\delta t V_{R_j} \\ -\frac{\delta t}{\left(R_{1a} C_{1a}\right)} k_f \left(R_{p1} i_{1_j} + k_1 v_{3_j}\right) \end{array} \end{array} \right] \tag{3.11}$$

With a slow start that ramps up the 2.5 V reference and 5 ms circuit time that covers one hundred cycles at the switching frequency, the circuit output does settle to the steady state. Figure 3.8a–e show the end results in pair. Each pair gives both turn-on transient and steady-state ten cycles.

Next, the type-II error amplifier is replaced by the corresponding digital filter. The analog type-II shows −26 db at 400 kHz. Digital sampling frequency is set at twice that frequency, and bilinear transform constant $C = 1.6 \times 10^6$, twice the sampling frequency. It leads to the following $H_{II}(z)$, (1.19), polynomial coefficients

$$a_0 = 69.576 \times 10^{-3} \quad a_1 = 429.176 \times 10^{-6} \quad a_2 = -69.147 \times 10^{-3}$$
$$b_1 = -1.961 \times 10^0 \quad b_2 = 960.903 \times 10^{-3}$$

The digital filter is implemented and it changes the schematic to Figure 3.9.

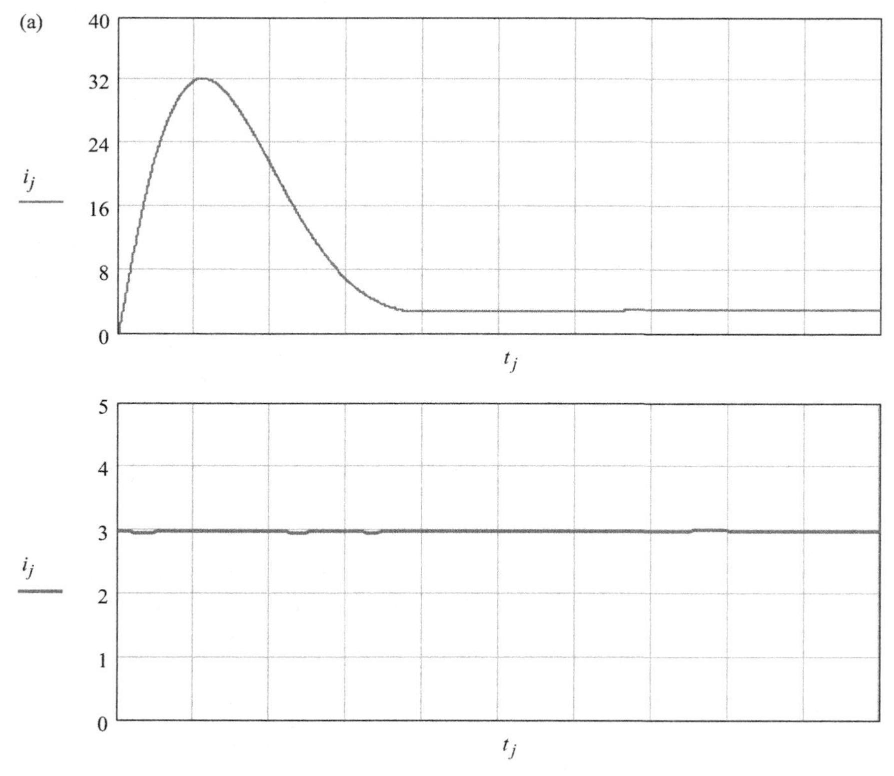

Figure 3.8 Figure 3.3, equation (3.11), time-domain (a) Input current; (b) primary winding/switch current; (c) error voltage and sawtooth; (d) output winding current; and (e) output voltage.

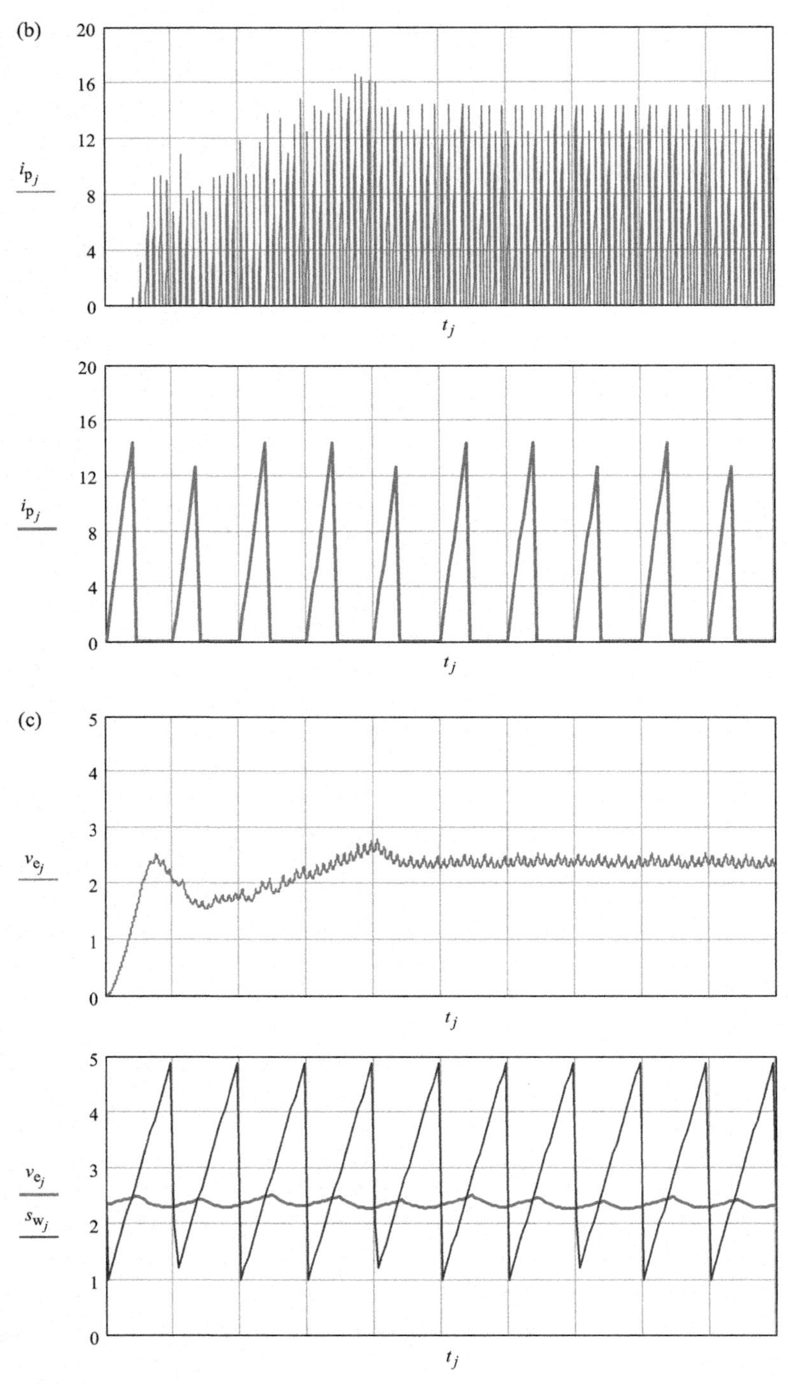

Figure 3.8 ***(cont.)***

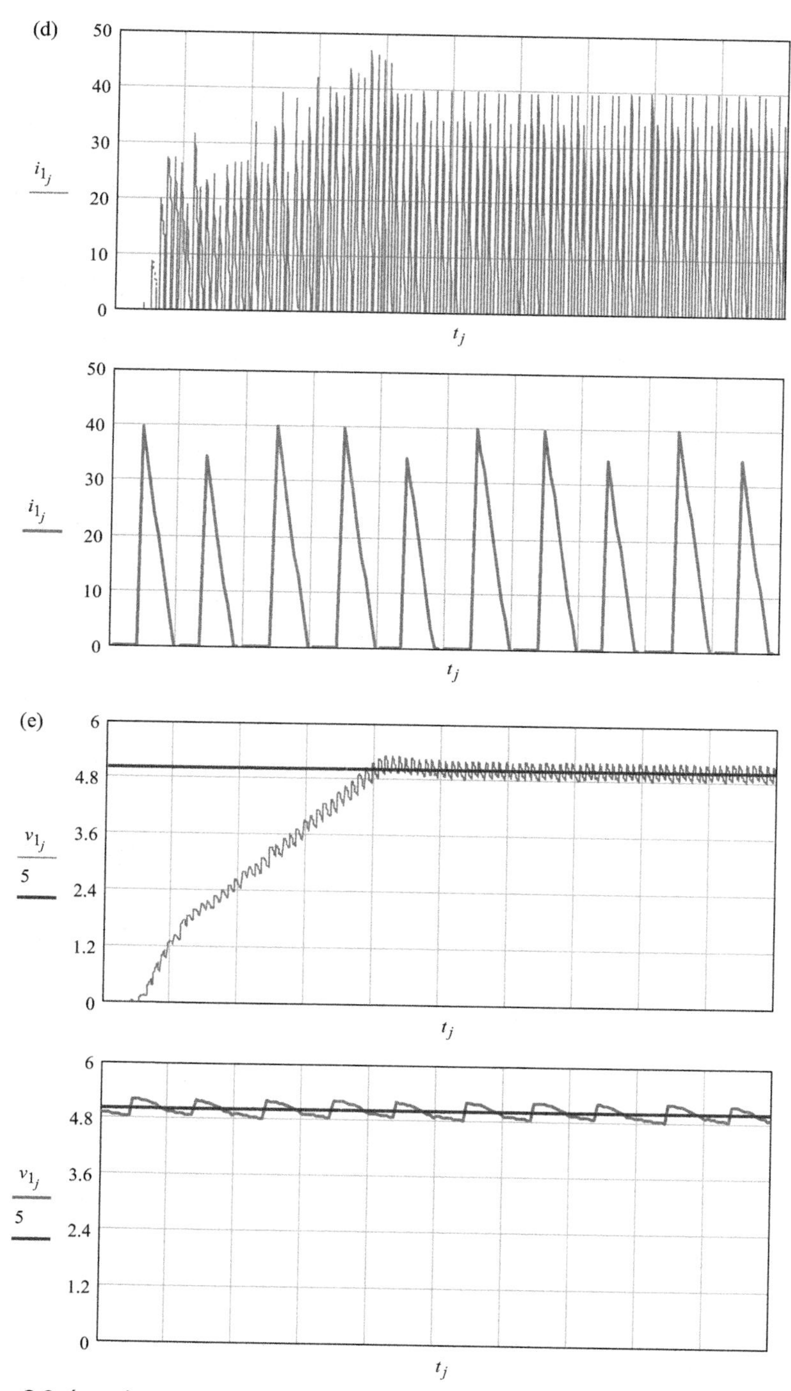

Figure 3.8 *(cont.)*

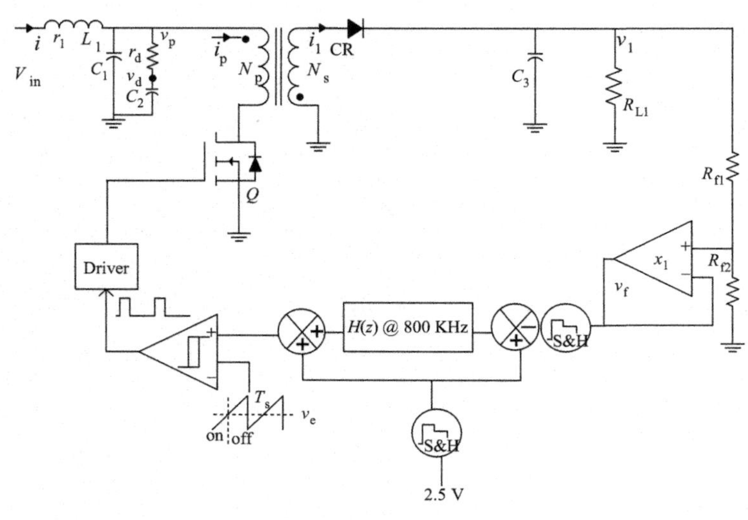

Figure 3.9 ***Flyback Converter in DCM with Digital Filter.***

The difference equation set for Figure 3.9 is easily derived by modifying the set, (3.11), for Figure 3.3; that is, replacing v_{ej} entry with y_j and adding intermediate digital variable w_j. The procedure results in equation set (3.12) and performance Figure 3.10.

$$
\begin{pmatrix} i_{j+1} \\ v_{p_{j+1}} \\ v_{d_{j+1}} \\ i_{p_{j+1}} \\ i_{1_{j+1}} \\ v_{3_{j+1}} \\ y_{j+1} \\ w_{j+1} \end{pmatrix}
=
\begin{bmatrix}
\left(1-\dfrac{\delta t}{L_1}\times r_1\right)i_j+\dfrac{\delta t}{L_1}\left(V_{in}-v_{p_j}\right) \\
\left[1-\dfrac{\delta t}{(C_1 r_d)}\right]v_{p_j}+\dfrac{\delta t}{C_1}\left(i_j-i_{p_j}+\dfrac{1}{r_d}v_{d_j}\right) \\
\left[1-\dfrac{\delta t}{(r_d C_2)}\right]v_{d_j}+\dfrac{\delta t}{(r_d C_2)}v_{p_j} \\
\text{if}\left[i_{p_j}<0,0,\text{if}\left[y_j>s_{w_j},\left[1-\dfrac{\delta t}{L_p}\left(r_p+R_{on}\right)\right]i_{p_j}+\dfrac{\delta t}{L_p}v_{p_j},0\right]\right] \\
\text{if}\left[i_{1_j}<0,0,\text{if}\left[y_j>s_{w_j},0,\sqrt{\dfrac{L_p}{L_{s1}}}i_{p_j}+\left[1-\dfrac{\delta t}{L_{s1}}\left(r_{s1}+R_{p1}\right)\right]i_{1_j}-\dfrac{\delta t}{L_{s1}}\left(V_{CR}+k_1 v_{3_j}\right)\right]\right] \\
\left[1+\dfrac{\delta t}{(r_3 C_3)}(k_1-1)\right]v_{3_j}+\dfrac{\delta t}{(r_3 C_3)}R_{p1}i_{1_j} \\
\text{if}\left[\begin{array}{l} y_j<0,0,\text{if}\left[\begin{array}{r} y_j>12,12,v_{r_j}+a^0\left[v_{r_j}-k_f\left(R_{p1}i_{1_j}+k_1 v_{3_j}\right)\right]-b1w_j-b2w_{j-1} \\ +a1w_j+a2w_{j-1}\end{array}\right]\end{array}\right] \\
\left[v_{r_j}-k_f\left(R_{p1}i_{1_j}+k_1 v_{3_j}\right)\right]-b1w_j-b2w_{j-1}
\end{bmatrix}
\tag{3.12}
$$

We shall also invoke MATLAB SIMULINK to reaffirm the converter performance. Figure 3.11 gives the simulation drawing with analog amplifier and results of a 5 ms real-time coverage.

Next, the simulation drawing with digital filter and results of a 5 ms real-time coverage is presented, Figure 3.12.

3.5 EXAMPLE – TWO OUTPUTS

Flyback converters serve well inexpensive applications in which low power and multiple outputs are needed, since a single storage inductor with multiple output windings can easily support the need. However, applications of this nature run into one major bottleneck. That is, given multiple outputs, only one, in general the one with the heaviest load, serves as the control loop's feedback source. As a result, only the main output selected as the

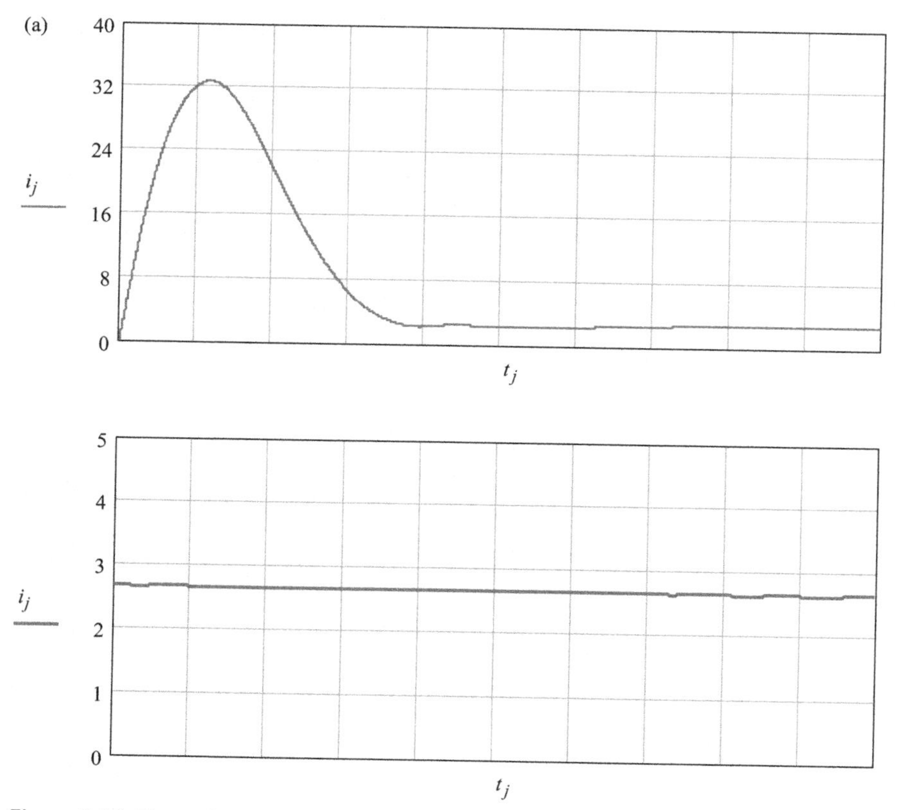

Figure 3.10 Figure 3.9, equation set (3.12), time-domain. (a) Input current; (b) primary winding/switch current; (c) error voltage and sawtooth; (d) secondary winding current; and (e) output voltage.

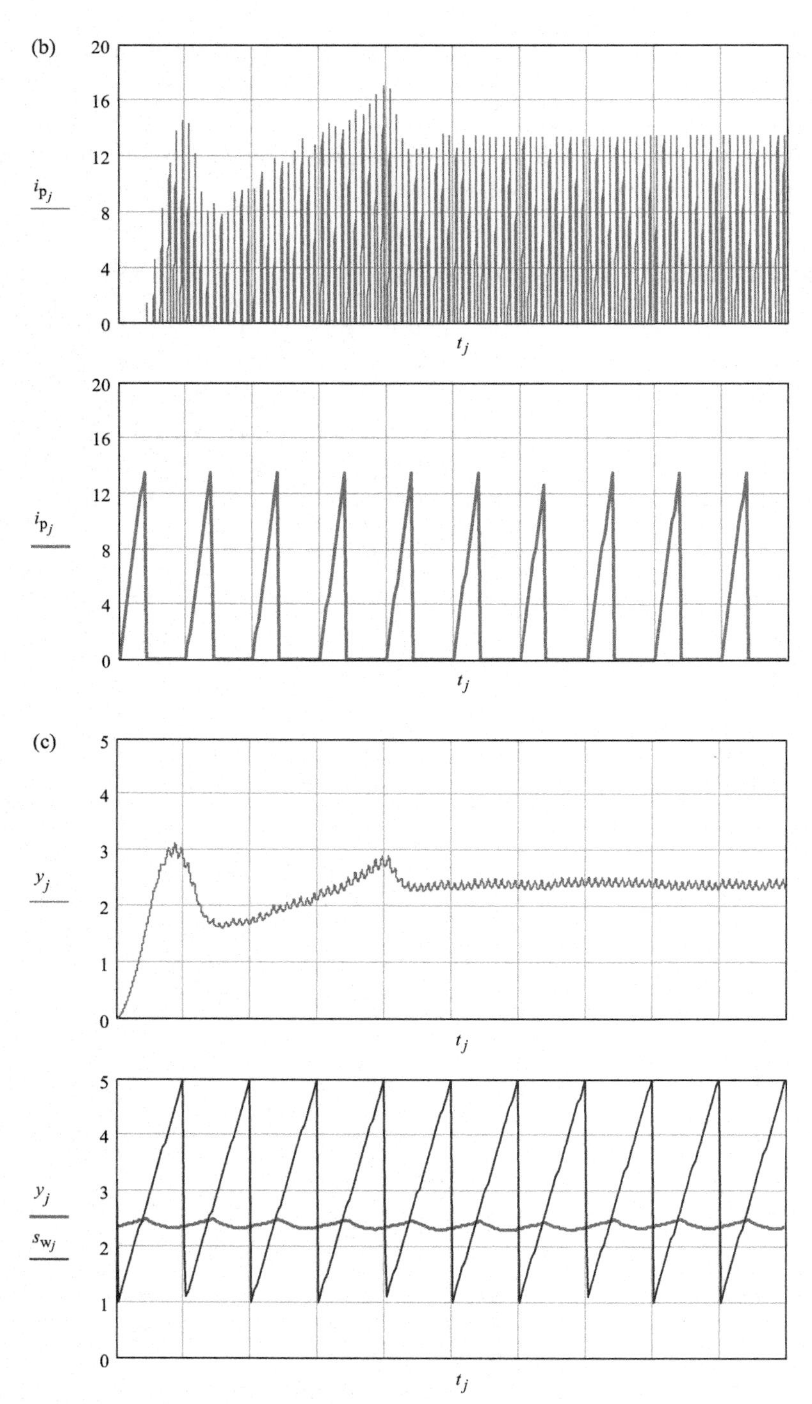

Figure 3.10 ***(cont.)***

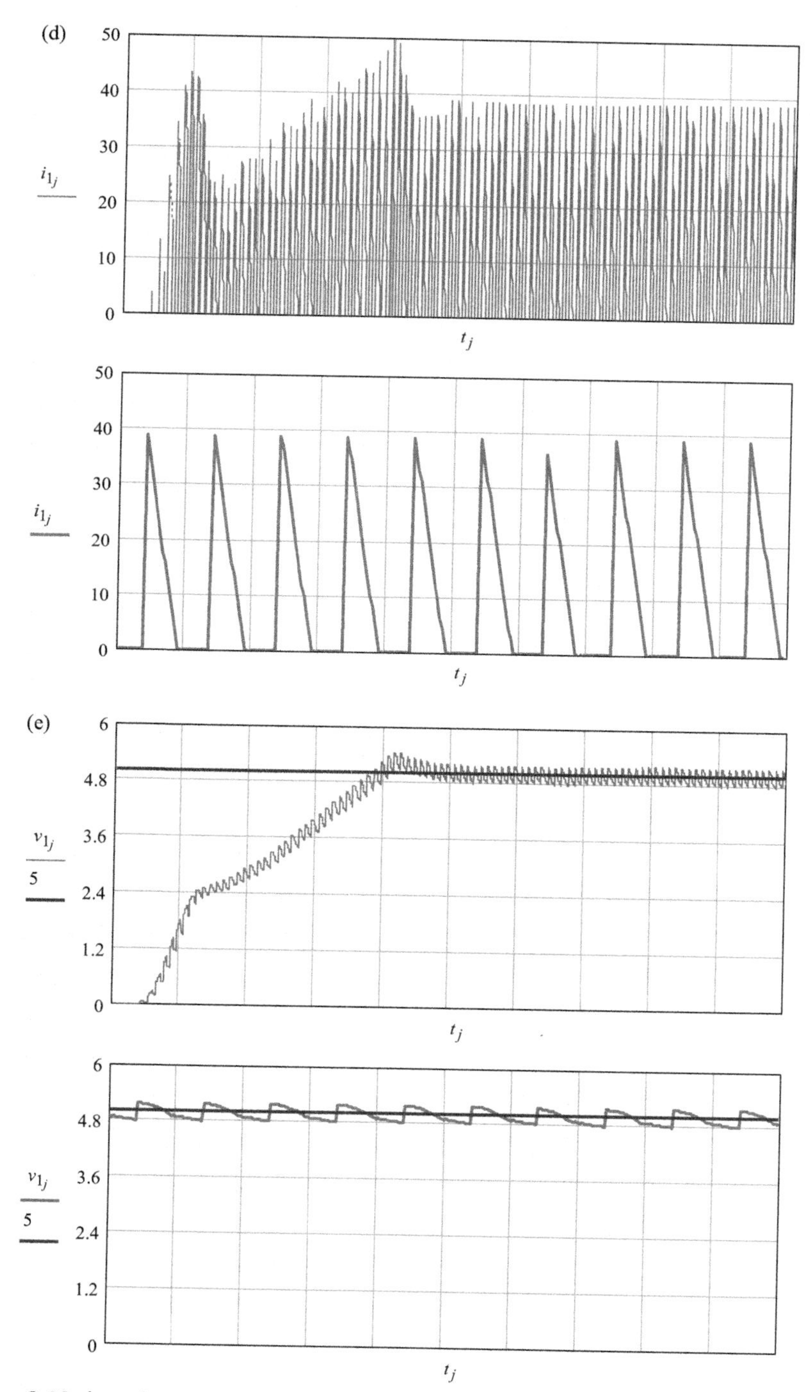

Figure 3.10 ***(cont.)***

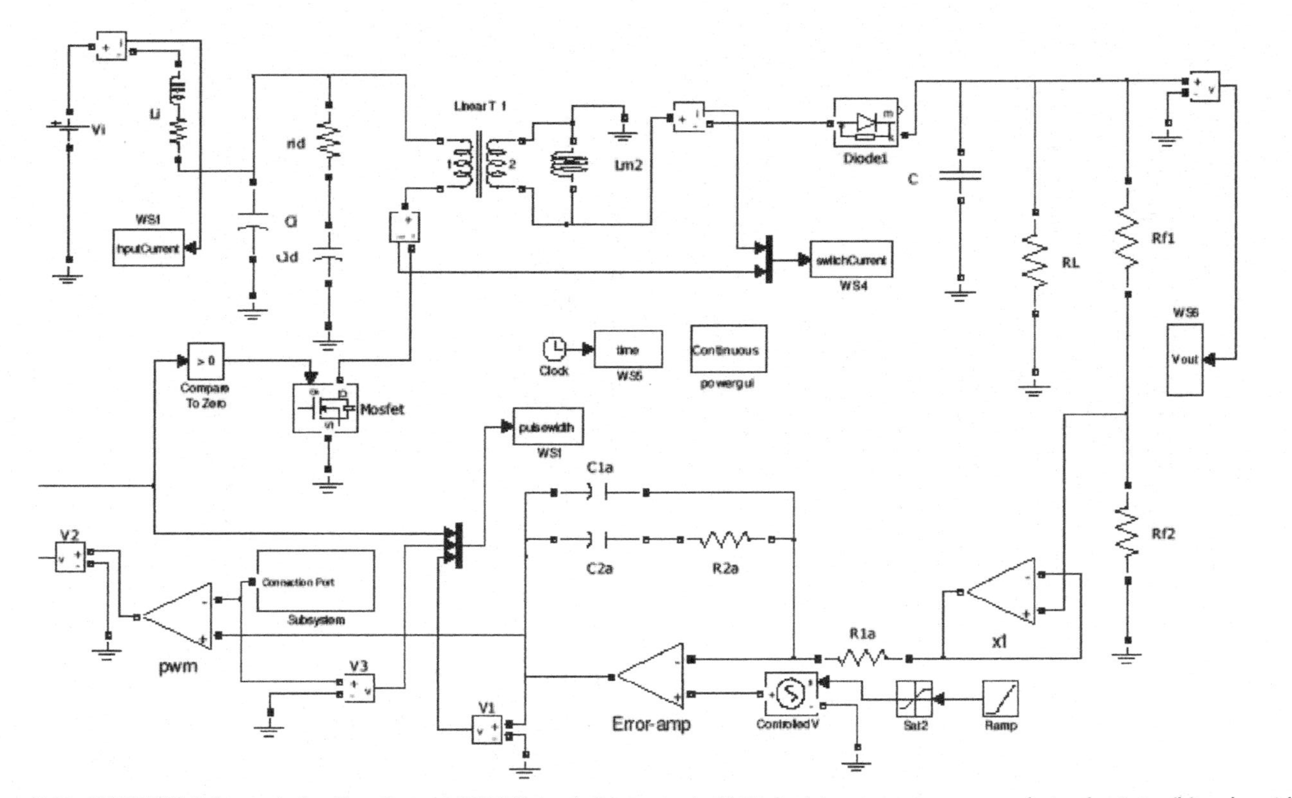

Figure 3.11 ***SIMULINK Schematic for One Output DCM Flyback Converter in V-Mode.*** (a) output; turn-on and steady state; (b) pulsewidth modulation, sawtooth clock, and error signal; steady state; and (c) inductor winding currents.

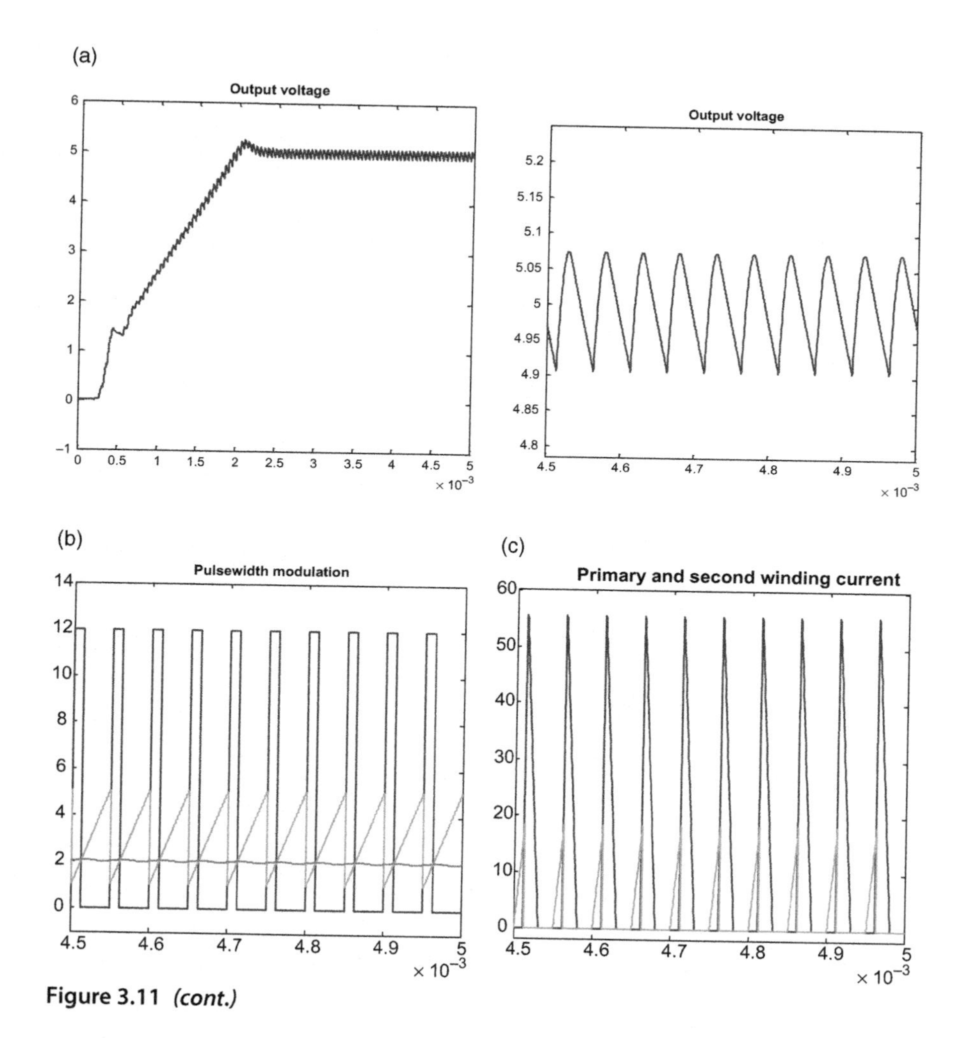

Figure 3.11 ***(cont.)***

feedback is regulated, while the rest are left free running and not regulated. With this example, the deficiency will be demonstrated.

Furthermore, and also to solidify the design process and boost design confidence, we attempt a new output configuration: a 3.3 V output up to 15 A to support digital circuit and a 12V output at 0.2 A to support analog, Figure 3.13.

The design procedure begins by assigning the primary on-duty cycle to be 0.4 and selecting the main output flyback duty cycle to be 0.45, and for the 12V/0.2 A output, a proprietary procedure yields $L_{s1} = 1.48\ \mu H$, $L_{s2} = 16\ \mu H$, $L_p = 20\ \mu H$, $C_1 = 6150\ \mu F$, and $C_2 = 100\ \mu F$. Here, because of two

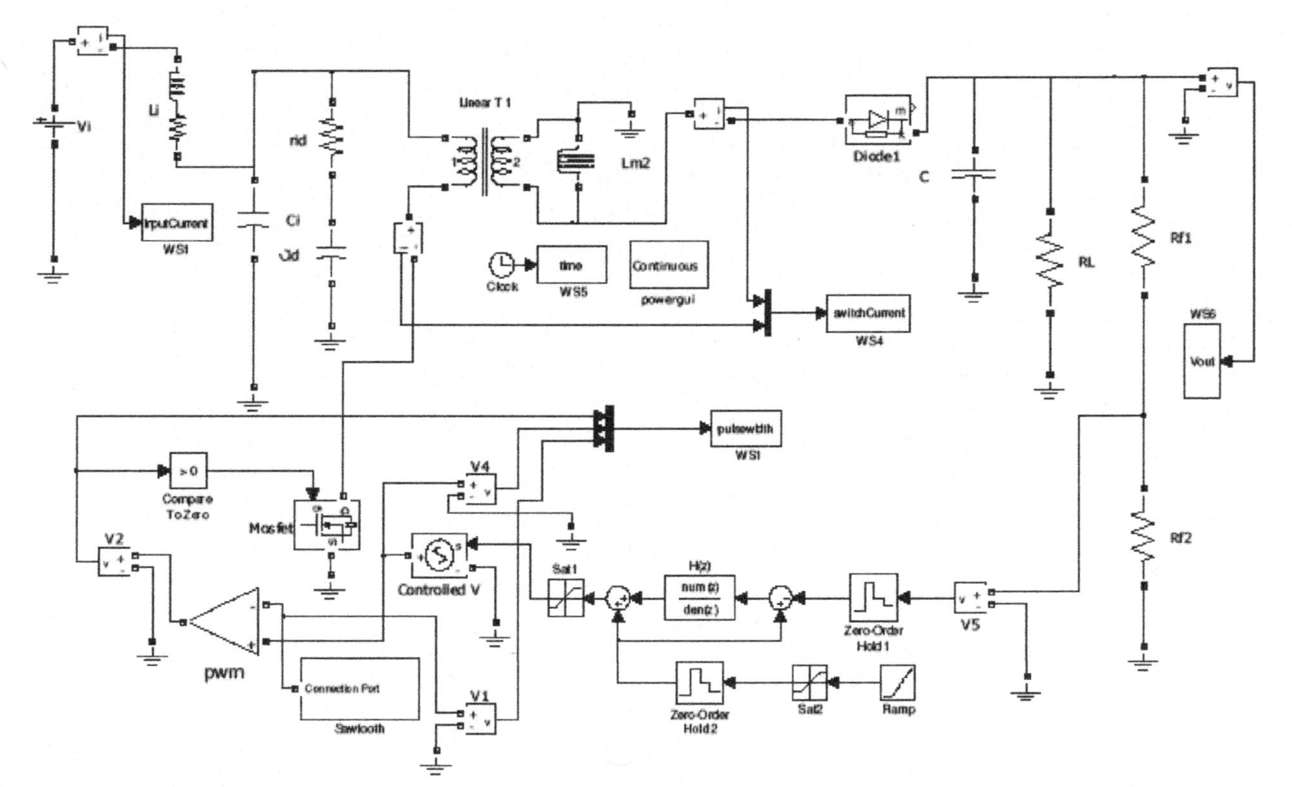

Figure 3.12 ***SIMULINK Drawing with Digital Filter.*** (a) output voltage; (b) PWM; and (c) inductor winding current.

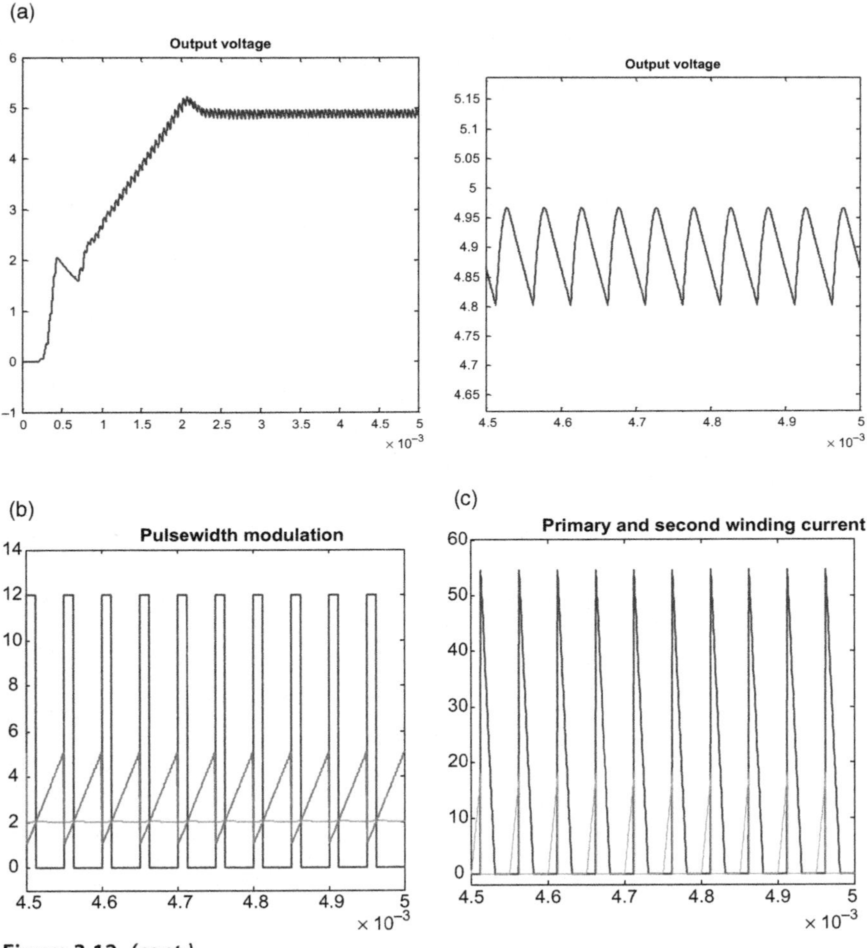

Figure 3.12 *(cont.)*

outputs, R and C load parameter values, used in (3.7) through (3.9), are to be evaluated by

$$R = \left[R_{L1}^{-1} + \left(\tfrac{L_{S1}}{L_{S2}} R_{L2} \right)^{-1} \right]^{-1}, \quad C = C_{L1} + \tfrac{L_{S2}}{L_{S1}} C_{L2} \tag{3.13}$$

In addition, and since there are two possible feedback routes, performance differences may be expected for the two configurations. We study first the configuration with the feedback taken from the 3.3V output, Figure 3.13.

Again, state-space averaged canonical model parameters depicted in Figure 3.4 are computed accordingly, and the power stage gain $G_{vd}(s)$ follows.

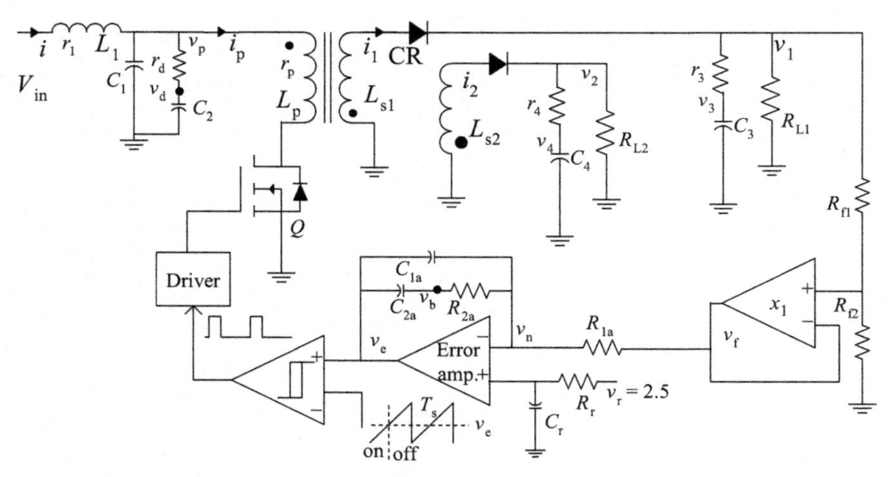

Figure 3.13 ***Two Outputs Flyback Converter with Voltage-Mode Control and with Feedback Taken from the Main Output, 3.3 V.***

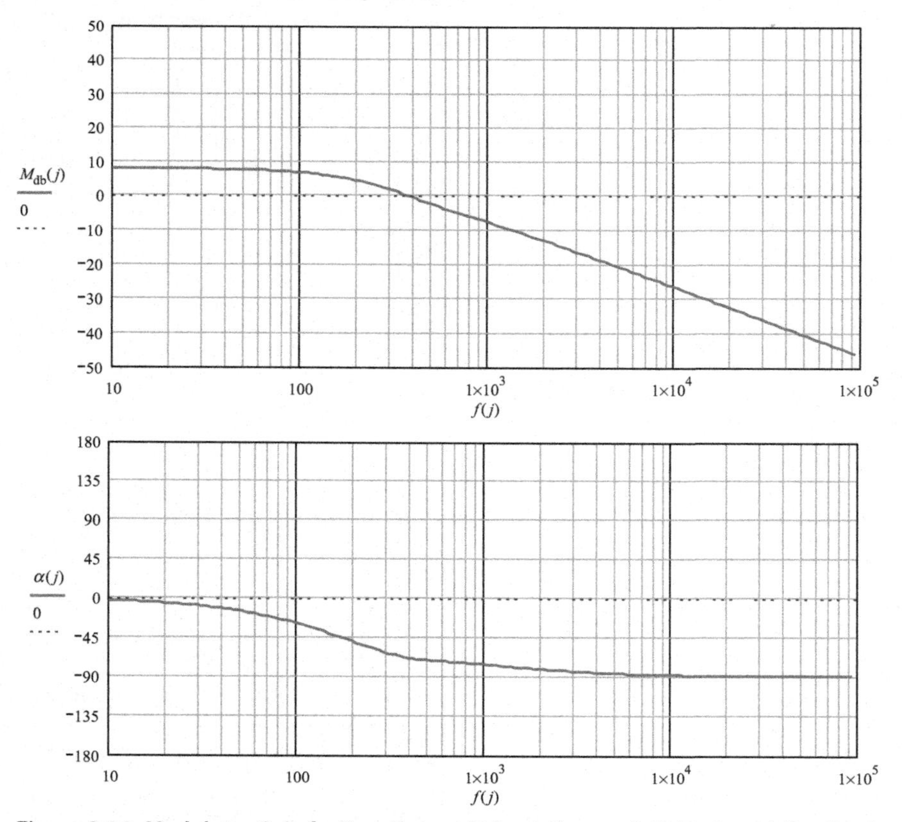

Figure 3.14 ***Modulator Gain for Two-Output Flyback Convert in V-Mode with Feedback Taken from 3.3 V.***

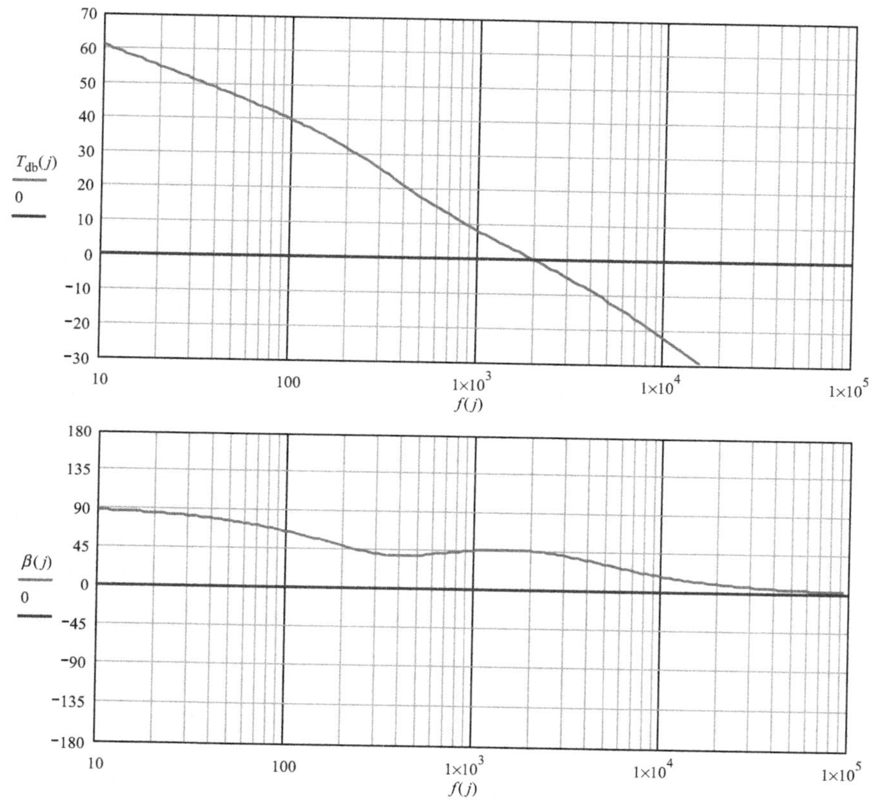

Figure 3.15 ***Loop Gain of a Flyback, Two-Output Converter in DCM with Main Output, 3.3 V, Feedback.***

With the same PWM gain and feedback ratio as given in Section 3.3, the modulator gain, equation form (3.10), of Figure 3.13 is obtained.

At frequency equal to 2 kHz, the modulator gain equals −13.12 db, and phase equals −81.45° (Figure 3.14). Passing both to type-II design (1.14), we obtain $R_{1a} = 2K$, $C_{1a} = 0.0044\ \mu F$, $R_{2a} = 12.15K$, and $C_{2a} = 0.013\ \mu F$. Combining the error amplifier with the modulator, the loop gain shows a 2 kHz crossover (Figure 3.15).

3.6 SIMULATION AND PERFORMANCE VERIFICATION – TWO OUTPUTS WITH FEEDBACK FROM THE MAIN

For confirming its time domain performance, we again invoke SIMULINK. Circuit of Figure 3.13 is converted to Figure 3.16, a SIMULINK analog model with the feedback taken from the main output.

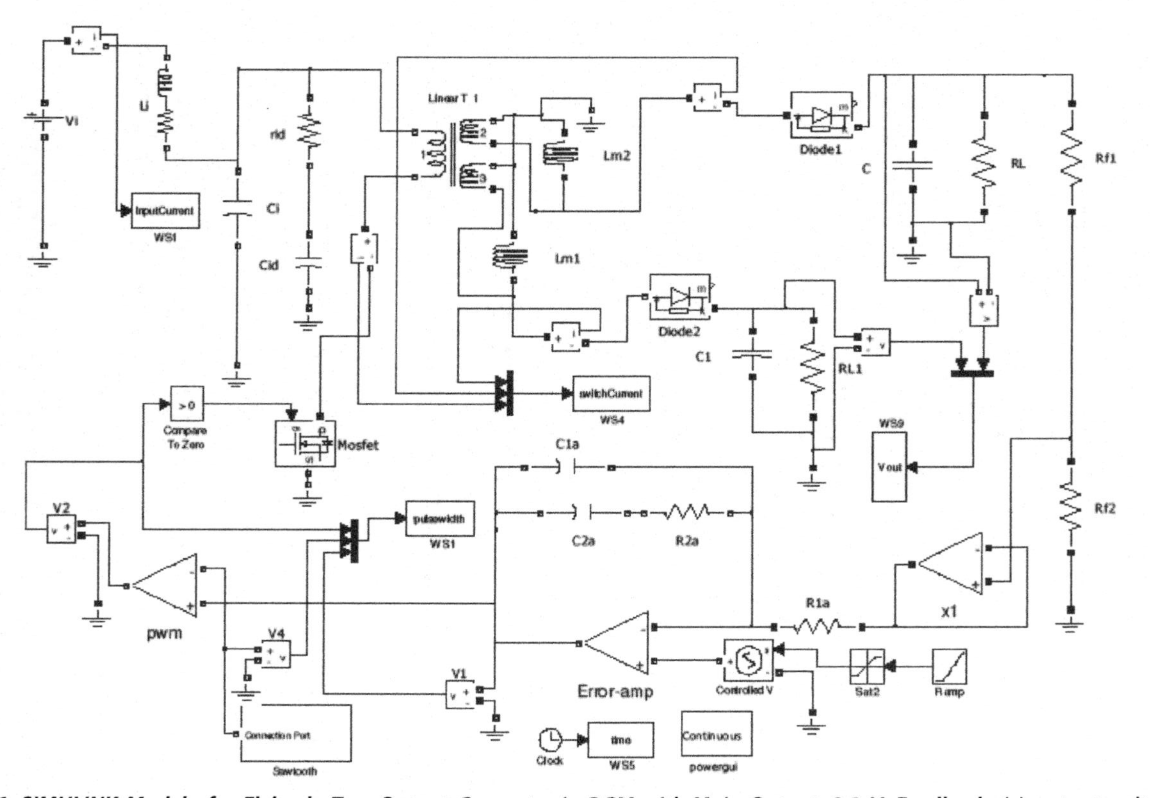

Figure 3.16 ***SIMULINK Model of a Flyback, Two-Output Converter in DCM with Main Output, 3.3 V, Feedback.*** (a) output voltages; and (b) pulsewidth modulation and switch currents.

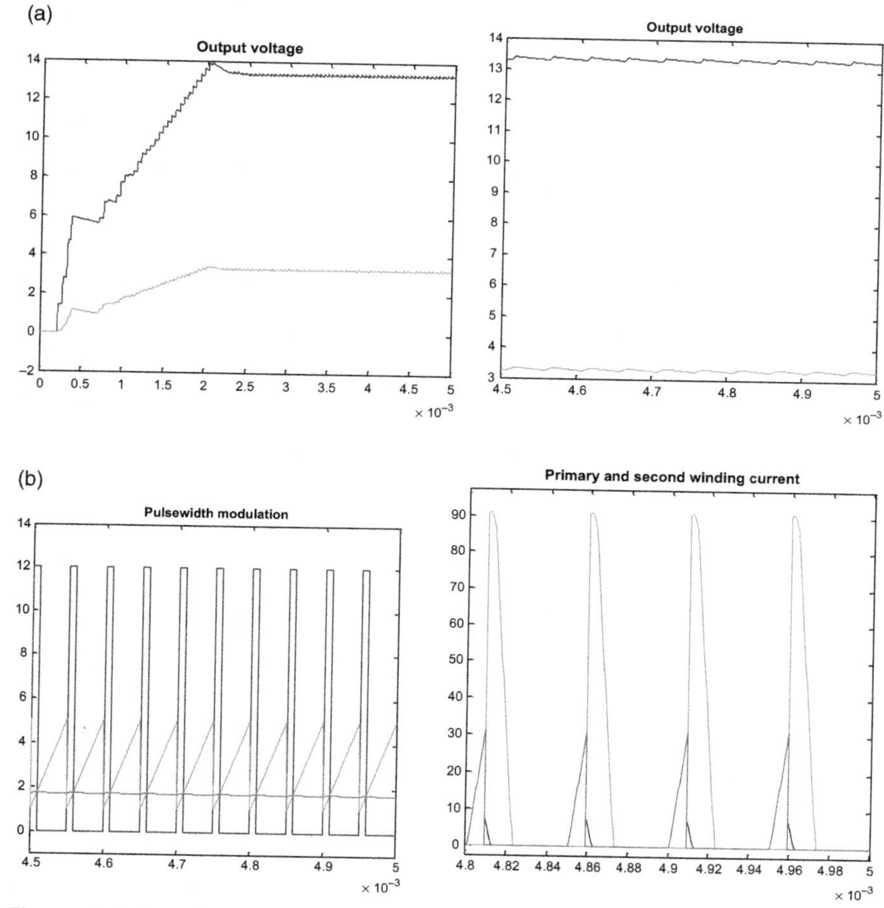

Figure 3.16 ***(cont.)***

Figure 3.16a,b give simulation results corresponding to a 5-ms real-time operation. Both the slow start transient and steady states are shown.

The simulation does show a regulated main output, while the other output is significantly off the mark.

Next, the type-II amplifier identified in the process leading to Figure 3.15 has the form, (1.17)

$$\frac{157.708\times10^{-6}\,s+1}{s\left(1.398\times10^{-9}\,s+34.822\times10^{-6}\right)} \tag{3.14}$$

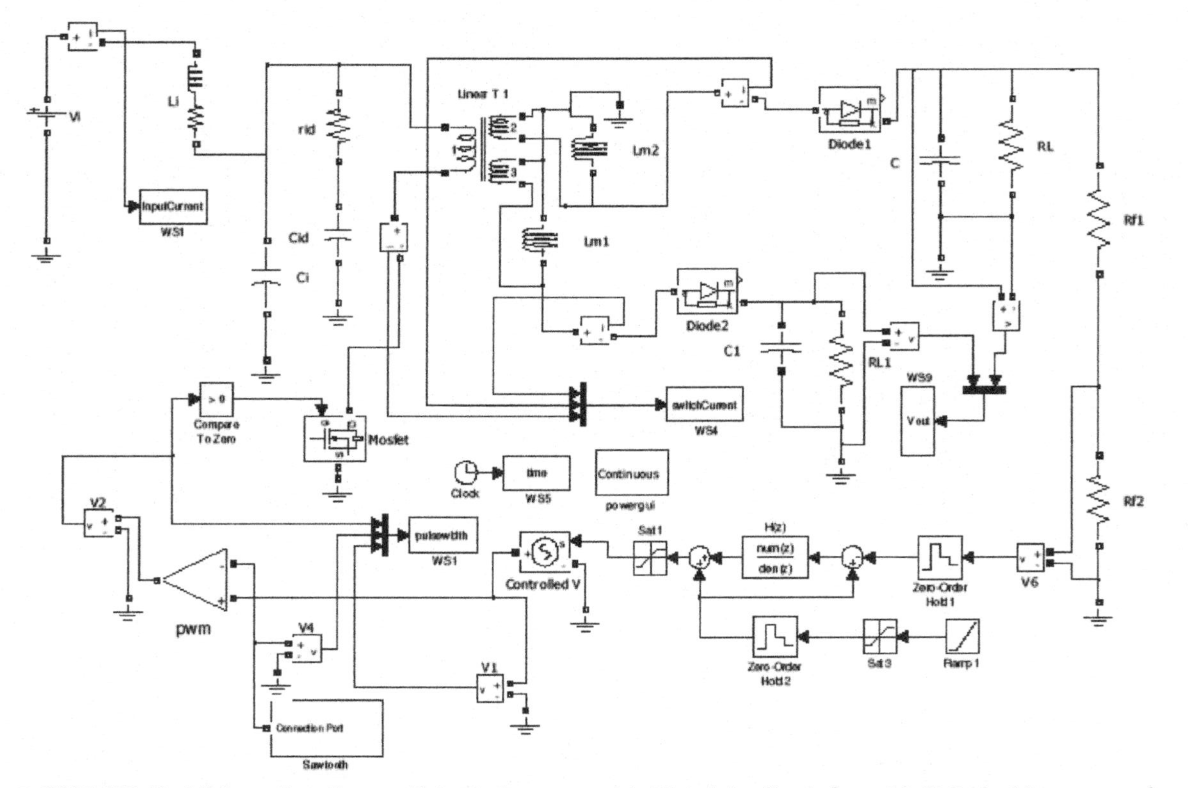

Figure 3.17 ***SIMULINK Model for a Two-Output Flyback Converter with Digital Feedback from 3.3 V Path.*** (a) output voltages; and (b) pulsewidth modulation and switch currents.

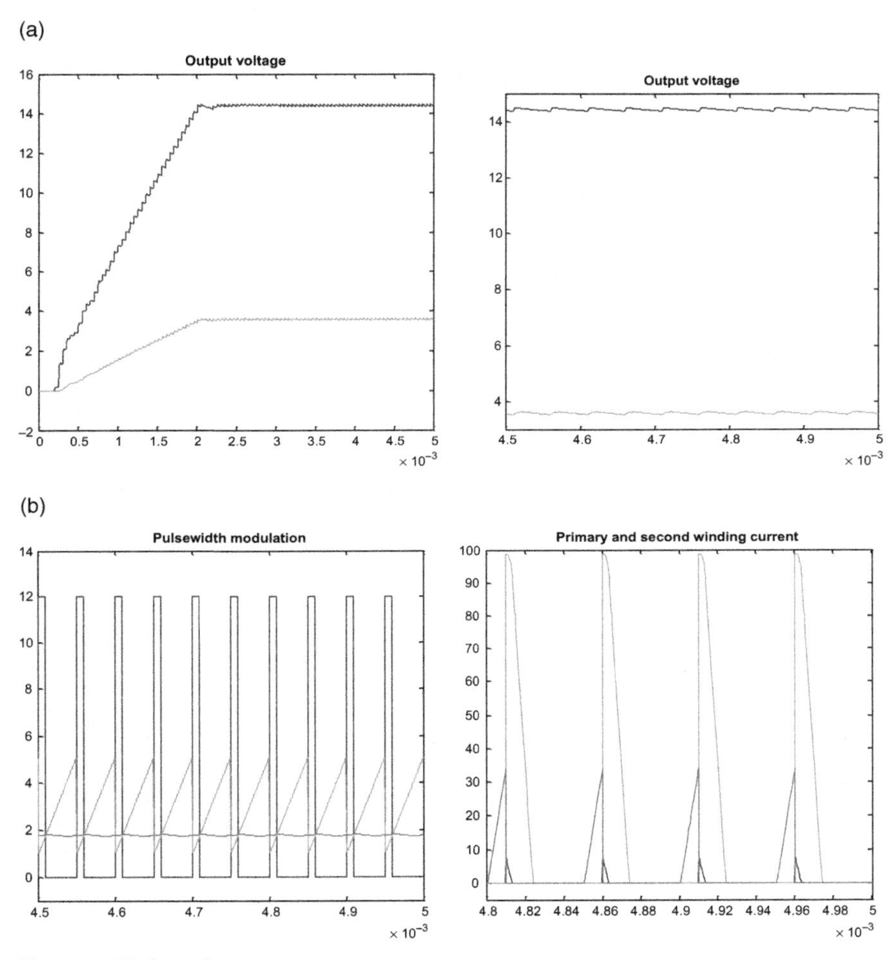

Figure 3.17 ***(cont.)***

It gives a digital counterpart,

$$\frac{a_0 + a_1 z^{-1} + a_2 z^{-2}}{1 + b_1 z^{-1} + b_2 z^{-2}}$$

$$a_0 = 69.689 \times 10^{-3} \quad a_1 = 550.176 \times 10^{-6} \quad a_2 = -69.139 \times 10^{-3}$$
$$b_1 = -1.969 \times 10^{0} \quad b_2 = 969.347 \times 10^{-3} \tag{3.15}$$

As a result, Figure 3.16 is next moved to the digital version, Figure 3.17. Figure 3.17a,b show its results. Again, 12V output is way off the mark since feedback is taken from 3.3V output.

3.7 TWO OUTPUTS WITH ALTERNATIVE FEEDBACK

As mentioned in Section 3.5, there is an alternative feedback path taken from 12 V output. Figure 3.19 gives the analog version of SIMULINK drawing.

Without duplicating the verbiage, the modulator gain, and a new one, is given (Figure 3.18a).

The modulator yields a gain of −8.8 db and a phase of −75° at 1 kHz. Again, for 45°-phase margin, a type-II amplifier is selected. Components surrounding the amplifier are identified: $R_{1a} = 2K$, $C_{1a} = 0.017\ \mu F$, $R_{2a} = 8.26K$, and $C_{2a} = 0.033\ \mu f$. Integrating the error amplifier with the modulator, the loop gain, Figure 3.18b, shows a 1 kHz crossover.

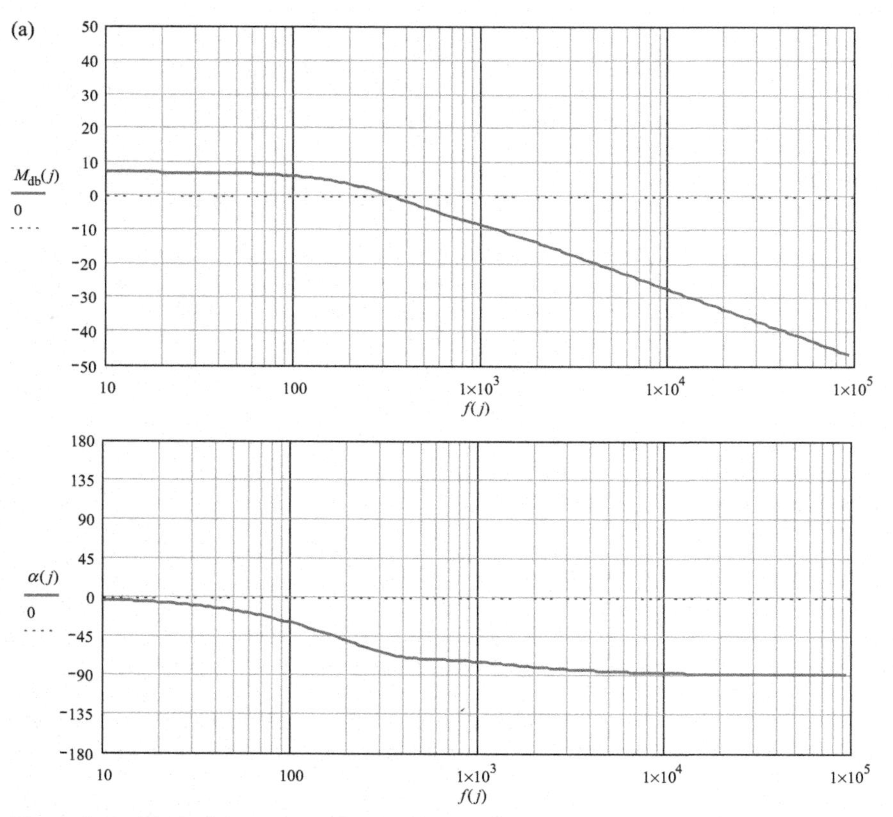

Figure 3.18 (a) Modulator gain of Figure 3.19 with alternative feedback; (b) loop gain of Figure 3.19 with alternative feedback.

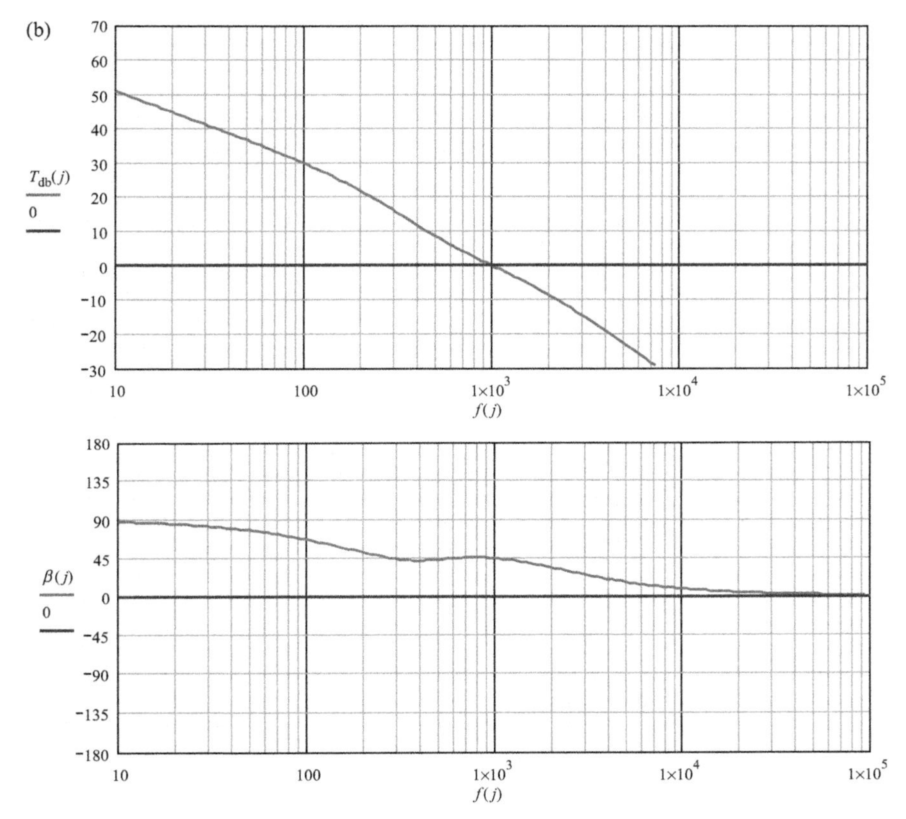

Figure 3.18 ***(cont.)***

Figure 3.19 is run for 5 ms coverage of real time. Figure 3.19a,b show the output voltages, pulsewidth modulation, and switch currents.

As expected, with the feedback closed on +12 V, this output is now well-regulated, while the main output is off a bit, less than the target of 3.3 V.

The type-II amplifier identified gives

$$\frac{275.664\times 10^{-6}\,s+1}{s\left(9.197\times 10^{-9}\,s+100.088\times 10^{-6}\right)} \tag{3.16}$$

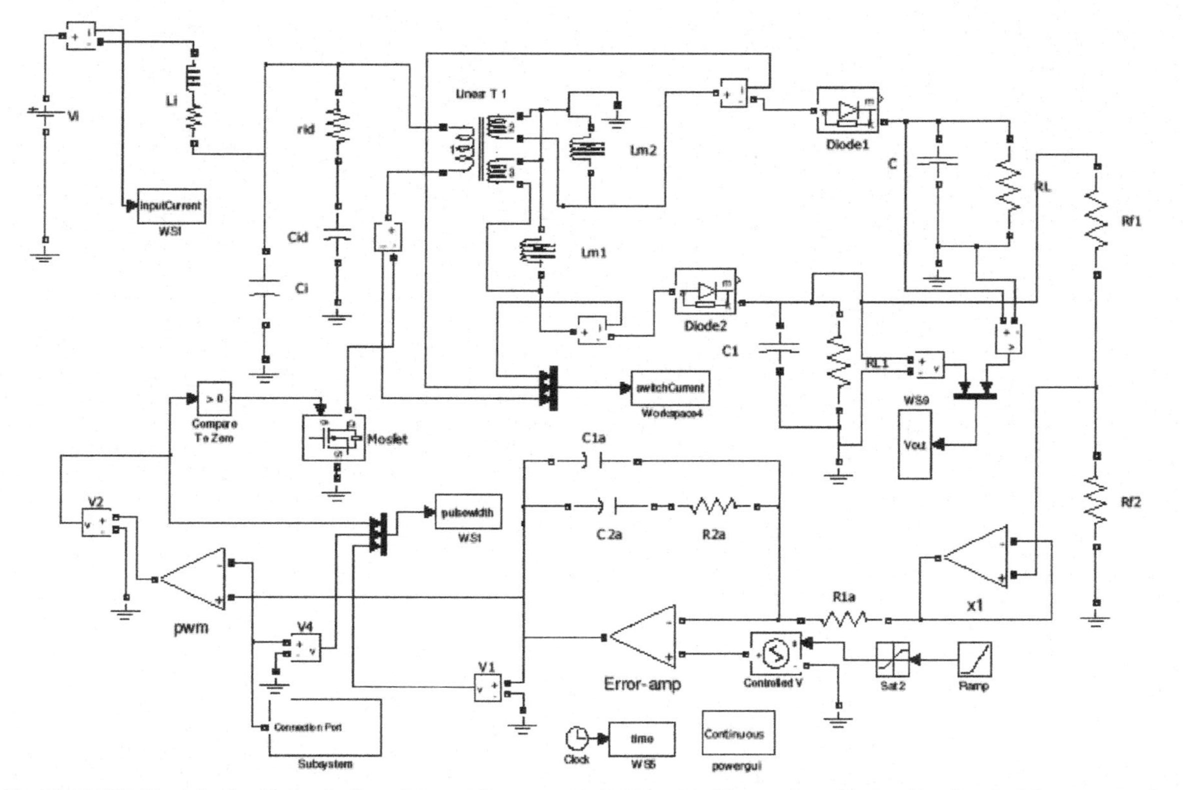

Figure 3.19 ***SIMULINK Model of a Flyback, Two-Output Converter in DCM with Alternative, Analog Feedback.*** (a) output voltages; and (b) pulsewidth modulation and switch current.

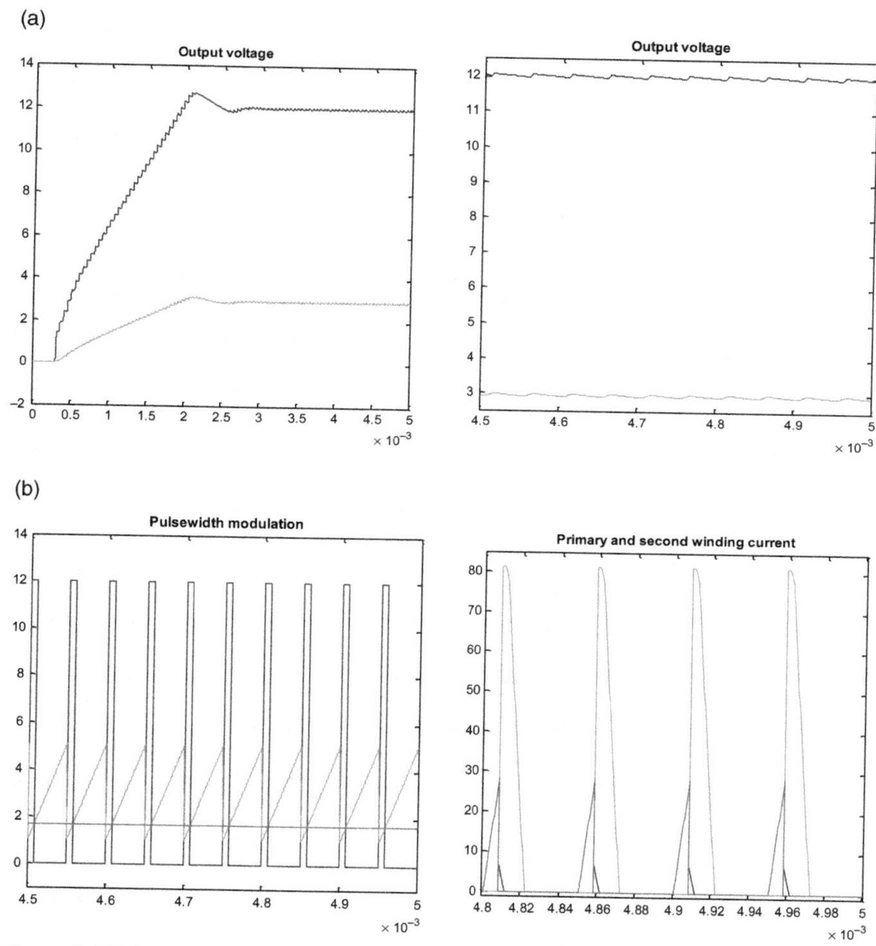

Figure 3.19 ***(cont.)***

The corresponding digital version is

$$\frac{a_0 + a_1 z^{-1} + a_2 z^{-2}}{1 + b_1 z^{-1} + b_2 z^{-2}}$$

$$a_0 = 73.611 \times 10^{-3} \quad a_1 = 1.323 \times 10^{-3} \quad a_2 = -72.288 \times 10^{-3} \tag{3.17}$$

$$b_1 = -1.947 \times 10^{0} \quad b_2 = 947.027 \times 10^{-3}$$

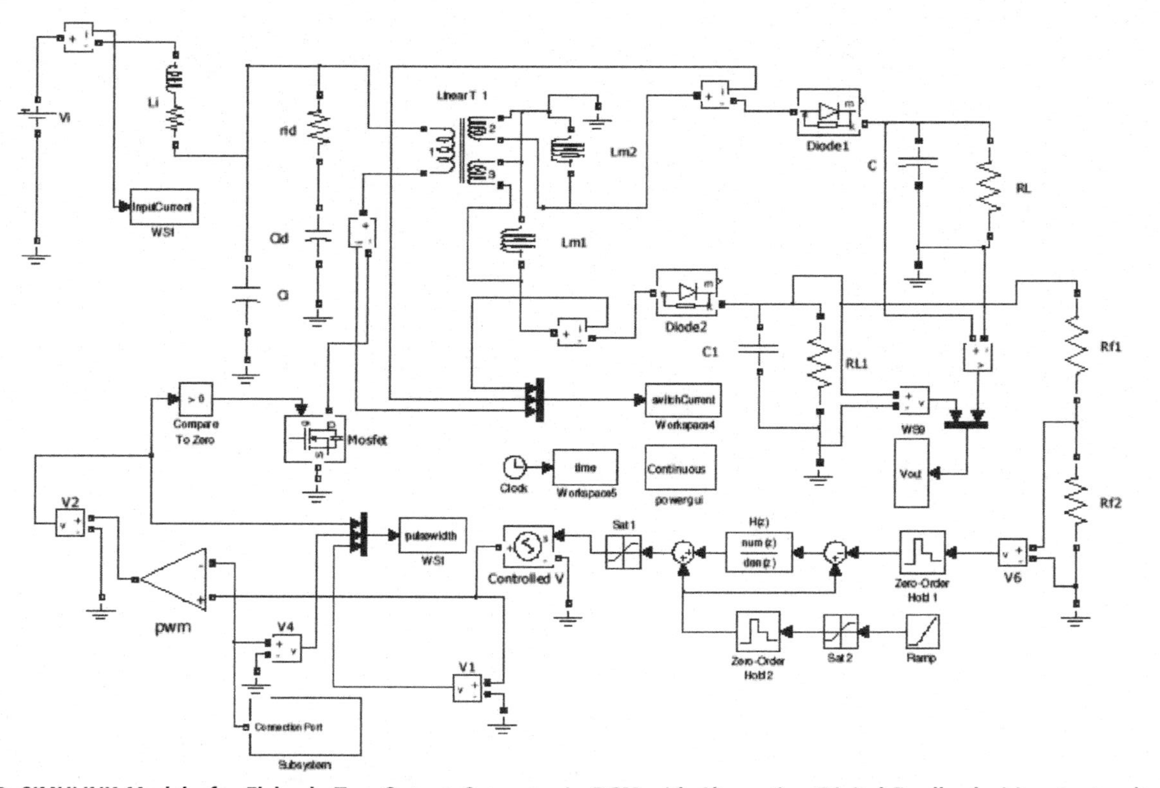

Figure 3.20 ***SIMULINK Model of a Flyback, Two-Output Converter in DCM with Alternative, Digital Feedback.*** (a) output voltages; and (b) pulsewidth modulation and switch current.

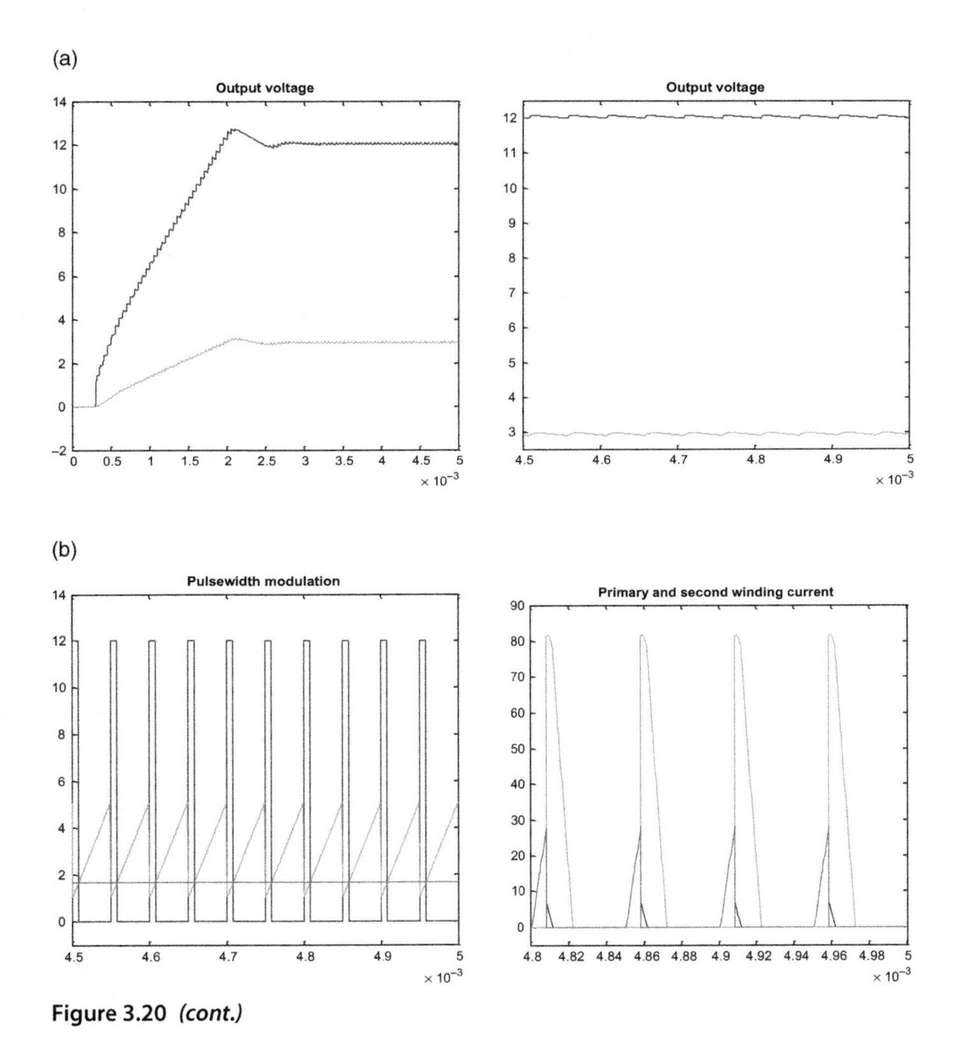

Figure 3.20 ***(cont.)***

In other words, the digital version of Figure 3.19 is Figure 3.20, and it gives performance Figure 3.20a,b.

CHAPTER 4

Flyback Converter with Current-Mode Control

4.1 CURRENT-MODE SCHEMATIC

To obtain peak current-mode control, the voltage-mode PWM in Figure 3.3 is replaced by a resistive current sensor R_s, a noninverting scaling amplifier (gain A_c), a comparator, and an RSFF (Figure 4.1).

Following the current sensing resistor scaling, a simple RC filter is introduced to take out high-frequency ringing and to prevent false triggering.

4.2 CURRENT-MODE PWM GAIN

Figure 3.2 gave both primary and secondary currents of a flyback converter with a single output in DCM. The top trace presents the primary current in an ideal ramp form without resistive effect. A more concise form, including winding resistance, sensing resistance, switch-on resistance, and inductance is easily derived as,

$$i_p(t) = \frac{V_{in}}{r_p + R_s + R_{on}}\left(1 - e^{-\frac{t}{\left[L_p/(r_p+R_s+R_{on})\right]}}\right) \tag{4.1}$$

Given a sensing resistor R_s and at a steady state, the current peak, $I_p(\mathrm{DT})\ R_sA_c$, intercepts the feedback error voltage v_e, T the witching period, and terminates a conduction period. The steady-state duty cycle is therefore determined by,

$$\frac{V_{in}}{r_p + R_s + R_{on}}\left(1 - e^{-\frac{D}{\left[L_p/(r_p+R_s+R_{on})\right]f}}\right)R_sA_c = v_e \tag{4.2}$$

Define an implicit function,

$$\begin{aligned} &f(V_{in}, v_e, D) \\ &= \left[V_{in}/(r_p + R_s + R_{on})\right]\left(1 - e^{-\frac{D}{\left[L_p/(r_p+R_s+R_{on})\right]f}}\right)R_sA_c - v_e \end{aligned} \tag{4.3}$$

Power Converters with Digital Filter Feedback Control
http://dx.doi.org/10.1016/B978-0-12-804298-4.00004-6

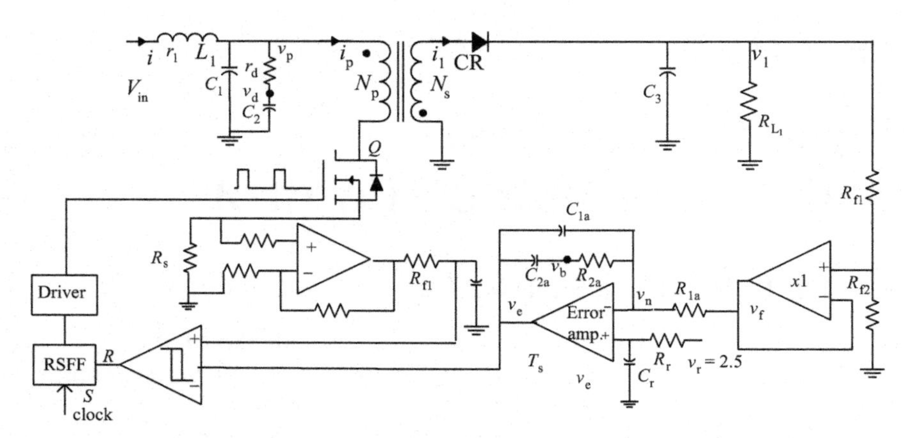

Figure 4.1 ***Single Output Flyback Converter with Current-Mode, Analog Feedback.***

The overall PWM contains two parts, the feedback, F_m and the feed-forward, F_g.

$$dD = \left(\frac{\partial D}{\partial v_e}\delta v_e\right) + \left(\frac{\partial D}{\partial V_{in}}\delta V_{in}\right) = F_m \delta v_e + F_g \delta V_{in} \tag{4.4}$$

in which the two gain factors are given by the Jacobian,

$$F_m = -\left(\frac{\partial f / \partial v_e}{\partial f / \partial D}\right), \quad F_g = -\left(\frac{\partial f / \partial V_{in}}{\partial f / \partial D}\right) \tag{4.5}$$

This implies a small signal block diagram, Figure 4.2, for the converter.

In most cases, the input supply is constant. As a result, the feedforward dynamics F_g and G_{vg} (sometimes designated as G_{vs} for nomenclature

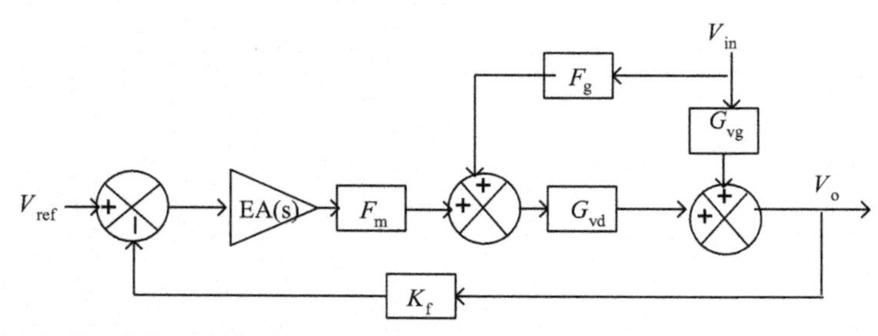

Figure 4.2 ***Small Signal Block Diagram.***

consideration, g for generator, and s for source) are not included in determining the modulator gain, which is,

$$M(s) = K_f F_m G_{vd}(s) \tag{4.6}$$

$G_{vd}(s)$ was given in (3.9); current mode control does not alter the power stage.

4.3 EXAMPLE

The same converter specification given in Section 3.3 Example shall be used here to illustrate the performance difference, which is actually better, between the current-mode and voltage-mode. To make the example more realistic, the filter surrounding the current sensor is also included. Consequently (4.3) is modified,

$$f(V_{in}, v_e, D) = \left(\frac{V_{in}}{r_p + R_s + R_{on}}\right)\left(1 - e^{-\frac{D}{[L_p/(r_p+R_s+R_{on})]f}}\right)\left(1 - e^{-(D/R_f C_f f)}\right) R_s A_c - v_e \tag{4.7}$$

In general, it is a good practice to set the steady state error voltage at about half of the source, for instance 12 V, supplying the error amplifier. Therefore, v_e shall be around 6 V. It is also known from the time domain study that primary peak current is about 14 A. Taking both into consideration, the current sensing resistor $R_s = 6/14$ may be selected. However, such a sensor drops excessively available source voltage. $R_s = 0.01$ is instead selected and followed by a scaling amplifier with a noninverting gain, $A_c = 43$. As for the filter, $R_f = 100$ and $C_f = 0.001\ \mu F$ were selected for the example. Then (4.5) gives $F_m = 2.850614$. Neither the feedback factor nor the power stage gain $G_{vd}(s)$ is changed. This leads to the current-mode modulator gain, as shown in Figure 4.3.

Compared with Figure 3.6, the voltage-mode, significant difference is observed. At 2 kHz, the current-mode modulator exhibits 10 db gain and $-77°$ phase. At the selected crossover frequency of 2 kHz and desired phase margin of 60°, the phase boost required is 47°. Using type-II design, all four components are obtained; $R_{1a} = 2K$, $C_{1a} = 0.05\ \mu F$, $R_{2a} = 748$, and $C_{2a} = 0.27\ \mu F$. The loop gain then follows, as seen in Figure 4.4.

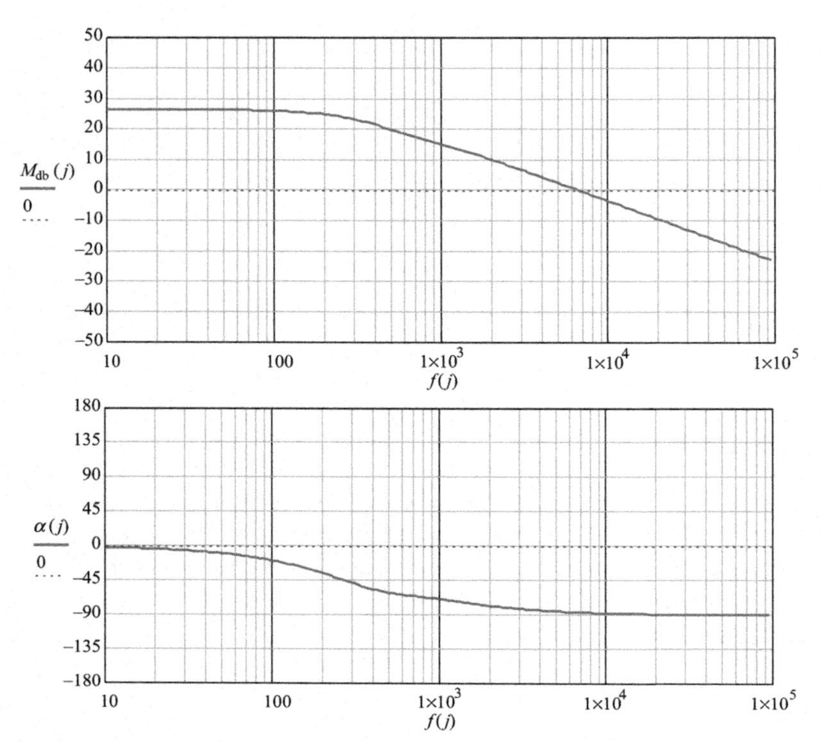

Figure 4.3 *Current-Mode Modulator Gain.*

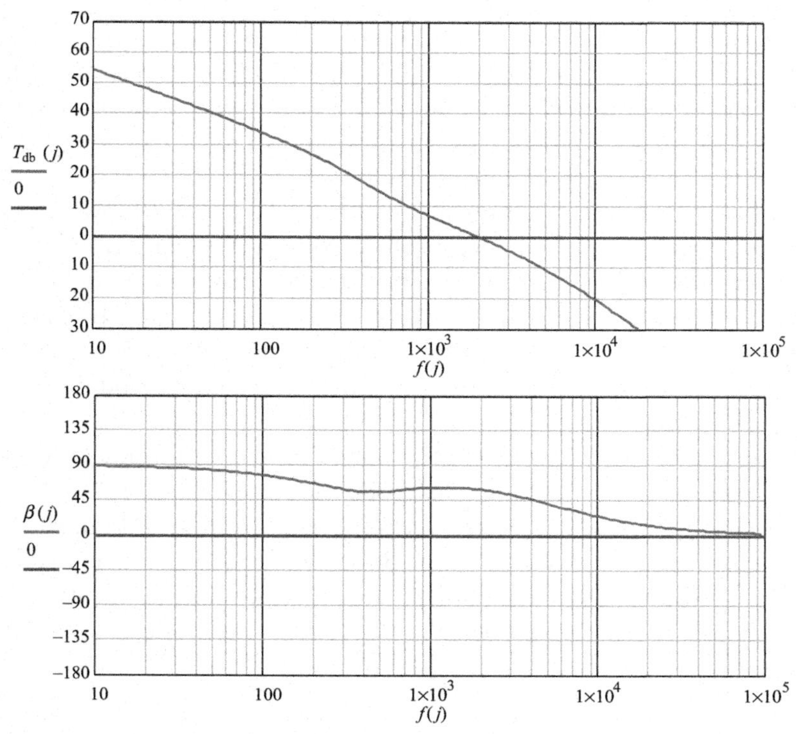

Figure 4.4 *Loop Gain with Current-Mode Control.*

4.4 SIMULATION AND PERFORMANCE VERIFICATION

Once more, the difference equation set describing Figure 4.1 is

$$\begin{pmatrix} q_{j+1} \\ i_{j+1} \\ v_{p_{j+1}} \\ v_{d_{j+1}} \\ i_{p_{j+1}} \\ i_{1_{j+1}} \\ v_{3_{j+1}} \\ v_{e_{j+1}} \\ v_{b_{j+1}} \end{pmatrix} := \begin{bmatrix} -\left[\left(\text{if}\left(R_s A_c i_{p_j} \ge v_{e_j},1,0\right)\right)v - \left[q_j v\left(\text{if}\left(c_{lk_j}=1,1,0\right)\right)\right]\right] \\ \left[1-\left((\delta t/L_1)r_1\right)\right]i_j + (\delta t/L_1)(V_{in}-v_{p_j}) \\ \left[1-\left(\delta t/(C_1 r_d)\right)\right]v_{p_j} + (\delta t/C_1)\left(i_j - i_{p_j} + (1/r_d)v_{d_j}\right) \\ \left[1-\left(\delta t/r_d C_2\right)\right]v_{d_j} + \left[\delta t/r_d C_2\right]v_{p_j} \\ \text{if}\left[i_{p_j}<0,0,\text{if}\left[q_j=1,\left[\left(1-(\delta t/L_p)\right)\left(r_p+R_s+R_{on}\right)\right]i_{p_j}+\dfrac{\delta t}{L_p}v_{p_j},0\right]\right] \\ \text{if}\left[i_{1_j}<0,0,\text{if}\left[q_j=1,0,\sqrt{(L_p/L_{s1})}i_{p_j}+\left[\left(1-(\delta t/L_{s1})\right)\left(r_{s1}+R_{p1}\right)\right]i_{1_j} \right.\right. \\ \left.\left. -(\delta t/L_{s1})\left(V_{CR}+k_1 v_{3_j}\right)\right]\right] \\ \left[1+\left(\delta t/(r_3 C_3)\right)(k_1-1)\right]v_{3_j} + \left(\delta t/r_3 C_3\right)R_{p1} i_{1_j} \\ \text{if}\left[v_{e_j}<0,0,\text{if}\left[v_{e_j}>12,12,v_{e_j}+\left[(\delta t/C_{1a})\left((1/R_{1a})+(1/R_{2a})\right)\text{VR}_j \right.\right.\right. \\ \left.\left.\left. -\left(\delta t/(R_{2a}C_{1a})\right)v_{b_j} - \left(\delta t/(R_{1a}C_{1a})\right)k_f\left(R_{p1}i_{1_j}+k_1 v_{3_j}\right)\right]\right]\right] \\ \left[1-\left[\left(1/(R_{2a}C_{2a})\right)+\left(1/(R_{2a}C_{1a})\right)\right]\delta t\right]v_{b_j} + \left[\left(1/(R_{2a}C_{2a})\right) \right. \\ \left. +(1/C_{1a})\left((1/R_{1a})+(1/R_{2a})\right)\right]\delta t\text{VR}_j - \left(\delta t/(R_{1a}C_{1a})\right)k_f\left(R_{p1}i_{1_j}+k_1 v_{3_j}\right) \end{bmatrix} \tag{4.8}$$

Here, v_3 stands for ideal C_3 voltage, excluding series parasitic resistance. Equation set (4.8) yields results as shown in Figure 4.5a–e.

The type-II error amplifier, identified previously, gives the magnitude profile as shown in Figure 4.5f.

It shows −25 db activity at 32 kHz. The digitizing sampling frequency is then made twice at 64 kHz. Therefore, the bilinear transform constant C = 128 kHz. With the constant selected, the corresponding $H(z)$ polynomial, a second order, is obtained with the following coefficients,

$$a_0 = 65.53\times10^{-3} \quad a_1 = 4.88\times10^{-3} \quad a_2 = -60.65\times10^{-3}$$
$$b_1 = -1.601\times10^{0} \quad b_2 = 600.985\times10^{-3}$$

Replacing the analog error amplifier in Figure 4.1 with a digital filter, Figure 4.6 shows the digital version of a single output flyback converter in DCM and with current-mode control.

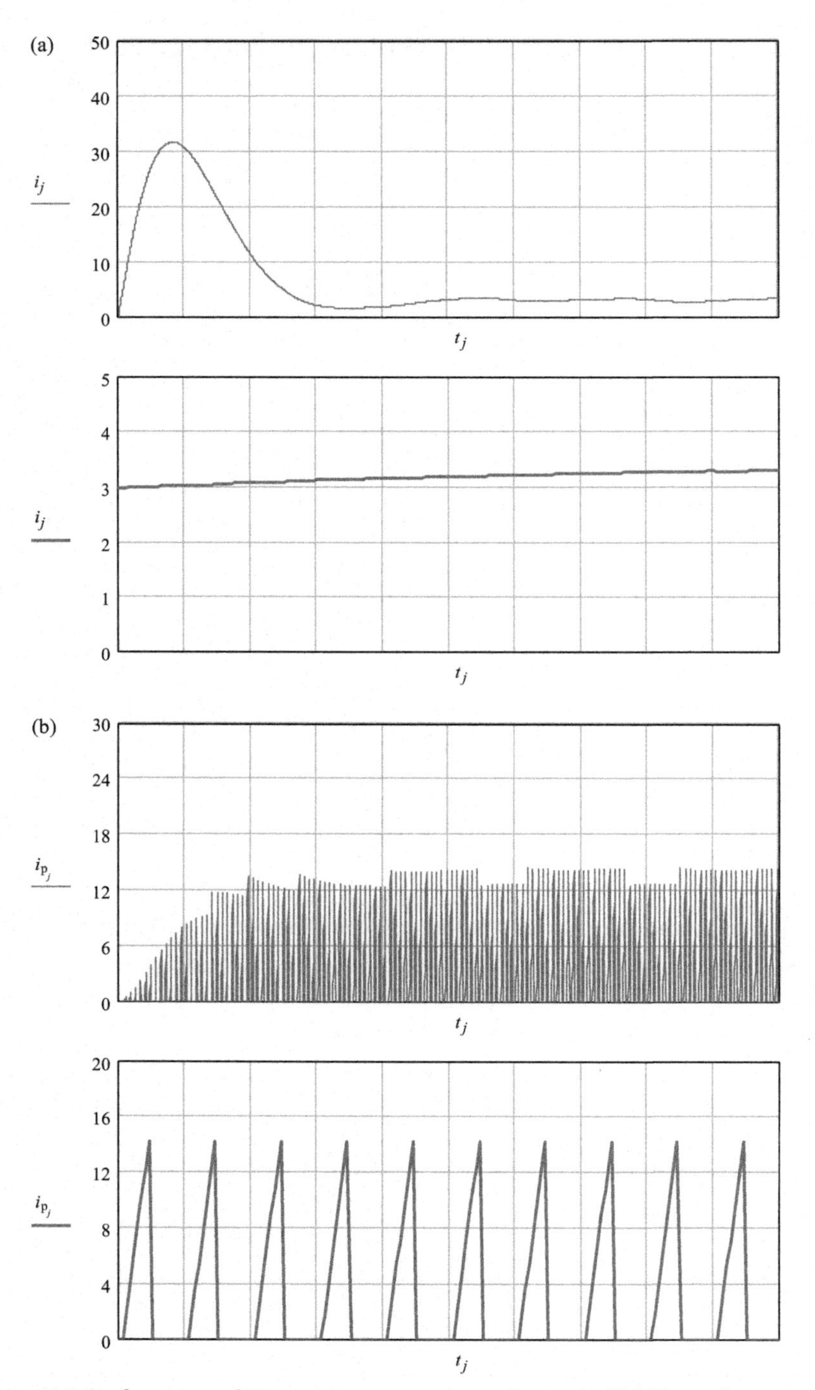

Figure 4.5 Performance of Figure 4.1 converter (equation set (4.8)) (a) Input line current, (b) primary winding/switch current, (c) error voltage and current sensing, (d) secondary winding current, (e) output voltage, and (f) type-II amplifier performance.

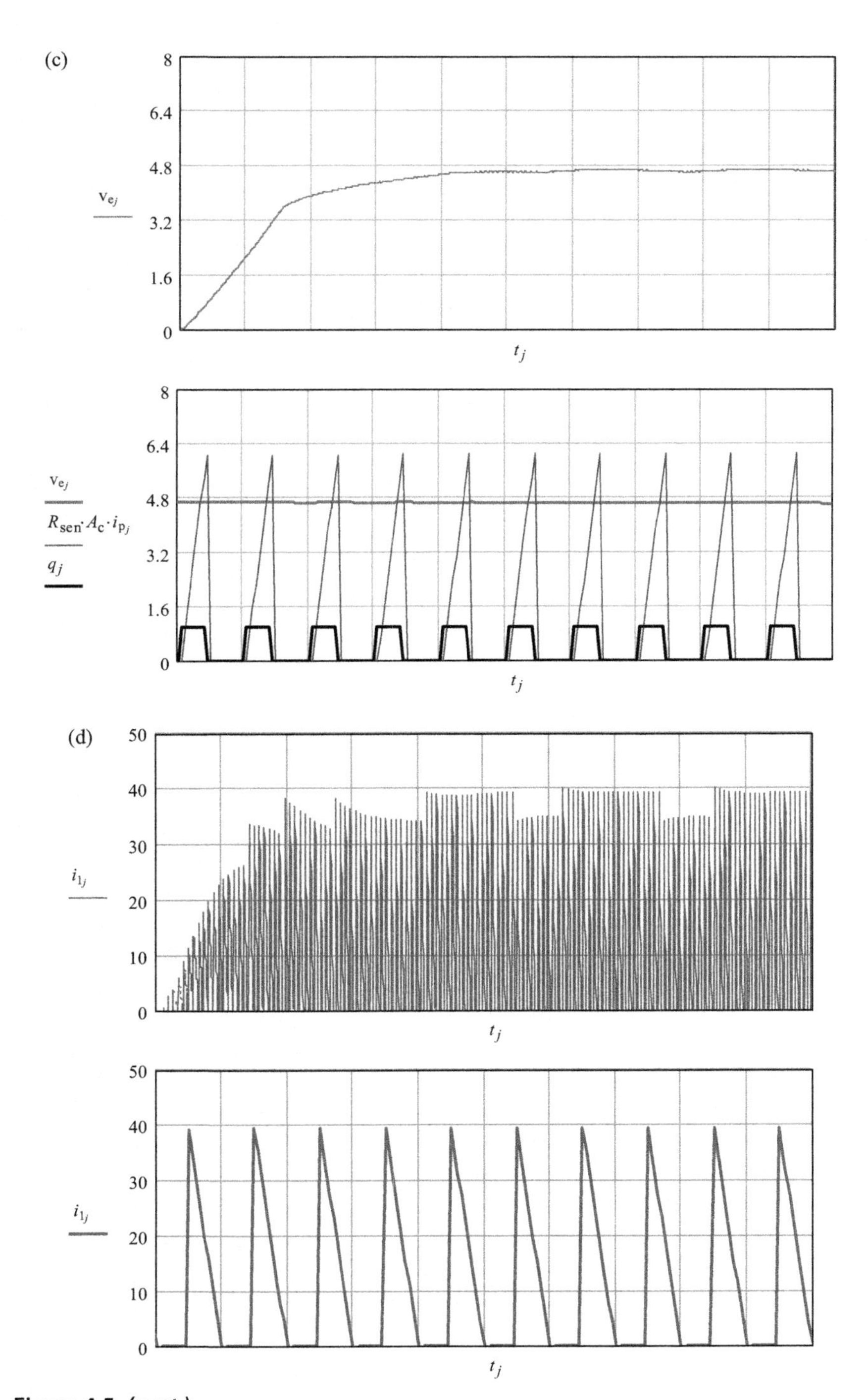

Figure 4.5 ***(cont.)***

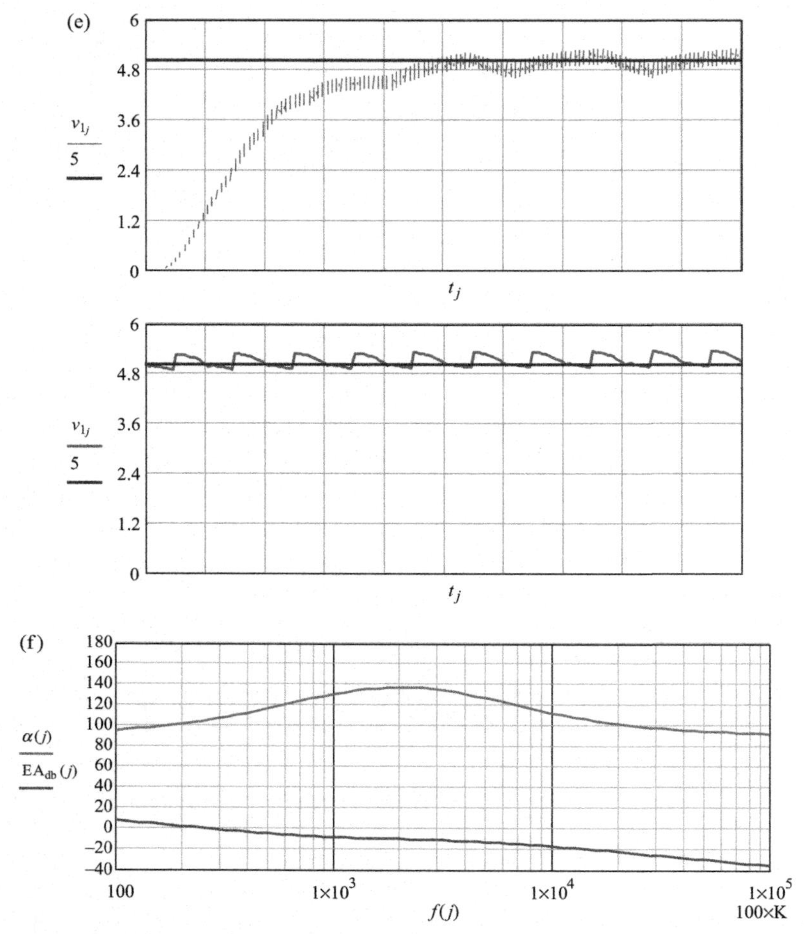

Figure 4.5 *(cont.)*

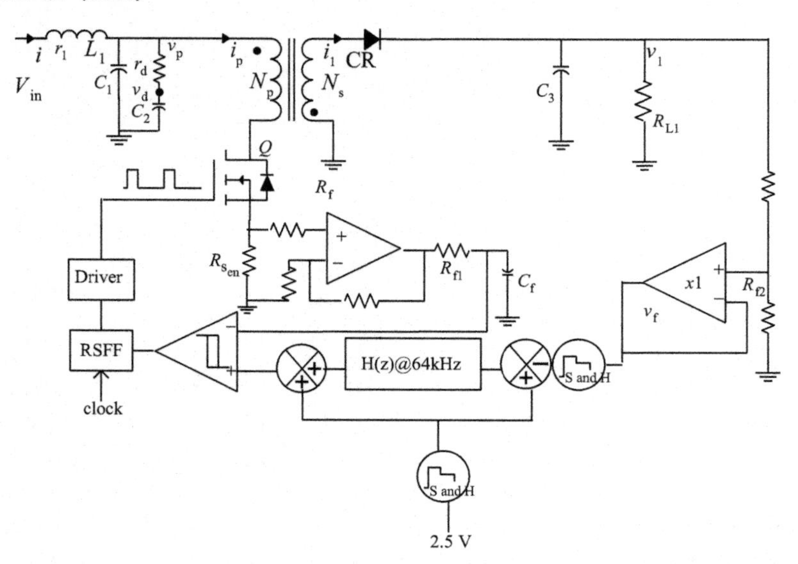

Figure 4.6 *Flyback Converter in DCM with Current-Mode Control and Digital Feedback.*

The difference equation set corresponding to Figure 4.6 is given as follows,

$$
\begin{pmatrix} q_{j+1} \\ i_{j+1} \\ v_{p_{j+1}} \\ v_{d_{j+1}} \\ i_{p_{j+1}} \\ i_{1_{j+1}} \\ v_{3_{j+1}} \\ \gamma_{j+1} \\ w_{j+1} \end{pmatrix} =
\begin{bmatrix}
-\left[\left(\text{if}\left(R_{s_{en}} i_{p_j} A_c \ge \gamma_j, 1, 0\right)\right)\nu - \left[q_j \nu\left(\text{if}\left(c_{lk_j} = 1, 1, 0\right)\right)\right]\right] \\
\left(1 - (\delta t / L_1) r_1\right) i_j + \left(\delta t / L_1\right)\left(V_{in} - v_{p_j}\right) \\
\left[1 - \left(\delta t / C_1 r_d\right)\right] v_{p_j} + (\delta t / C_1)\left(i_j - i_{p_j} + (1 / r_d) v_{d_j}\right) \\
\left[1 - \left(\delta t / r_d C_2\right)\right] v_{d_j} + \left(\delta t / r_d C_2\right) v_{p_j} \\
\text{if}\left[i_{p_j} < 0, 0, \text{if}\left[q_j = 1, \left[1 - (\delta t / L_p)\left(r_p + R_{s_{en}} + R_{on}\right)\right] i_{p_j} + (\delta t / L_p) v_{p_j}, 0\right]\right] \\
\text{if}\left[i1_j < 0, 0, \text{if}\left[q_j = 1, 0, \sqrt{(L_p / L_{s1})} i_{p_j} + \left[1 - (\delta t / L_{s1})\left(r_{s1} + R_{p1}\right)\right] \mathrm{i}_{1_j} \right.\right. \\
\left.\left. - (\delta t / L_{s1})\left(V_{CR} + k_1 v_{3_j}\right)\right]\right] \\
\left[1 + (\delta t / (r_3 C_3)(k_1 - 1)\right] v_{3_j} + (\delta t / r_3 C_3) R_{p1} i_{1_j} \\
\text{if}\left[\gamma_j < 0, 0, \text{if}\left[\gamma_j > 12, 12, \mathrm{VR}_j + a_0\left[\left[\mathrm{VR}_j - k_f\left(R_{p1} i_{1_j} + k_1 v_{3_j}\right)\right] \right.\right.\right. \\
\left.\left.\left. - b_1 w_j - b_2 w_{j-1}\right] + a_1 w_j + a_2 w_{j-1}\right]\right] \\
\left[\mathrm{VR}_j - k_f\left(R_{p1} i1_j + k_1 v_{3_j}\right)\right] - b_1 w_j + b_2 w_{j-1}
\end{bmatrix}
$$

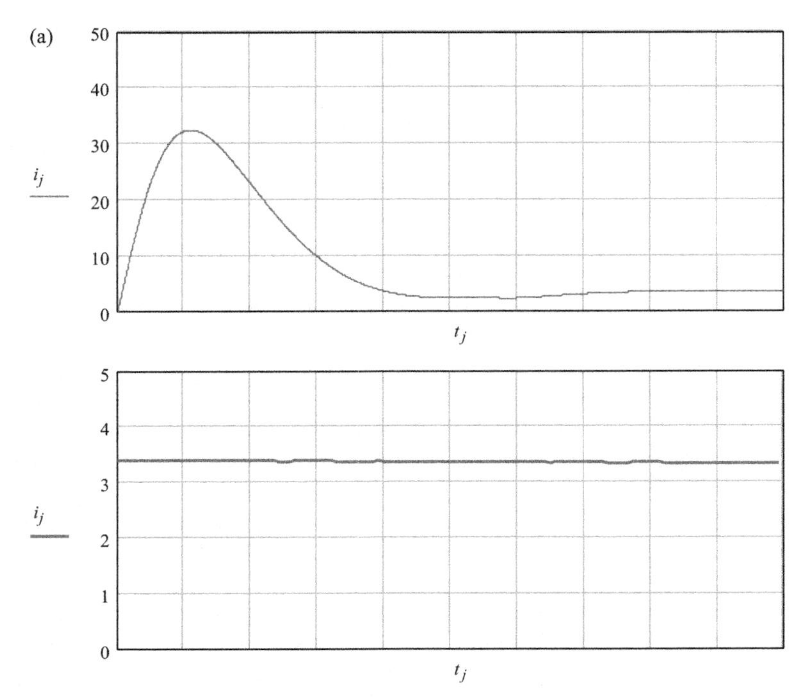

Figure 4.7 Performance of Figure 4.6 circuit (a) Input current, (b) primary winding/switch current, (c) error voltage and current sensing, (d) secondary winding current, and (e) output voltage.

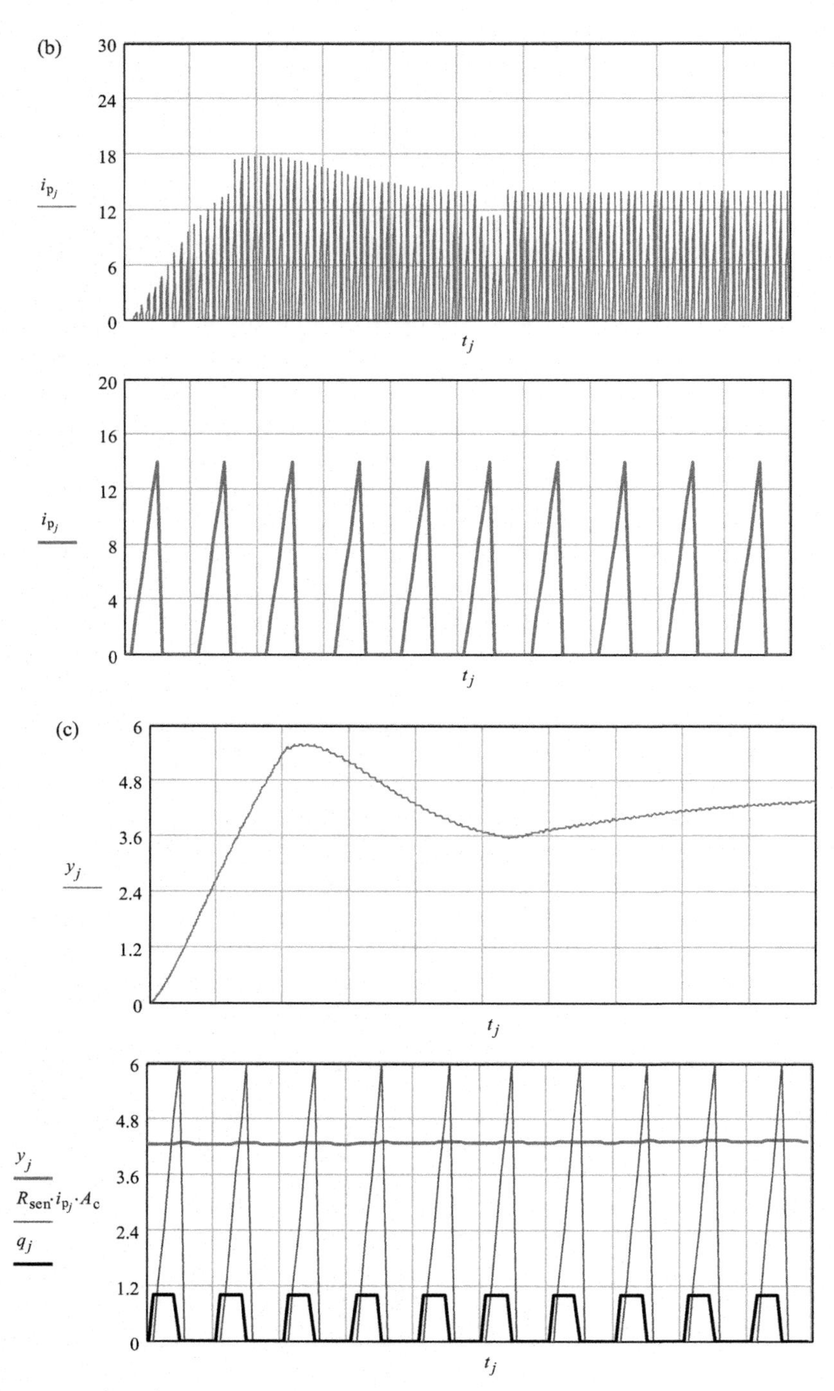

Figure 4.7 ***(cont.)***

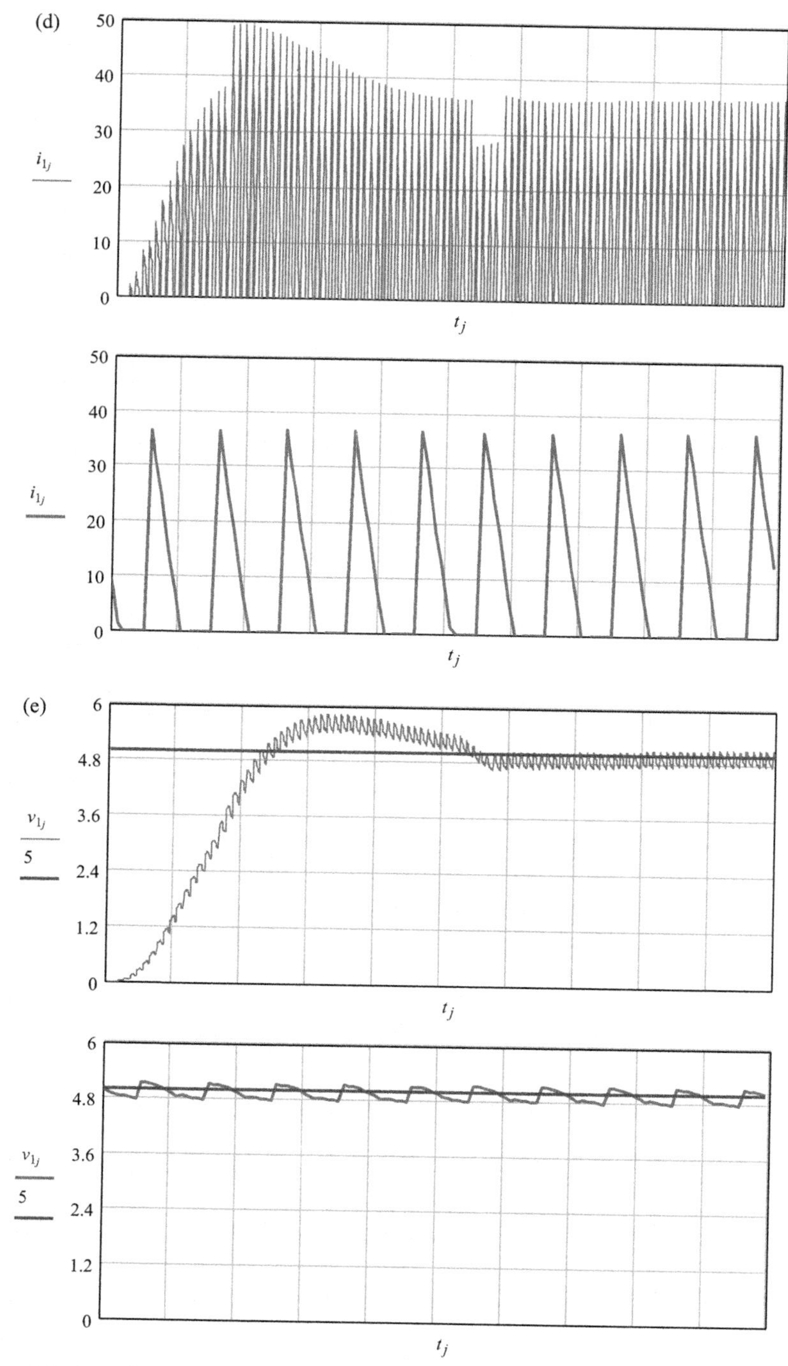

Figure 4.7 *(cont.)*

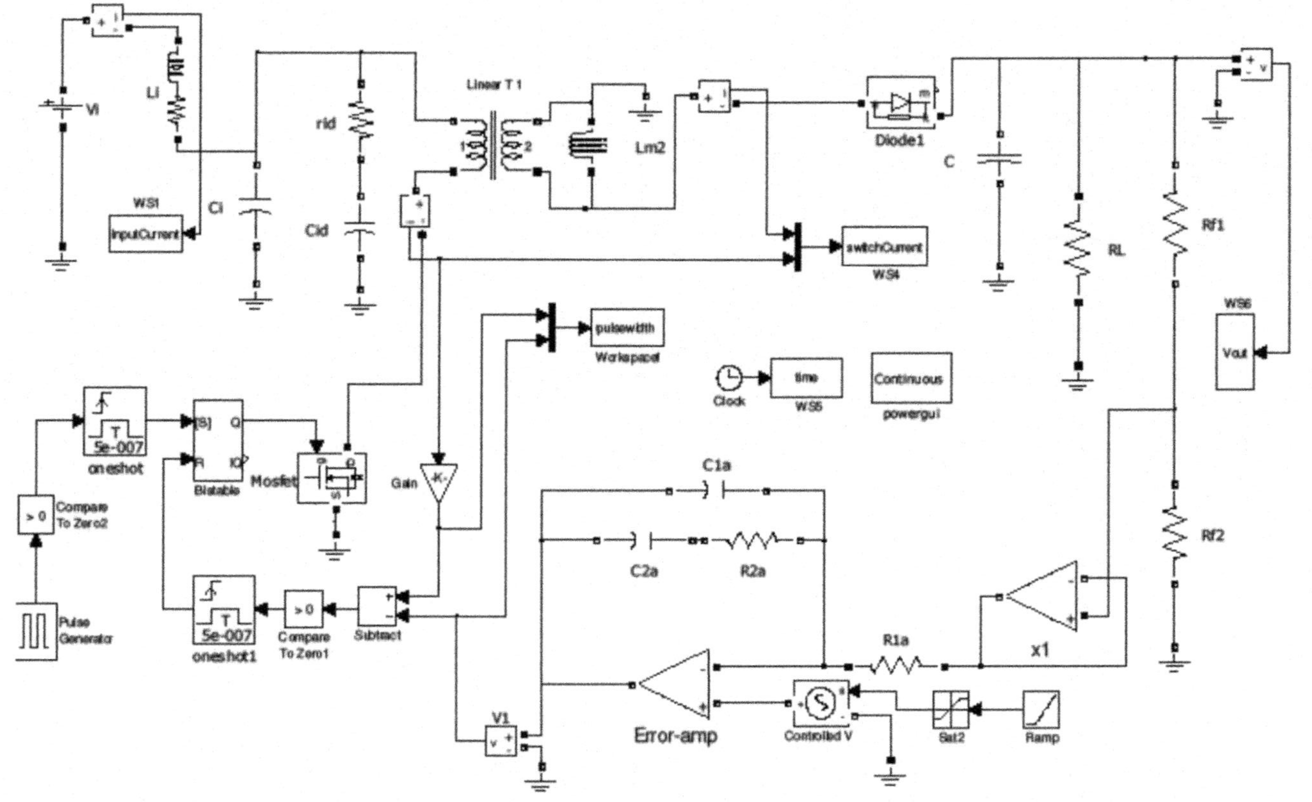

Figure 4.8 ***SIMULINK Model of One Output Flyback in DCM with Current Mode and Analog Feedback.*** (a) output voltage, and (b) pulse-width modulation (current feedback intercepts error signal) and switch current.

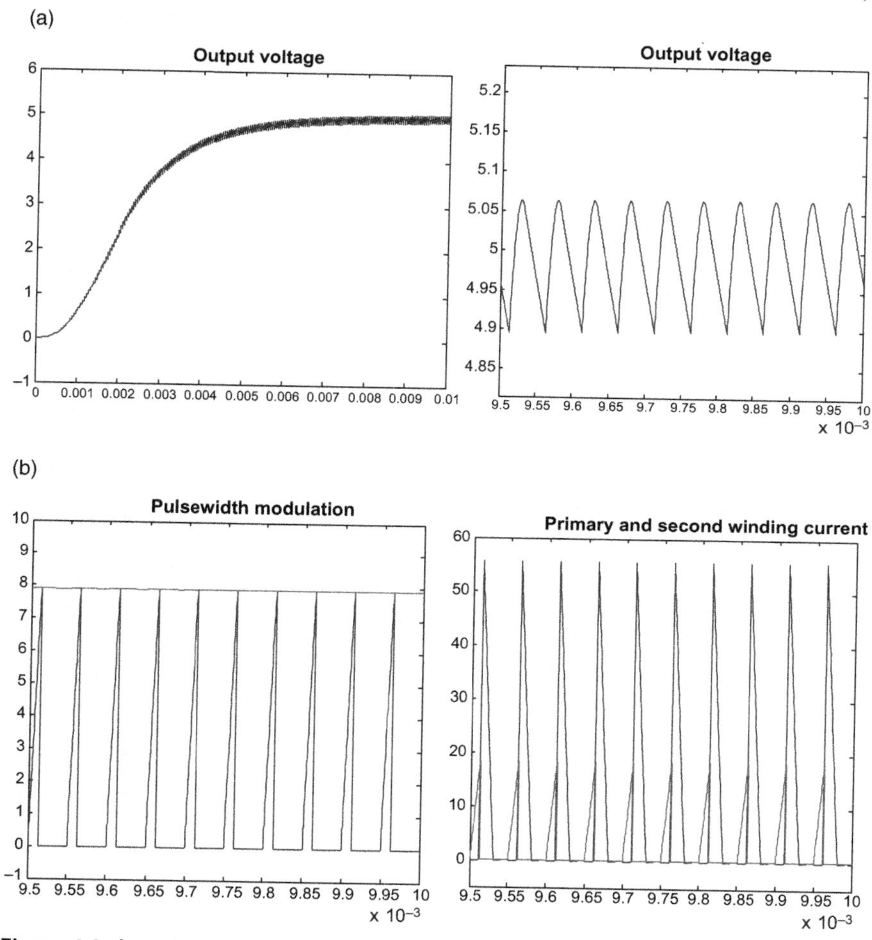

Figure 4.8 ***(cont.)***

Next, the difference-equation-based simulations given above, Figure 4.5 and Figure 4.7, are replaced by one of the graphic-approach, SIMULINK models, and analog control version, Figure 4.8, as well as digital feedback version, Figure 4.9.

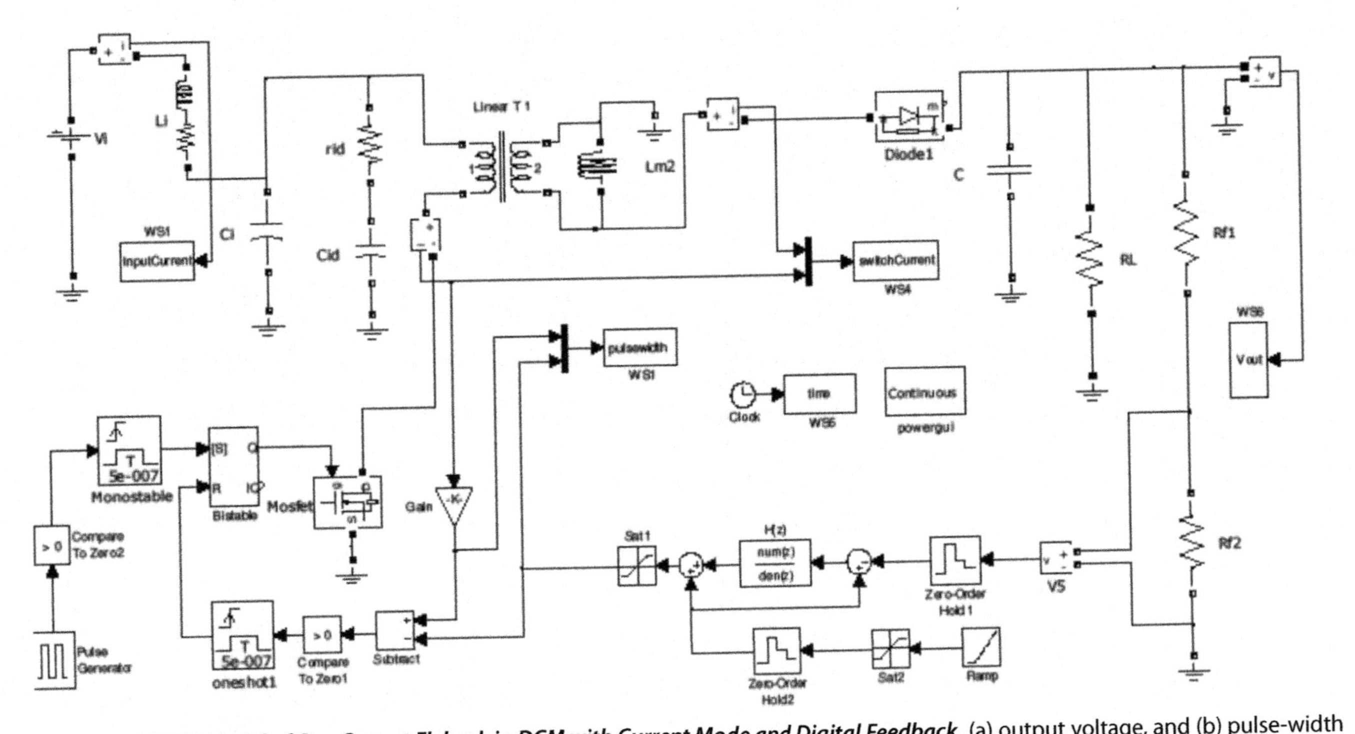

Figure 4.9 ***SIMULINK Model of One Output Flyback in DCM with Current Mode and Digital Feedback.*** (a) output voltage, and (b) pulse-width modulation (current feedback intercepts error signal) and switch current.

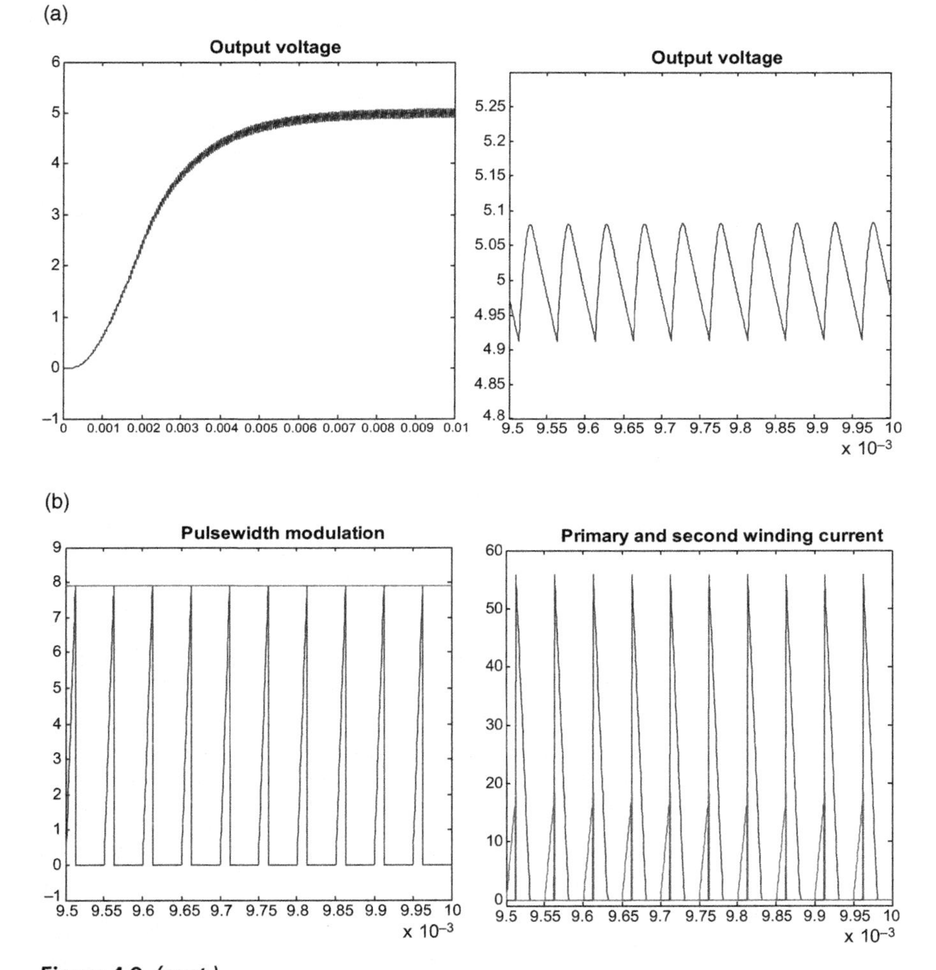

Figure 4.9 *(cont.)*

PART III

Linear Regulator and LED Array Driver

Linear regulators, particularly the precision ones, play some critical role in precision sensors, professional instruments, and medical equipments, which set very stringent requirements for dealing with either expensive systems or life objects. Digital improvements on the conventional linear regulators may sound slightly overkill. But the future need of integrating laboratory or clinical environment with the web shall justify it.

LED is beginning to force even government agencies to take action in setting new regulatory policy for energy consideration. As such, constant current drivers are covered in this part.

CHAPTER 5

Linear Regulator

Switch-mode power converters presented so far generate noisy environments because of incessant ON/OFF switching of power semiconductors at high frequency and at high current rate, *di/dt*. Such EMI-rich background is highly undesirable for safety considerations called for in medical and other precision equipments. Linear regulators, either bipolar transistor or unipolar MOSFET, are then preferred in such applications.

5.1 BIPOLAR LINEAR REGULATOR

The linear regulator works by sampling its output with a simple voltage divider. The output sample is then compared against a command reference utilizing a high gain analog IC. The comparison generates an error voltage, V_2, which drives a current-controlled NPN transistor Q_1. The NPN transistor is driven in a linear fashion and generates a control current I_{C1} that consists of two parts. One part, through R_5, biases the Q_2 emitter-base junction. The other, through the Q_2 base terminal, regulates the series element Q_2 itself. Depending on the loading condition, the output moves with it. Heavier load decreases the output and increases the error voltage. A higher error drives higher base current for the series pass transistor. Consequently, a higher collector current I_{C2} is delivered and pulls up the output voltage. When the load decreases, all signal movements get reversed. By doing so, the output is regulated.

5.2 DERIVATION OF MODULATOR GAIN

Following the same procedure, Figure 5.1 is first placed in its small signal equivalent circuit form, Figure 5.2.

For the purpose of developing modulator gain, and loop gain, eventually the equivalent circuit loop is broken at the error voltage node, and an external low-level test signal, v_i, is injected. Both transistors are replaced by the common-emitter *h*-parameters. In this case, the simpler version considering only current gain, h_{fe}, and input impedance, h_{ie}, is used. The transistor output feedback factor, h_{re}, and output conductance, h_{oe}, are omitted. Several

Power Converters with Digital Filter Feedback Control
http://dx.doi.org/10.1016/B978-0-12-804298-4.00005-8

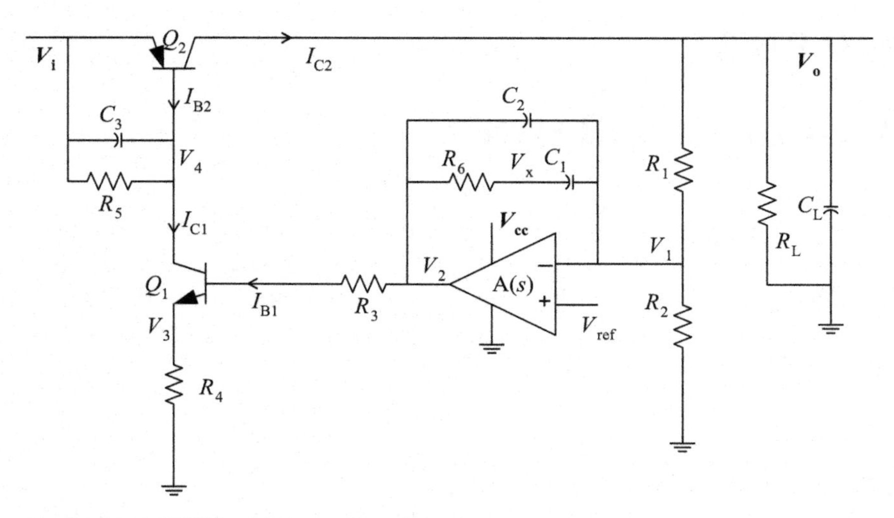

Figure 5.1 *A PNP Bipolar Linear Regulator.*

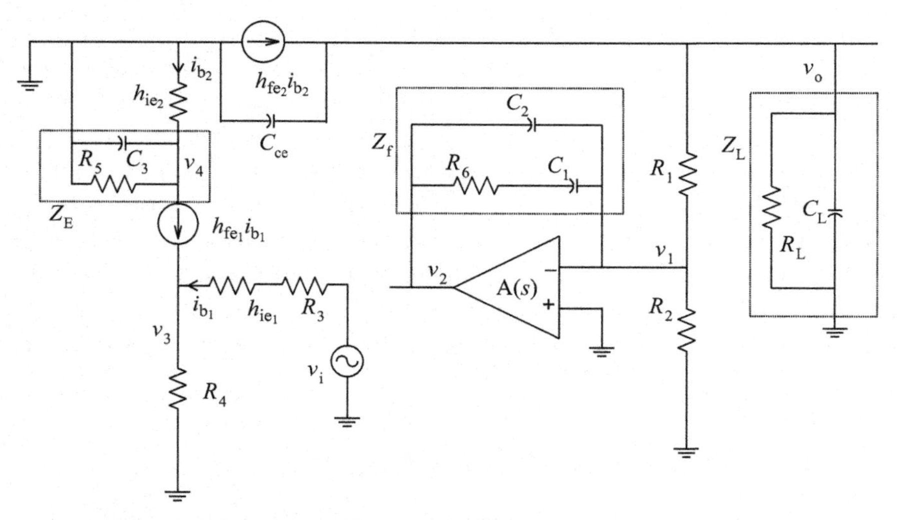

Figure 5.2 *Small Signal Equivalent Circuit for Figure 5.1.*

variables are assigned as independent: v_3, v_4, and v_o. At node v_3, the following Kirchhoff's current law (KCL) equation is established.

$$\left(\frac{1+h_{fe_1}}{R_3+h_{ie_1}}+\frac{1}{R_4}\right)v_3=\frac{1+h_{fe_1}}{R_3+h_{ie_1}}v_i \tag{5.1}$$

And, at node v_4,

$$\frac{h_{fe1}}{R_3 + h_{ie_1}} v_3 - \left(\frac{1}{h_{ie_2}} + \frac{1}{Z_E(s)} \right) v_4 = \frac{h_{fe_1}}{R_3 + h_{ie_1}} v_i \tag{5.2}$$

at the output node,

$$\frac{h_{fe_2}}{h_{ie_2}} v_4 + \left(\frac{1}{R_1 + R_2} + C_{ce} s + \frac{1}{Z_L(s)} \right) v_o = 0 \tag{5.3}$$

The above three equations give the output as,

$$v_o = \frac{\begin{vmatrix} \frac{1+h_{fe_1}}{R_3 + h_{ie_1}} + \frac{1}{R_4} & 0 & \frac{1+h_{fe_1}}{R_3 + h_{ie_1}} v_i \\ \frac{h_{fe_1}}{R_3 + h_{ie_1}} & -\left(\frac{1}{h_{ie_2}} + \frac{1}{Z_E(s)} \right) & \frac{h_{fe_1}}{R_3 + h_{ie_1}} v_i \\ 0 & \frac{h_{fe_2}}{h_{ie_2}} & 0 \end{vmatrix}}{\begin{vmatrix} \frac{1+h_{fe_1}}{R_3 + h_{ie_1}} + \frac{1}{R_4} & 0 & 0 \\ \frac{h_{fe_1}}{R_3 + h_{ie_1}} & -\left(\frac{1}{h_{ie_2}} + \frac{1}{Z_E(s)} \right) & 0 \\ 0 & \frac{h_{fe_2}}{h_{ie_2}} & \frac{1}{R_1 + R_2} + C_{ce} s + \frac{1}{Z_L(s)} \end{vmatrix}} \tag{5.4}$$

The modulator gain, from the test signal injection to the error amplifier input, is then given as,

$$M(s)=\frac{\begin{vmatrix} \frac{1+h_{fe_1}}{R_3+h_{ie_1}}+\frac{1}{R_4} & 0 & \frac{1+h_{fe_1}}{R_3+h_{ie_1}} \\ \frac{h_{fe_1}}{R_3+h_{ie_1}} & -\left(\frac{1}{h_{ie_2}}+\frac{1}{Z_E(s)}\right) & \frac{h_{fe_1}}{R_3+h_{ie_1}} \\ 0 & \frac{h_{fe_2}}{h_{ie_2}} & 0 \end{vmatrix}\frac{R_2}{R_1+R_2}}{\begin{vmatrix} \frac{1+h_{fe_1}}{R_3+h_{ie_1}}+\frac{1}{R_4} & 0 & 0 \\ \frac{h_{fe_1}}{R_3+h_{ie_1}} & -\left(\frac{1}{h_{ie_2}}+\frac{1}{Z_E(s)}\right) & 0 \\ 0 & \frac{h_{fe_2}}{h_{ie_2}} & \frac{1}{R_1+R_2}+C_{ce}s+\frac{1}{Z_L(s)} \end{vmatrix}} \tag{5.5}$$

5.3 EXAMPLE – BIPOLAR LINEAR REGULATOR

We shall again use an example to show that the general process applied for switch-mode power converter is just as applicable for linear regulator.

Briefly, the operating conditions, all component values, device parameters are given. V_i = 5 V, V_o = 3.3 V, V_{ref} = 2.5 V, loading R_L = 1.65, h_{fe} = 100, h_{ie} = 30, R_1 = 4.9K, R_2 = 15.3K, R_3 = 4.9K, R_4 = 163, R_5 = 10K, C_L = 22 μF, and C_3 = 0.001 μF. (5.5) gives the modulator gain, as shown in Figure 5.3.

At a selected crossover frequency of 20 kHz, the plot shows −17.5 db gain and −77° phase. If 60° phase margin is desired at the crossover frequency, a type-II amplifier with C_1 = 613 pf, C_2 = 112 pf, and R_6 = 32.9K is obtained. It gives a loop gain, as shown in Figure 5.4.

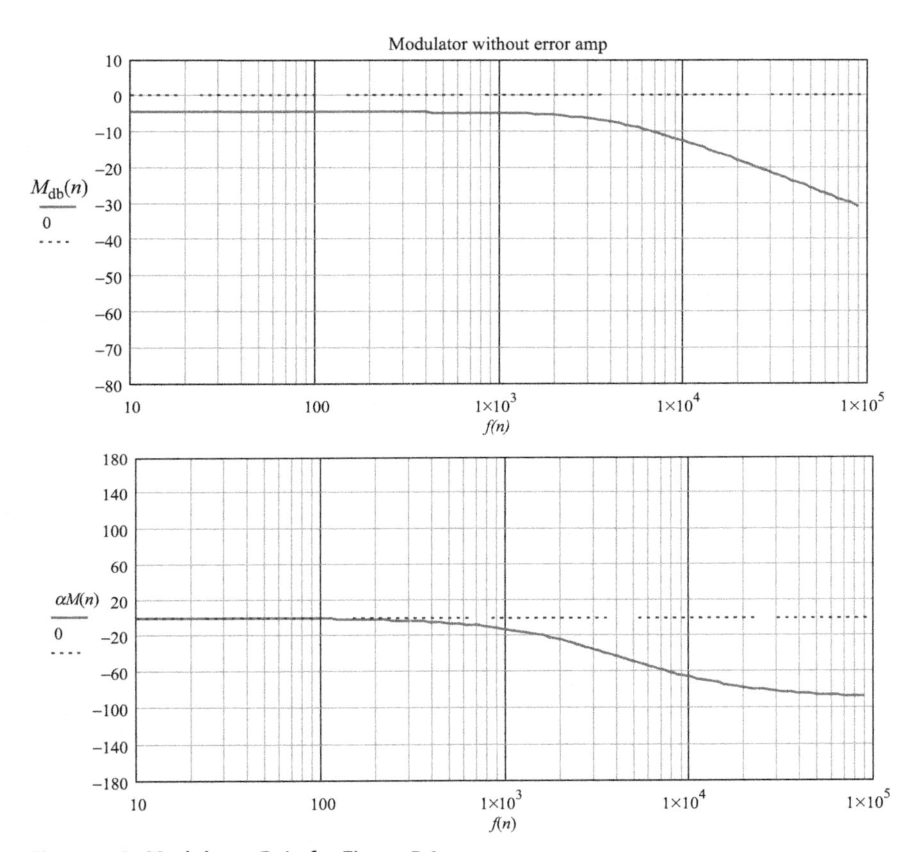

Figure 5.3 ***Modulator Gain for Figure 5.1.***

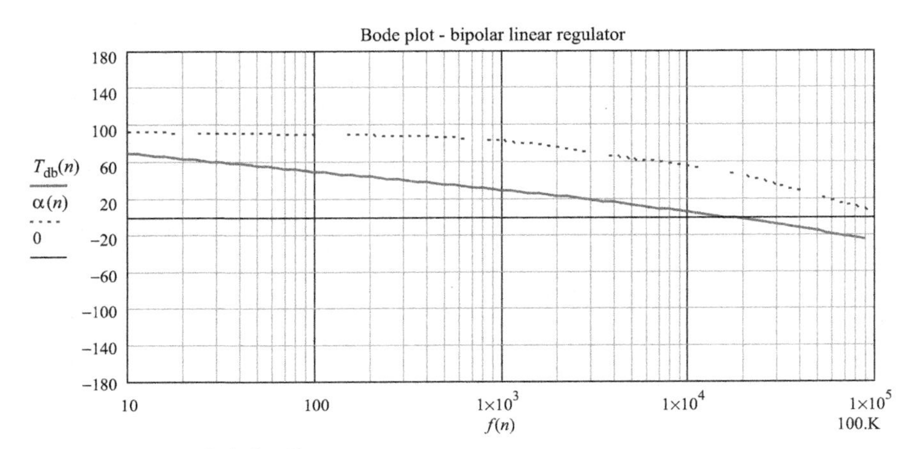

Figure 5.4 ***Loop Gain for Figure 5.1.***

The analog error amplifier identified here gets transformed into a digital one with bilinear transformation and a transform constant, $C = 24$ MHz. Then the following polynomial coefficients are obtained.

$$a_0 = 0.099 \quad a_1 = 4.058 \times 10^{-4} \quad a_2 = -0.098$$
$$b_1 = -1.974 \quad b_2 = 0.974$$

5.4 BIPOLAR LINEAR REGULATOR IN TIME DOMAIN

Prior to setting up the difference equations describing the regulator given in Figure 5.1, a few words regarding the difference existing between the switch-mode converter and linear regulator are warranted.

Power transistors in SMPS operate in non-linear, ON/OFF fashion. They are treated as switches with a small on-state resistance when turned on, or an open circuit when turned off. This behavior is easily incorporated into the dynamic equations used in all previous four chapters.

Power transistors in the linear regulator operate instead in the linear region of their operating curves. That being said, it does not mean that the device operation characteristics can be expressed in forms suitable for incorporating a difference equation set. Operation in the linear region simply means that it is limited to a small region in which linear system concepts are in force, or, as stated from another angle, the system is not expected to be so perturbed such that it moves out of the small region once it had moved in.

In the case of linear regulators with bipolar transistors, base-emitter junction current/voltage, exponentially linked, is the dominant parameter that set the stage for linear operation. As an exponential function, it cannot be brought into any first order differential equation and yields a simple solution.

Therefore, the base-emitter junction voltages of both Q_1 and Q_2 in Figure 5.1 are treated as independent parameters with preset values subjected to iterative and educational tweaking if so required.

Now, we go forward and set up the Q_2 collector current.

$$I_{C_2} = h_{FE_2}\left(I_{C_1} - \frac{V_{be_2}}{R_5}\right) = h_{FE_2}\left(\frac{h_{FE_1}\left(v_e - V_{be_1}\right)}{R_3 + \left(1 + h_{FE_1}\right)R_4} - \frac{V_{be_2}}{R_5}\right) \tag{5.6}$$

and, output voltage,

$$C_L \frac{dv_o}{dt} = I_{C_2} - \frac{v_o}{R_L} \tag{5.7}$$

The error voltage, v_e, and local feedback, v_b, at R_6 and C_1 junction, had been treated in the last chapter. They will not be repeated here. The overall difference equation set is therefore given as, ($R_{1a} = R_1/R_2$, $R_{2a} = R_6$, $C_{1a} = C_2$, $C_{2a} = C_1$, $k_f = R_2/(R_1 + R_2)$)

$$\begin{pmatrix} v_{o_{j+1}} \\ v_{e_{j+1}} \\ v_{b_{j+1}} \end{pmatrix} = \begin{bmatrix} \left(1 - \frac{\delta t}{R_L C_L}\right) v_{o_j} + \frac{\delta t}{C_L} h_{FE_2} \left[\frac{h_{FE_1}\left(v_{e_j} - V_{be_1}\right)}{R_3 + \left(1 + h_{FE_1}\right) R_4} - \frac{V_{be_2}}{R_5} \right] \\ \text{if}\left[v_{e_j} < 0, 0, \text{if}\left[v_{e_j} > 12, 12, v_{e_j} + \left[\begin{matrix} \frac{\delta t}{C_{1a}}\left(\frac{1}{R_{1a}} + \frac{1}{R_{2a}}\right) V_{ref} \\ -\frac{\delta t}{\left(R_{2a} C_{1a}\right)} v_{b_j} - \frac{\delta t}{\left(R_{1a} C_{1a}\right)} k_f v_{o_j} \end{matrix} \right] \right] \right] \\ \left[1 - \left[\frac{1}{\left(R_{2a} C_{2a}\right)} + \frac{1}{\left(R_{2a} C_{1a}\right)}\right] \delta t \right] v_{b_j} + \left[\frac{1}{\left(R_{2a} C_{2a}\right)} + \frac{1}{C_{1a}}\left(\frac{1}{R_{1a}} + \frac{1}{R_{2a}}\right)\right] \\ \delta t V_{ref} - \frac{\delta t}{\left(R_{1a} C_{1a}\right)} k_f v_{o_j} \end{bmatrix} \tag{5.8}$$

The dynamic equation set, (5.8), gives Figure 5.5.

Readers may have noticed that the error voltage is approaching 5 V. That is the reason the error amplifier is powered by a separate source, V_{cc} for instance, at 12 V, rather than V_i. This headache is encountered often in linear regulators.

Next, the analog error amplifier is replaced by its digital equivalent and this action results in changing equation (5.8) to (5.9). (5.9) then yields Figure 5.6.

$$\begin{pmatrix} v_{o_{j+1}} \\ y_{j+1} \\ w_{j+1} \end{pmatrix} = \begin{bmatrix} \left(1 - \frac{\delta t}{R_L C_L}\right) v_{o_j} + \frac{\delta t}{C_L} h_{FE_2} \left[\frac{h_{FE_1}\left(y_j - V_{be_1}\right)}{R_3 + \left(1 + h_{FE_1}\right) R_4} - \frac{V_{be_2}}{R_5} \right] \\ \text{if}\left[y_j < 0, 0, \text{if}\left[\begin{matrix} y_j > 12, 12, V_{ref} + a_0 \left[\left(V_{ref} - k_f v_{o_j}\right) - b_1 w_j - b_2 w_{j-1}\right] \\ + a_1 w_j + a_2 w_{j-1} \end{matrix} \right] \right] \\ \left(V_{ref} - k_f v_{o_j}\right) - b_1 w_j - b_2 w_{j-1} \end{bmatrix} \tag{5.9}$$

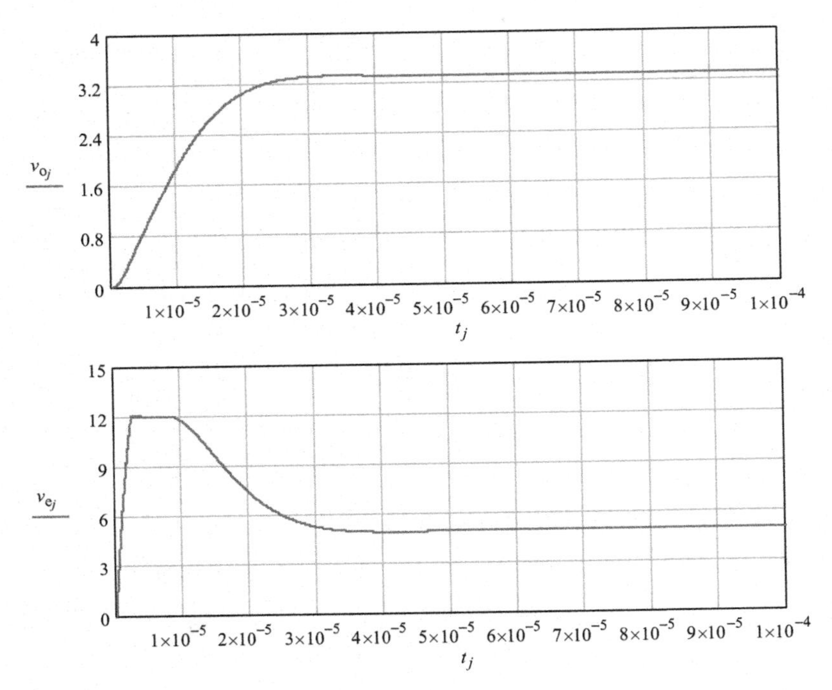

Figure 5.5 ***Bipolar Linear Regulator Performance.***

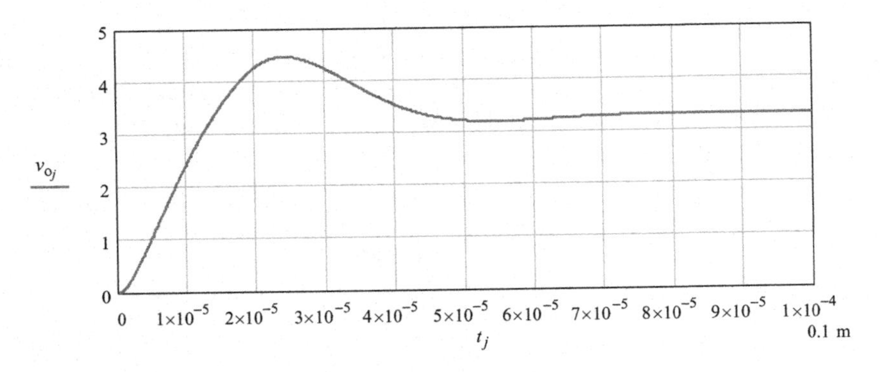

Figure 5.6 ***Bipolar Linear Regulator with Digital Filter.***

5.5 MOSFET LINEAR REGULATOR

Bipolar transistors are current-controlled current device. This is self-evident by the appearance of transistor current gains, h_{FE}, and current quantities, voltage/resistance, in the first equation of (5.8) and (5.9).

By contrast, unipolar MOSFETs are voltage-controlled devices. The key parameter that differentiates MOSFET from bipolar is the transconductance,

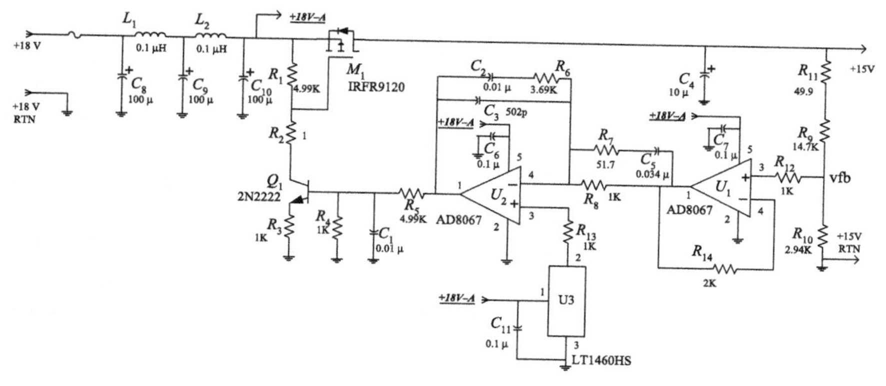

Figure 5.7 ***Linear Regulator with P-Channel MOSFET.***

labeled g_m in the device data sheet, signifying output current controlled by input voltage.

In this section, an actual linear regulator with p-channel MOSFET that is used in a phased-array radar power system is presented in its entirety, Figure 5.7, including device identification and part values.

Figure 5.7 is redrawn and parts renumbered to simplify mathematical formulation. The redrawn figure, Figure 5.11, is then placed in small signal form, Figure 5.8, with the loop open at the error amplifier output node.

There are seven node variable;, v_s, v_g, v_d, v_{fb}, v_b, v_e, and v_c. At each node, a KCL equation is derived and placed in the order given accordingly. The modulator gain is then expressed as $M(s) = v_{fb}/v_{in}$, with the denominator $D(s)$ and the numerator $N_{fb}(s)$.

$$N_{fb}(s) = \begin{bmatrix} \frac{1}{Z_s(s)} + \frac{1}{Z_g(s)} + C_{ds}s + g_m & \frac{-1}{Z_g(s)} - g_m & -(C_{ds}s) & 0 & 0 & 0 & 0 \\ \frac{1}{Z_g(s)} & \frac{-1}{Z_g(s)} - C_{gs}s - \frac{1}{R_2} & C_{gs}s & 0 & 0 & 0 & \frac{1}{R_2} \\ C_{ds}s + g_m & -g_m + C_{gd}s & -C_{ds}s - C_{gd}s - \frac{1}{ZL(s)} - \frac{1}{R_7} & \frac{1}{R_7} & 0 & 0 & 0 \\ 0 & 0 & \frac{1}{R_7} & \frac{-1}{R_7} - \frac{1}{R_8} & 0 & 0 & 0 \\ 0 & 0 & 0 & 0 & \frac{-1}{R_4} - \frac{1}{Z_b(s)} - \frac{1}{h_{ie}} & \frac{1}{h_{ie}} & 0 \\ 0 & 0 & 0 & 0 & \frac{(1+h_{fe})}{h_{ie}} & \frac{-(1+h_{fe})}{h_{ie}} - \frac{1}{R_e} & 0 \\ 0 & \frac{1}{R_2} & 0 & 0 & -\frac{h_{fe}}{h_{ie}} & \frac{h_{fe}}{h_{ie}} & -\frac{1}{R_2} \end{bmatrix} \tag{5.10}$$

$$N_{\mathrm{fb}}(s) = \begin{bmatrix} \frac{1}{Z_s(s)} + \frac{1}{Z_g(s)} + C_{ds}s + g_m & \frac{-1}{Z_g(s)} - g_m & -(C_{ds}s) & 0 & 0 & 0 & 0 \\ \frac{1}{Z_g(s)} & \frac{-1}{Z_g(s)} - C_{gd}s - \frac{1}{R_2} & C_{gd}s & 0 & 0 & 0 & \frac{1}{R_2} \\ C_{ds}s + g_m & -g_m + C_{gd}s & -C_{ds}s - C_{gd}s - \frac{1}{ZL(s)} - \frac{1}{R_7} & 0 & 0 & 0 & 0 \\ 0 & 0 & \frac{1}{R_7} & 0 & 0 & 0 & 0 \\ 0 & 0 & 0 & -\frac{1}{R_4} & \frac{-1}{R_4} - \frac{1}{Z_b(s)} - \frac{1}{h_{ie}} & \frac{1}{h_{ie}} & 0 \\ 0 & 0 & 0 & 0 & \frac{(1+h_{fe})}{h_{ie}} & \frac{-(1+h_{fe})}{h_{ie}} - \frac{1}{R_e} & 0 \\ 0 & \frac{1}{R_2} & 0 & 0 & -\frac{h_{fe}}{h_{ie}} & \frac{h_{fe}}{h_{ie}} & -\frac{1}{R_2} \end{bmatrix} \tag{5.11}$$

5.6 EXAMPLE – MOSFET LINEAR REGULATOR

From International Rectifier's data sheet, M_1 of Figure 5.7 gives $C_{iss} = 270$ pF, $C_{oss} = 170$ pF, $C_{rss} = 31$ pF, $C_{gd} = C_{rss}$, $C_{ds} = C_{oss} - C_{gd}$, and $g_m = 1$ at the expected operating point. For Q_2, $h_{ie} = 100$, $h_{fe} = 100$. The rest (Figure 5.8) is $R_1 = 5$K, $R_2 = 1$, $R_e = 1$K, $R_3 = 1$K, $C_b = 0.01$ μF, $R_4 = 5$K, $R_8 = 3.01$K, $R_7 = 15$K, load $R_L = 10$, $C_3 = 10$ μF, $L_1 = L_2 = 0.33$ μH, $C_8 = C_9 = C_{10} = 100$ μF. The component set yields a modulator gain Figure 5.9.

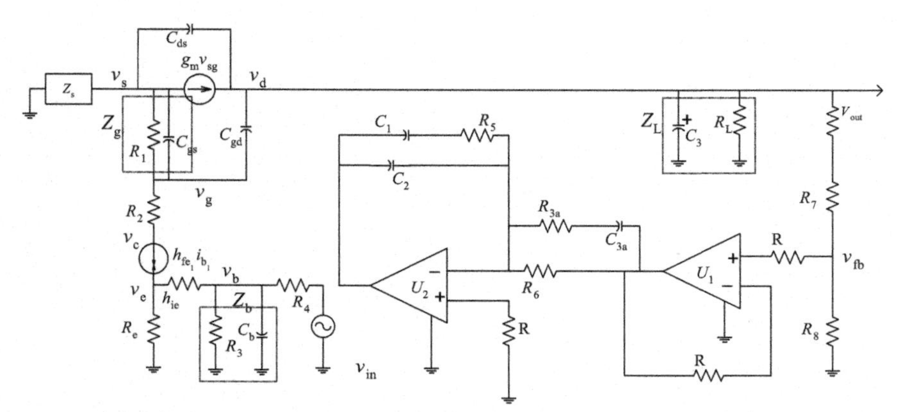

Figure 5.8 ***MOSFET Regulator in Small Signal Form for Modulator Gain Derivation.***

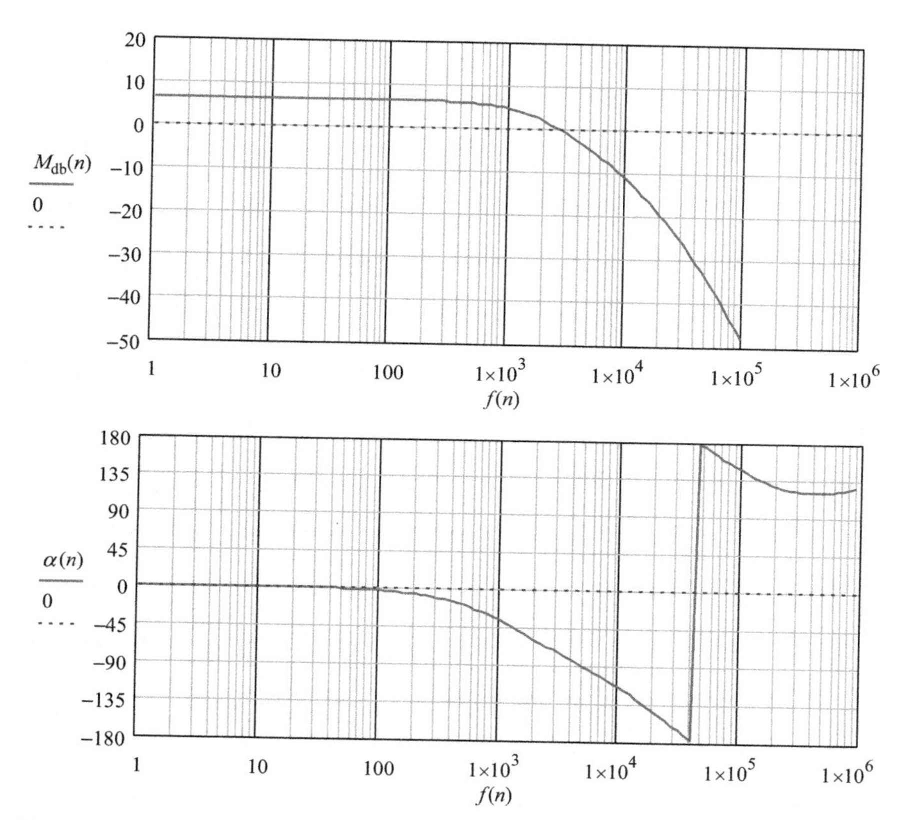

Figure 5.9 ***Modulator Gain of MOSFET Regulator.***

For 20 kHz crossover, the modulator gain/phase plots read −19.5 db and −144°. The excessive phase lag calls for a type-III amplifier. Section 1.15 gives us $R_6 = 1\text{K}$, $R_{3a} = 134$, $C_{3a} = 0.02\ \mu\text{F}$, $R_5 = 3.65\text{K}$, $C_1 = 0.006\ \mu\text{F}$, and $C_2 = 840$ pF. The resulting loop gain plot looks like Figure 5.10.

The actual measurement, Figure 5.10b, matches well the theoretical result given here, Figure 5.10a. Prior to time domain studies, the type-III analog error amplifier is transformed to digital. The following z-transform polynomial coefficients are obtained.

$$a_0 = 0.124 \quad a_1 = -0.124 \quad a_2 = -0.124 \quad a_3 = 0.124$$

$$b_1 = -2.982 \quad b_1 = 2.964 \quad b_3 = -0.982$$

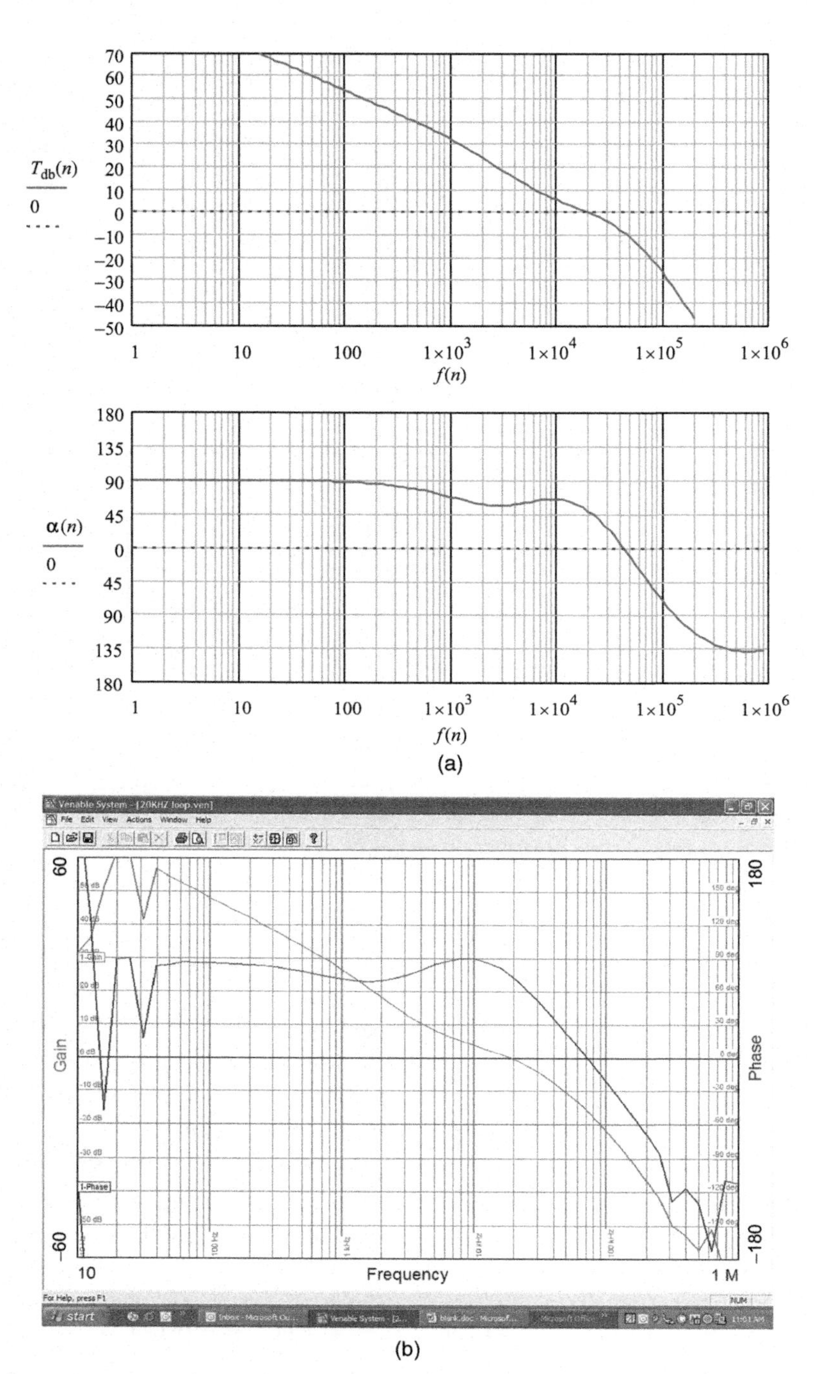

Figure 5.10 (a) Loop gain of MOFSET regulator, and (b) loop gain, actual measurement.

5.7 MOSFET LINEAR REGULATOR IN TIME DOMAIN

Based on Figure 5.11, the difference equation set for the regulator with analog control is formulated, (5.12). Here, the MOSFET is modeled as a controlled current source with square law properties; that is, drain current $I_d = \alpha(V_{gs} - V_{th})^2$, α being curve fitting parameter for MOSFET, V_{th} the gate-to-source threshold voltage, about 4 V. Figure 5.12 shows a slow start and an output settling to steady state in less than 0.5 ms.

$$\begin{pmatrix} i_{2_{j+1}} \\ i_{3_{j+1}} \\ v_{1_{j+1}} \\ v_{2_{j+1}} \\ v_{s_{j+1}} \\ v_{o_{j+1}} \\ v_{a_{j+1}} \\ v_{b_{j+1}} \\ v_{e_{j+1}} \end{pmatrix} = \begin{bmatrix} \left(1-\frac{r\delta t}{L_1}\right) i_{2_j} + \frac{\delta t}{L_1} v_{1_j} - \frac{\delta t}{L_1} v_{2_j} \\ \left(1-\frac{r\delta t}{L_2}\right) i_{3_j} + \frac{\delta t}{L_2} v_{2_j} - \frac{\delta t}{L_2} v_{s_j} \\ \left(1-\frac{\delta t}{R_g C_8}\right) v_{1_j} - \frac{\delta t}{C_8} i_{2_j} + \frac{\delta t}{R_g C_8} V_{in} \\ v_{2_j} + \frac{\delta t}{C_9}\left(i_{2_j} - i_{3_j}\right) \\ v_{s_j} + \frac{\delta t}{C_{10}}\left[i_{3_j} - \alpha\left[R_1 \frac{h_{FE_1}\left(v_{e_j} - V_{be_1}\right)}{R_4 + \left(1+h_{FE_1}\right)R_e} - V_{th}\right]^2 - \frac{h_{FE_1}\left(v_{e_j} - V_{be_1}\right)}{R_4 + \left(1+h_{FE_1}\right)R_e}\right] \\ \left(1-\frac{\delta t}{R_L C_3}\right) v_{o_j} + \frac{\delta t}{C_3}\alpha\left[R_1 \frac{h_{FE_1}\left(\gamma_j - V_{be_1}\right)}{R_4 + \left(1+h_{FE_1}\right)R_e} - V_{th}\right]^2 \\ v_{a_j} + \frac{\delta t}{C_{3a}} \frac{k_f v_{o_j} - v_{a_j}}{R_{3a}} \\ v_{b_j} + \delta t\left(\frac{v_{r_j} - v_{b_j}}{R_5 C_2} - \frac{k_f v_{o_j} - v_{r_j}}{R_6 C_2} - \frac{k_f v_{o_j} - v_{a_j}}{R_{3a} C_2} + \frac{v_{r_j} - v_{b_j}}{R_5 C_1}\right) \\ \text{if}\left[v_{e_j} < 0, 0, \text{if}\left[v_{e_j} > 15, 15, v_{e_j} + \frac{\delta t}{C_2}\left(\frac{v_{r_j} - v_{b_j}}{R_5} - \frac{k_f v_{o_j} - v_{r_j}}{R_6} - \frac{k_f v_{o_j} - v_{a_j}}{R_{3a}}\right)\right]\right] \end{bmatrix} \tag{5.12}$$

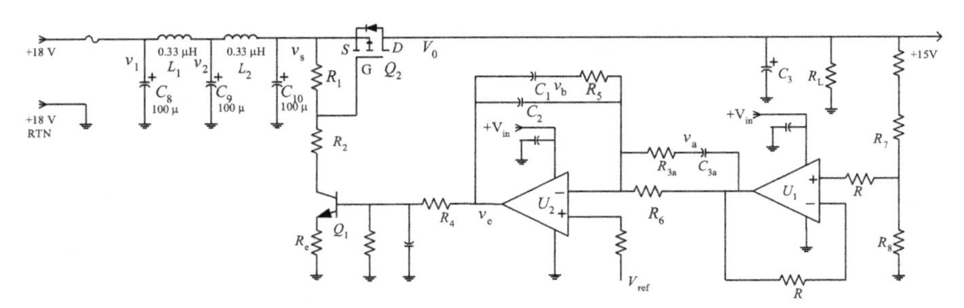

Figure 5.11 ***MOSFET Regulator Redrawn and Part Renamed for Analysis.***

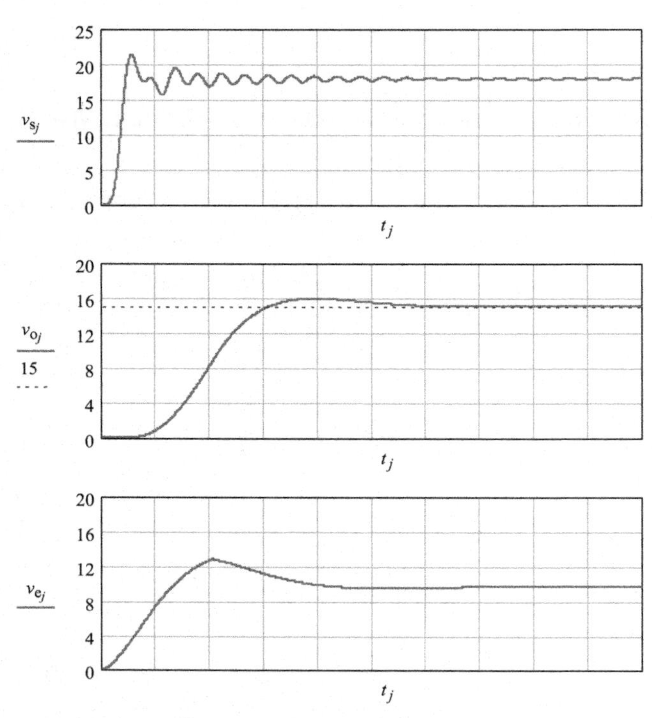

Figure 5.12 ***MOSFET Linear Regulator Performance with Analog Control.*** Top, MOSFET source – second, output; bottom, error voltage.

Replacing the analog error amplifier with its digital counterpart, the difference equation set, (5.12), changes to (5.13).

$$
\begin{pmatrix} i_{2_{j+1}} \\ i_{3_{j+1}} \\ v_{1_{j+1}} \\ v_{2_{j+1}} \\ v_{s_{j+1}} \\ v_{o_{j+1}} \\ y_{j+1} \\ w_{j+1} \end{pmatrix} :=
\begin{bmatrix}
\left(1-\frac{r\delta t}{L_1}\right)i_{2_j}+\frac{\delta t}{L_1}v_{1_j}-\frac{\delta t}{L_1}v_{2_j} \\
\left(1-\frac{r\delta t}{L_2}\right)i_{3_j}+\frac{\delta t}{\mathrm{L}_2}v_{2_j}-\frac{\delta t}{L_2}v_{s_j} \\
\left(1-\frac{\delta t}{R_g C_8}\right)v_{1_j}-\frac{\delta t}{C_8}i_{2_j}+\frac{\delta t}{R_g C_8}V_{in} \\
v_{2_j}+\frac{\delta t}{C_9}\left(i_{2_j}-i_{3_j}\right) \\
v_{s_j}+\frac{\delta t}{C_{10}}\left[i_{3_j}-\alpha\left[R_1\frac{h_{FE_1}\left(y_j-V_{be_1}\right)}{R_4+\left(1+h_{FE_1}\right)R_e}-V_{th}\right]^2-\frac{h_{FE_1}\left(y_j-V_{be_1}\right)}{R_4+\left(1+h_{FE_1}\right)R_e}\right] \\
\left(1-\frac{\delta t}{R_L C_3}\right)v_{o_j}+\frac{\delta t}{C_3}\alpha\left[R_1\frac{h_{FE_1}\left(\mathrm{y}_j-V_{be_1}\right)}{R_4+\left(1+h_{FE_1}\right)R_e}-V_{th}\right]^2 \\
\text{if}\left[y_j<0,0,\text{if}\left[y_j>12,12,v_{r_j}+a_0\left[\left(v_{r_j}-k_f vo_j\right)-b_1w_j-b_2w_{j-1}-b_3w_{j-2}\right.\right.\right. \\
\left.\left.\left. +a_1w_j+a_2w_{j-1}+a_3w_{j-2}\right]\right]\right] \\
\left(v_{r_j}-k_f v_{o_j}\right)-b_1w_j-b_2w_{j-1}-b_3w_{j-2}
\end{bmatrix}
\tag{5.13}
$$

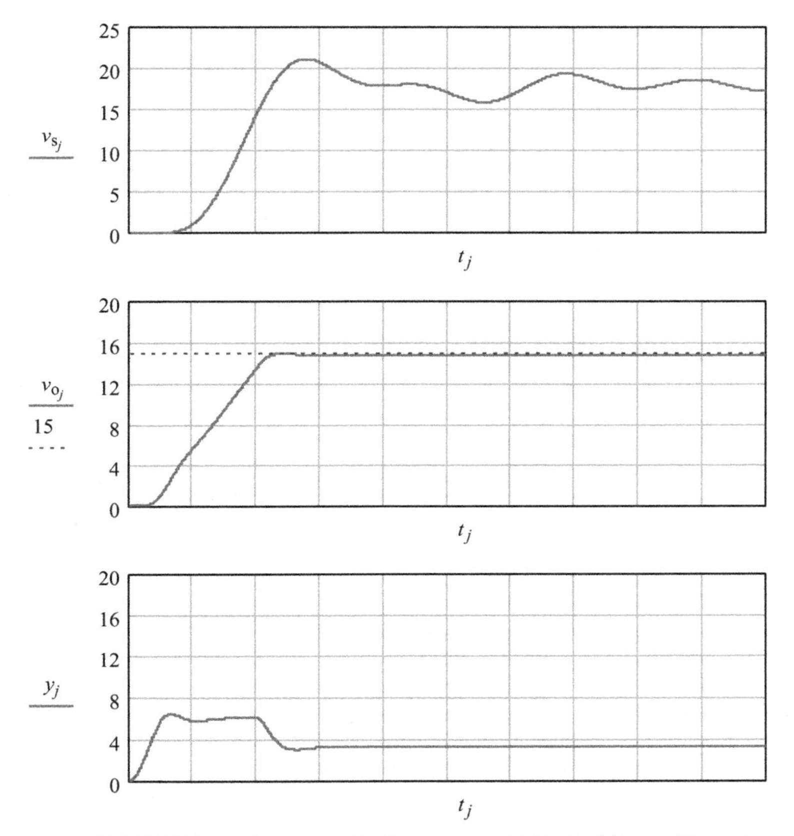

Figure 5.13 ***MOSFET Linear Regulator Performance with Digital Control for 0.1 ms.***

The MOSFET regulator with digital filter seems to perform better, Figure 5.13.

It is quite interesting to see that the control loop comes into regulation as soon as the MOSFET source supply reaches 18 V.

CHAPTER 6

LED Driver

Light-emitting diodes (LEDs) are making huge strides in lighting revolution. It is all about efficiency and reliability. A conventional incandescent, 100 W light bulb gives about 1200 lm (Lumen) and has an average of 1000 h (hours) of operation life. By contrast, a modern LED, for instance Philips LUXEON Rebel ES or Cree Xlamp XM-L, gives more than 100 lm at a forward current of 0.35 A, and a forward voltage of about 2.8 V. That is 100 lm/W for LED, compared with 12 lm/W for the former. In addition, LEDs enjoy very long operation life, easily exceeding 40,000–50,000 h.

Being intrinsically a semiconductor diode, LED is driven only in one direction by the DC source, instead of AC. Existing AC supply must be processed to be suitable for driving LED. It turns out that while in theory, LEDs have the advantage of a long operation life, the power converters, often the AC–DC, actually have worse reliability and become the weakest link.

With that in mind, this chapter treats power converters specifically designed for driving LEDs in a single string or in an array form.

6.1 LED MODEL

LEDs are basically solid-state diodes. The most important characteristic information is their forward current versus forward voltage curve, resembling a segment of an exponential function. For instance, Cree's LED mentioned earlier gives the following performance curve, Figure 6.1; tolerance and thermal drift are not included.

There are several ways to model the device. The better one, in this writer's opinion, is the exponential model in the form of $i_F(v_F) = a\exp(bv_F) + c$ with three unknown parameters, a, b, and c yet to be determined.

With a software tool like MathCAD, the three parameters can be easily found numerically. For this example, $a = 9.66\times10^{-3}$, $b = 1.818$, and $c = -1.157$ are obtained, and it yields Figure 6.2 exponential model.

6.2 DRIVING LED LOAD

Figures 6.1 and 6.2 show the forward current rising sharply as a function of the forward voltage within a narrow range, equivalent to a high sensitivity, 0.2–2 A, a factor of 10, from 2.7 V to 3.2 V, a mere 0.5 V swing. Therefore,

Power Converters with Digital Filter Feedback Control
http://dx.doi.org/10.1016/B978-0-12-804298-4.00006-X

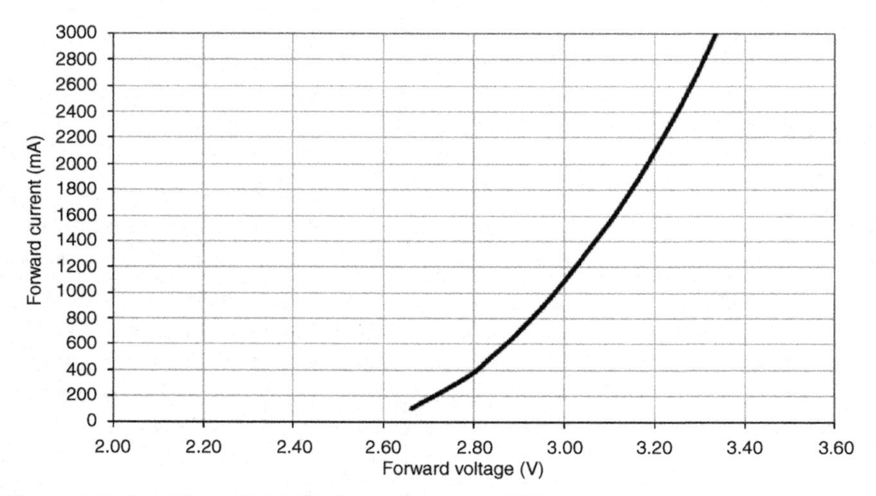

Figure 6.1 ***Cree Xlamp XM-L I/V Curve, Typical at 25°C.***

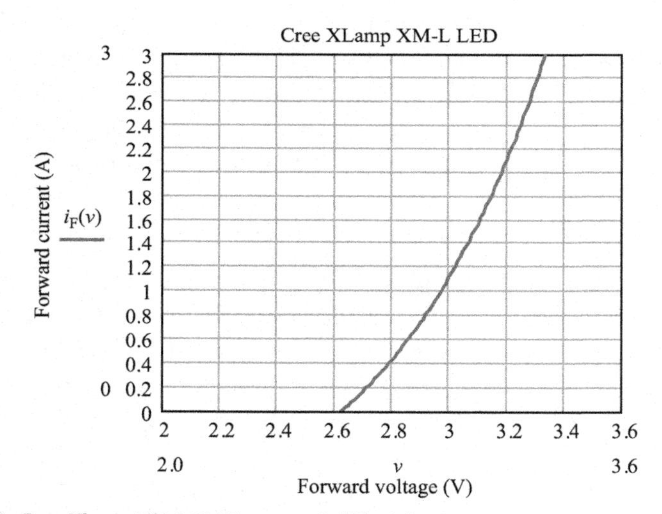

Figure 6.2 ***Cree Xlamp XM-L I/V Exponential Model.***

if driven directly by a DC voltage source with imperfect output regulation, the forward current will develop a large swing. Moreover, it is not a good practice to apply directly a finite voltage to a string of diodes in series, since the diode string, when conducting, sustains a clamping voltage that is not likely equal exactly to the driving source. Assume the source voltage is higher than that of the diode string, the differential voltage without a balance, or limiting, resistor may cause damage. This understanding of the

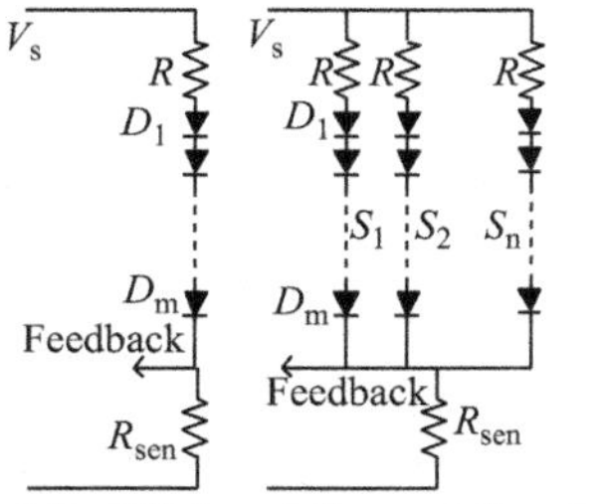

Figure 6.3 ***LED Load Current Sensing; A Single String and an Array.***

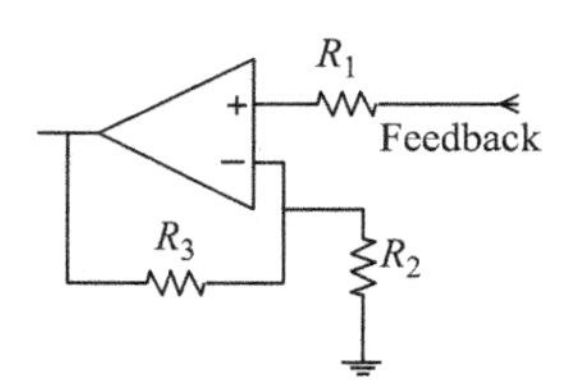

Figure 6.4 ***Current Sensing Preamplifier.***

nature of diodes prompts engineers to drive LEDs with the current source, or more appropriately, the constant current source.

So far, all converter presented are voltage regulators. They require modifications to work as current-regulated supplies. In the case of voltage supply, output voltage is sampled, fed back, and compared with a voltage reference. Instinctively, for current sourcing supply, output current shall be sampled and fed back to control the desired current supply. This is exactly what is done for LED driver. Figure 6.3 gives two configurations, one for a single string with m LEDs, the other for an array with n strings. Since R_{sen} is in series with the LED load, it shall be a very low value resistor, 10–20 mΩ, so not to waste drive source. However, even for an array yielding tens of thousand lumens, the total drive current is still in the range of a few amperes at the most. That means the current sensor drop is quite low, perhaps less than 50 mV. Treating such a low voltage as a control parameter near ground level runs the risk of interference by ground noise. It is a good practice and a necessity to raise the feedback higher. This is easily done by incorporating a noninverting gain buffer, Figure 6.4. The circuit has the equivalent effect of raising the current sensing resistor value by a factor of $(1 + R_3/R_2)$ without waste.

In addition to the current sensing resistor, balancing resistors are sprinkled with each load string. Sizing those resistors such that they meet the requirement while still maintaining desired overall efficiency takes some efforts. We shall use the exponential model to show how it is done.

We start with a single diode in series with a resistor, and the combination is powered by a voltage source, Figure 6.5. Using the exponential model, a KVL equation around the circuit loop gives

$$R_x I_F + V_F = R_x(a e^{bV_F} + c) + V_F = V_s \tag{6.1}$$

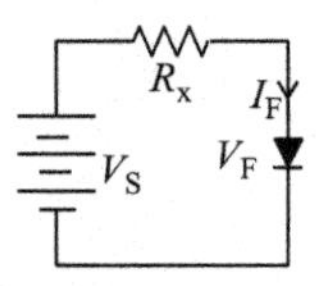

Figure 6.5 ***Single LED Plus Resistor Circuit.***

That is,

$$R_x = \frac{V_s - V_F}{a e^{bV_F} + c} \tag{6.2}$$

Given a device with known model parameters and a desired current, the corresponding forward voltage and supply source determine the series resistance accordingly.

There is another way to find the resistance value. It is a graphical approach. (6.1) is rewritten in the following form

$$\frac{V_s - V_F}{R_x} = I_F = a e^{bV_F} + c \tag{6.3}$$

The right hand side of (6.3) is the diode I_F–V_F curve. The left hand side gives a linear conductance line. Let's use Philips LUXEON Rebel ES LED ($a = 0.014$, $b = 1.604$, $c = -0.974$) as an example to describe the graphical procedure. Assume a supply of 3.2 V and a desired current of 0.5 A, two points on the (V, I) plane are identified and placed: the open source (open circuit) point (3.2 V, 0) and the point (2.92 V, 0.5 A) at which the 0.5 A horizontal line intercepts the LED I_F–V_F curve. The slant line connecting both points is the conductance line required, Figure 6.6. It gives a resistance (3.2–2.92)/0.5 = 0.56 Ω.

When it comes to commercial or industrial lighting, configurations in array forms like Figure 6.3 are preferred, since just a few devices do not offer sufficient lumen for large space. In that case, a graphical approach would not handle it effectively. Instead, the numerical method works better. To represent the array current i_A, the exponential model is modified for a ($m \times n$) array, in which drops across balance resistor and current sensing resistor are, in general, small compared with driving source V_s.

$$i_A = n\left(a e^{b(V_s/m)} + c\right) \tag{6.4}$$

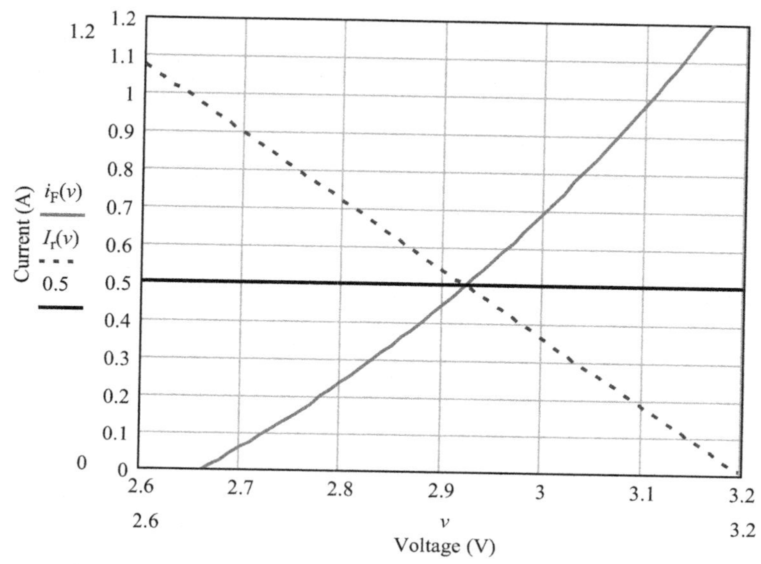

Figure 6.6 ***Graphical Approach Determining the Series Resistance.***

6.3 A TYPICAL INDUSTRIAL LED DRIVER STRUCTURE

As indicated in Section 6.2, LEDs shall be driven with DC current supply. Existing electrical utility power of 50–60 Hz and 120–240 VAC is incompatible and needs to be processed for lighting applications employing solid-state LEDs. In order to preserve the high-efficiency LED offers, power processors converting AC to DC must also perform efficiently. The need renders the switch-mode power supply (SMPS) the exclusive converter of choice. However, complications arise when SMPS is called upon to interface with the AC source. Basically, SMPS expects DC source input. Or, put it in other words, SMPS is fundamentally a DC–DC power processor. Therefore, a front-end stage is required to bridge the gap. The simplest form it comes in is a full-wave bridge rectifier loaded with a large smoothing capacitor. The combination does deliver an acceptable DC source to the DC–DC converter that follows, but with a highly undesirable effects—a pulsating line input current that notoriously violates telecommunication regulations. Over the past years, a better front-end, power-factor-correction (PFC) has effectively replaced the simple full-wave rectifier. This brings us to a typical, industrial LED driver, Figure 6.7, with voltage-mode control; transformer reset winding is not shown, but understood.

Figure 6.7 is a two-stage driver. The first is the PFC that makes the AC input current in phase with the AC source to meet the regulatory compliance

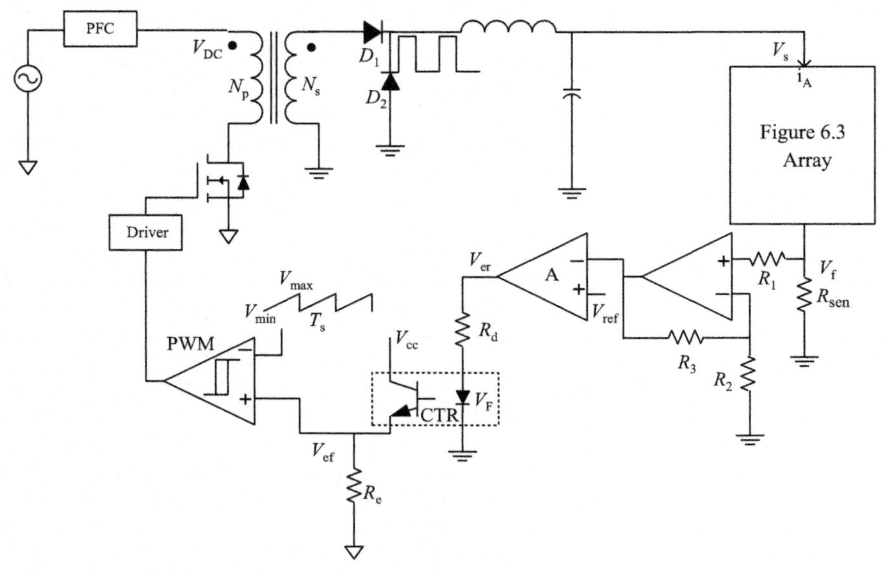

Figure 6.7 ***A Typical Commercial LED Driver Structure.***

requirement. Most existing commercial PFC design is based on boost topology that takes AC input ranging from 100 VAC to 277 VAC (universal AC compatible) and converts it to about 440 VDC (277√2 + 10%) at V_{DC} node, feeding the isolation transformer. Since the topic is not the focus of this writing, it shall not be covered.

The second stage is the DC–DC converter that takes the input V_{DC} and converts it to an output voltage, V_s. The output voltage is not regulated. Instead, it is a drifting voltage that must match, or support, the LED load current, which is regulated. Briefly, this is how it works.

Starting at the current feedback, it gives

$$V_f = i_A R_{sen} \tag{6.5}$$

The error amplifier output is

$$V_{er} = A\left[V_{ref} - V_f\left(1 + \tfrac{R_3}{R_2}\right)\right] \tag{6.6}$$

An optical isolator with current transfer ratio (CTR) maintains the isolation barrier and gives

$$V_{ef} = \frac{V_{er} - V_F}{R_d} \mathrm{CTR}\, R_e \tag{6.7}$$

Given a switching period T_s and a maximum duty cycle of 95% by design, and corresponding to a carrier swing V_{min} to V_{max}, the cyclic, PWM ramp repetition is expressed as

$$V_{ramp}(t) = V_{min} + \frac{V_{max} - V_{min}}{0.95T_s} t \tag{6.8}$$

The steady state duty cycle is determined by the ramp intercepting the error voltage.

$$V_{ramp}(DT_s) = V_{min} + \frac{V_{max} - V_{min}}{0.95T_s} DT_s = V_{ef} \tag{6.9}$$

Given a duty cycle, the power stage yields an output voltage

$$\frac{V_{DC}}{N_p} N_s D = V_s \tag{6.10}$$

With patience, and in theory, all six equations can be consolidated into a single, close-loop equation, but it is a transcendental, implicit equation with an exponential term that prevents us from solving it. We therefore leave it as it is. Readers are to be reminded that the output voltage variable V_s appears in both sides of the equation.

$$\frac{V_{DC}}{N_p} N_s \frac{0.95\left(\left[\left(A\left(V_{ref} - n\left(a\,e^{b(Vs/m)} + c\right) \cdot R_{sen}\left(1 + (R_3/R_2)\right)\right) - V_F\right) / \left(R_d\right)\right] \mathrm{CTR}\, R_e - V_{min}\right)}{V_{max} - V_{min}} = V_s \tag{6.11}$$

6.4 AN LED ARRAY DRIVER WITH VOLTAGE-MODE CONTROL

We now expand Figure 6.7 and present a more realistic driver schematic, Figure 6.8. Since the second stage is the focus, more circuit details, including local housekeeping supply, are shown. For the example, the same Philips LUXEON Rebel ES LEDs, Figure 6.6, are configured as a 13 × 6 array: 13 in a series string, 6 strings in parallel. The overall string voltage is expected to be about 40 V, while the six string current about 3 A, 0.5 A/string. For overvoltage protection, in case a string opens, multiple zener diodes, ZV_{ov}, clamping in series across the load structure, provide shutdown when it is

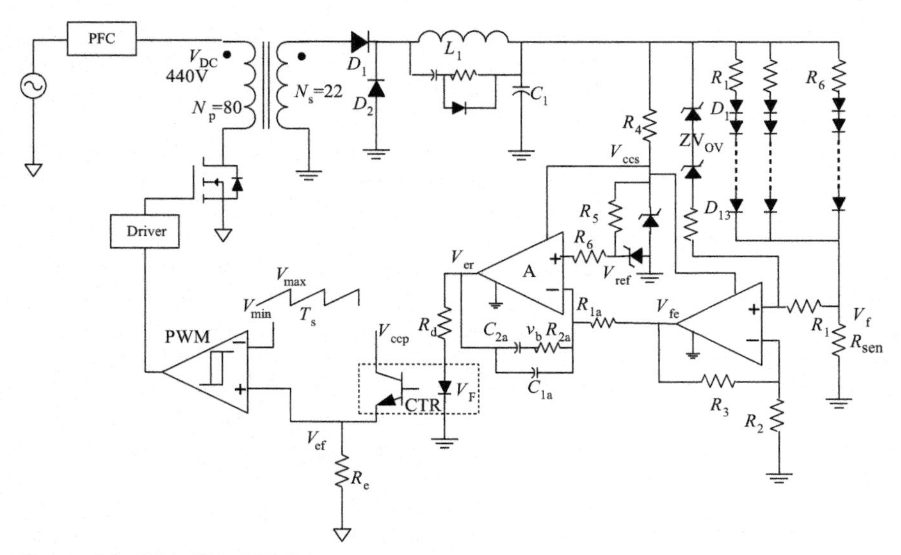

Figure 6.8 ***A Realistic LED Driver.***

activated. R_4 and a zener diode provide local supply, V_{ccs}, while R_5 and another zener provide precision reference, $V_{ref} = 2.5\,V$. Snubber is also shown across the output filter inductor L_1, ~500 μH. C_1 equals ~20 μF. Its design is more a trial-and-error effort requiring less analytical support, and shall not be covered here. R_{sen} is chosen as 0.01 Ω, a low value to minimize non-productive dissipative drop.

R_2 and R_3 are chosen such that at the desired drive current 3 A, the current feedback signal is matching V_{ref}, 2.5V. If R_1 and R_2 are assigned 1K, R_3 turns out to be 82.5K, a standard E96, 1% part. A typical optocoupler 4N35 (emitter V_{F_max} = 1.5 V @10 mA, CTR = 100%), or 4N36, 4N37 is also selected. Since a general purpose, analog IC operational amplifier is usually designed with limited output drive capability, for instance less than 10 mA, 2 mA output current is chosen in this case, and, considering the dynamic range, steady state V_{er} is intentionally set to about half, 7 V, of the IC supply, V_{ccs} ~15 V. Together, an upper bound, 2.75K, for R_d is chosen. If a 100-kHz sawtooth is selected with 3 V swing (V_{min} = 1 V, V_{max} = 4 V), a 2 V effective error voltage, V_{ef}, will yield a 33% duty cycle that also corresponds to about 40 V [=(440/80)22 × 0.33] output drive for the LED load. It also implies a 1K resistor for R_e. Since PWM IC and MOSFET gate driver are readily available, we will not discuss details surrounding both. Next, we begin the process finding the modulator gain; V_{fe}/V_{er}.

There is first an error voltage isolation and translation using an optical coupler.

$$V_{\mathrm{ef}} = \frac{V_{\mathrm{er}} - V_{\mathrm{F}}}{R_{\mathrm{d}}}\mathrm{CTR}\,R_{\mathrm{e}} \qquad G_{\mathrm{opto}} = \frac{dV_{\mathrm{ef}}}{dV_{\mathrm{er}}} = \frac{\mathrm{CTR}\,R_{\mathrm{e}}}{R_{\mathrm{d}}} \tag{6.12}$$

(6.9) produces a steady state duty cycle function and PWM gain

$$D = \frac{0.95(V_{\mathrm{ef}} - V_{\mathrm{min}})}{V_{\mathrm{max}} - V_{\mathrm{min}}} \qquad G_{\mathrm{pwm}} = \frac{dD}{dV_{\mathrm{ef}}} = \frac{0.95}{V_{\mathrm{max}} - V_{\mathrm{min}}} \tag{6.13}$$

The development of power stage gain involves more efforts. It begins with the DC gain

$$V_{\mathrm{ODC}} = \frac{V_{\mathrm{DC}}}{N_{\mathrm{p}}}N_{\mathrm{s}}D - V_{\mathrm{rectifier}} \qquad \frac{dV_{\mathrm{ODC}}}{dD} = \frac{V_{\mathrm{DC}}N_{\mathrm{s}}}{N_{\mathrm{p}}} \tag{6.14}$$

Then there is an AC gain, including the output filter loaded by the LED array. This step requires the functional expression of the LED array's conductance. From (6.4), we have

$$G_{\mathrm{LED}} = \frac{di_{\mathrm{A}}}{dV_{\mathrm{o}}} = na\frac{b}{m}\mathrm{e}^{b(V_{\mathrm{o}}/m)} \tag{6.15}$$

Then, the filter transfer function is expressed

$$H(s) = \frac{\left(C_1 s + G_{\mathrm{LED}}\right)^{-1}}{r_{\mathrm{L}} + L_1 s + \left(C_1 s + G_{\mathrm{LED}}\right)^{-1}} \tag{6.16}$$

With that, the power stage gain equals

$$G_{\mathrm{pwr}} = \frac{dV_{\mathrm{fe}}}{dD} = \frac{V_{\mathrm{DC}}N_{\mathrm{s}}}{N_{\mathrm{p}}}H(s)G_{\mathrm{LED}}R_{\mathrm{sen}}\left(1 + \frac{R_3}{R_2}\right) \tag{6.17}$$

As a result, the modulator gain is given as

$$\begin{aligned} M(s) &= G_{\mathrm{opto}}G_{\mathrm{pwm}}G_{\mathrm{pwr}} \\ &= G_{\mathrm{opto}}G_{\mathrm{pwm}}\frac{V_{\mathrm{DC}}N_{\mathrm{s}}}{N_{\mathrm{p}}}H(s)G_{\mathrm{LED}}R_{\mathrm{sen}}\left(1 + \frac{R_3}{R_2}\right) \end{aligned} \tag{6.18}$$

With all components, device parameters, and operating conditions, the example gives a modulator gain (Figure 6.9).

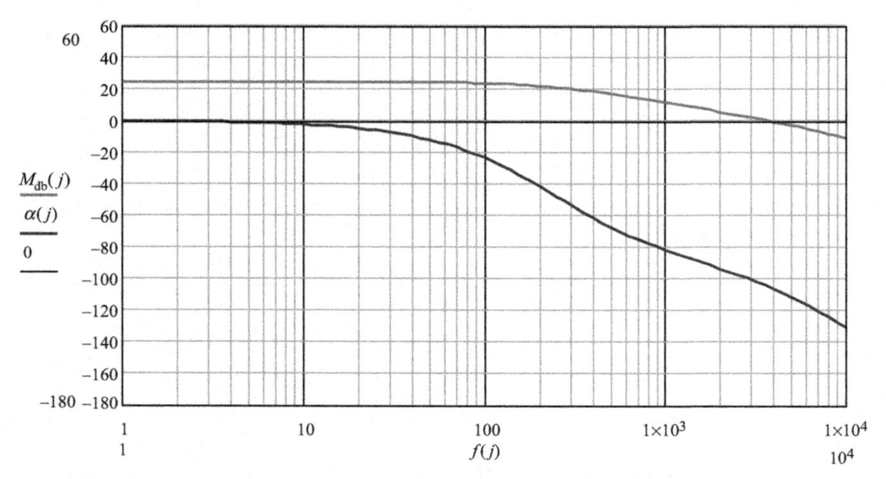

Figure 6.9 ***Modulator Gain of Figure 6.8's DC–DC Section.***

At 2 kHz, the modulator gain is 5.35 db, the phase −93°. If the phase margin desired at the crossover frequency of 2 kHz is 45°, phase boost required equals 48° and results in a pole/zero separation factor k = 2.605 if type-II amplifier is used. A type-III shall also work, but we treat type-II first. The resulting type-II has the following components, based on equation (1.14): R_{1a} = 2K, R_{2a} = 1.27K, C_{1a} = 0.028 μF, and C_{2a} = 0.16 μF.

By incorporating the error amplifier identified, it results in a loop gain (Figure 6.10).

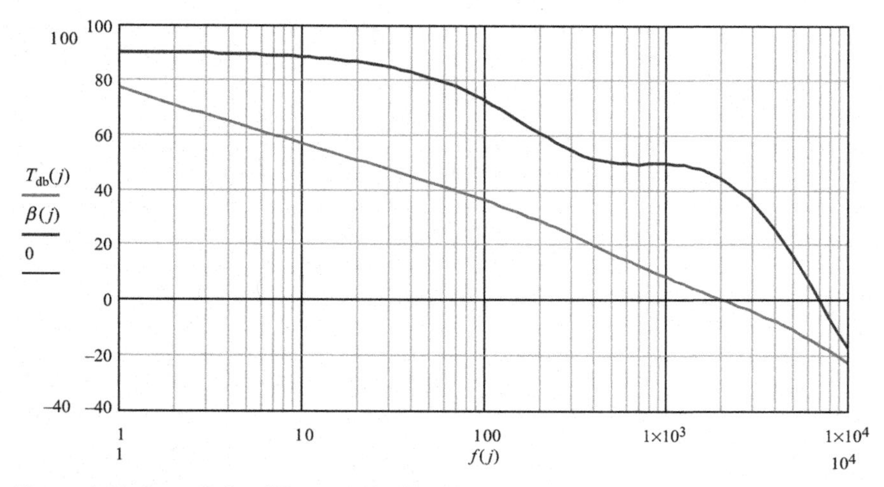

Figure 6.10 ***Loop Gain of Figure 6.8's DC–DC Section.***

With all parts and operating condition settled, Figure 6.8 yields the difference equation set (inductor current, I; output voltage, v_o; R_{2a} and C_{2a} node voltage, v_b; and error amplifier output voltage, v_e).

$$\begin{pmatrix} i_{j+1} \\ v_{9+1} \\ v_{b_{j+1}} \\ v_{e_{j+1}} \end{pmatrix} = \begin{bmatrix} \text{if}\left[i_j < 0, 0, i_j + \dfrac{\delta t}{L_1}\left[\text{if}\left[\dfrac{v_{e_j} - 1.2}{R_d}\text{CTR}R_e > s_{w_j}, \dfrac{N_s}{N_P}\left[V_{DC} - R_{on}\left(\dfrac{N_s}{N_P} i_j \right) \right] - V_d, -V_d \right] - \left(r_L i_j + v_{o_j} \right) \right] \right] \\ v_{o_j} + \dfrac{\delta t}{C_1}\left[i_j - n_p\left(a e^{b(v_{o_j}/m_s)} + c \right) \right] \\ \left[1 - \left[\dfrac{1}{(R_{2a}C_{2a})} + \dfrac{1}{(R_{2a}C_{1a})} \right]\delta t \right] v_{b_j} + \left[\dfrac{1}{(R_{2a}C_{2a})} + \dfrac{1}{C_{1a}}\left(\dfrac{1}{R_{1a}} + \dfrac{1}{R_{2a}} \right) \right]\delta t \cdot v_{r_j} - \dfrac{\delta t}{(R_{1a}C_{1a})}\left[n_p\left(a e^{b(v_{o_j}/m_s)} + c \right) R_{sen}\left(1 + \dfrac{R_3}{R_2} \right) \right] \\ \text{if}\left[v_{e_j} < 0, 0, \text{if}\left[v_{e_j} > 12, 12, v_{e_j} + \left[\dfrac{\delta t}{C_{1a}}\left(\dfrac{1}{R_{1a}} + \dfrac{1}{R_{2a}} \right) v_{r_j} - \dfrac{\delta t}{(R_{2a}C_{1a})} v_{b_j} - \dfrac{\delta t}{(R_{1a}C_{1a})}\left[n_p\left(a e^{b(v_{o_j}/m_s)} + c \right) R_{sen}\left(1 + \dfrac{R_3}{R_2} \right) \right] \right] \right] \right] \end{bmatrix} \quad (6.19a)$$

Several auxiliary equations are also derived; D_1 current, i_s; LED array current, i_A; switch current, i_p; D_2 current, i_d; D_1/D_2 cathode voltage, v_s; and the effective error voltage, v_{ef}, feeding the PWM block.

$$\begin{aligned} i_{s_j} &= \text{if}\left(\frac{v_{e_j} - 1.2}{R_d}\text{CTR}R_e > s_{w_j}, i_j, 0 \right) \\ i_{A_j} &= n_p\left(a e^{b(v_{o_j}/m_s)} + c \right) \\ i_{p_j} &= \frac{N_s}{N_P} i_{s_j} \\ i_{d_j} &= i_j - i_{s_j} \\ v_{s_j} &= \text{if}\left[\frac{v_{e_j} - 1.2}{R_d}\text{CTR}R_e > s_{w_j}, \frac{N_s}{N_P}\left[V_{DC} - R_{on} i_{p_j} \right] - V_d, -V_d \right] \\ v_{ef_j} &= \frac{\text{v}_{e_j} - 1.2}{R_d}\text{CTR}R_e \end{aligned} \quad (6.19b)$$

By conducting iterative simulation over 200 switching periods and by slowly starting the 2.5 V reference voltage, Figure 6.11a–g show an extremely well-behaving LED array driver.

We now move to the next task converting the analog control to digital. It shall be noted that the conversion effort will focus only on the error amplifier. The R_{sen} scaling amplifier and the optocoupler circuits still remain in the analog domain.

The type-II error amplifier identified earlier gives the following frequency responses, Figure 6.12.

Since the converter is operating at 100 kHz, a sampling frequency at 15 times of the frequency is selected, that is, 1.5 MHz. The z-domain bilinear transformation constant C shall then equal 3 MHz. The choice gives the $H_{II}(z)$ coefficients, (1.17–1.19) as

$$a_0 = 5.84\times10^{-3} \quad a_1 = 18.75\times10^{-6} \quad a_2 = -5.821\times10^{-3}$$

$$b_1 = -1.978\times10^{0} \quad b_2 = 978.411\times10^{-3}$$

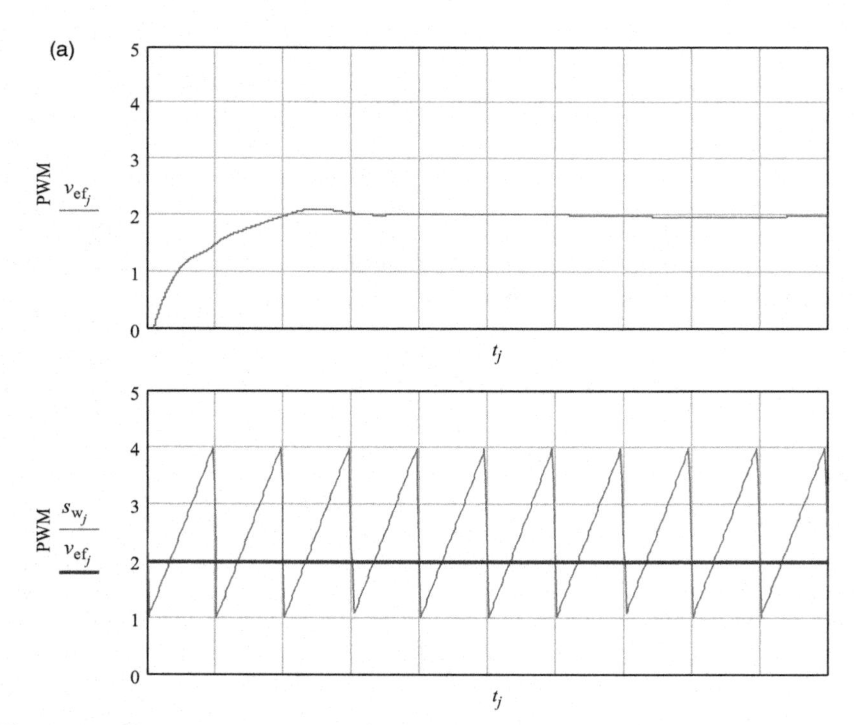

Figure 6.11 Equation set (6.19) performance (a) Error feedback at PWM block input, (b) Error amplifier output, (c) D_1/D_2 common cathode voltage, (d) D_1 current, (e) D_2 current, (f) Inductor current, (g) Output voltage, and (h) LED array current.

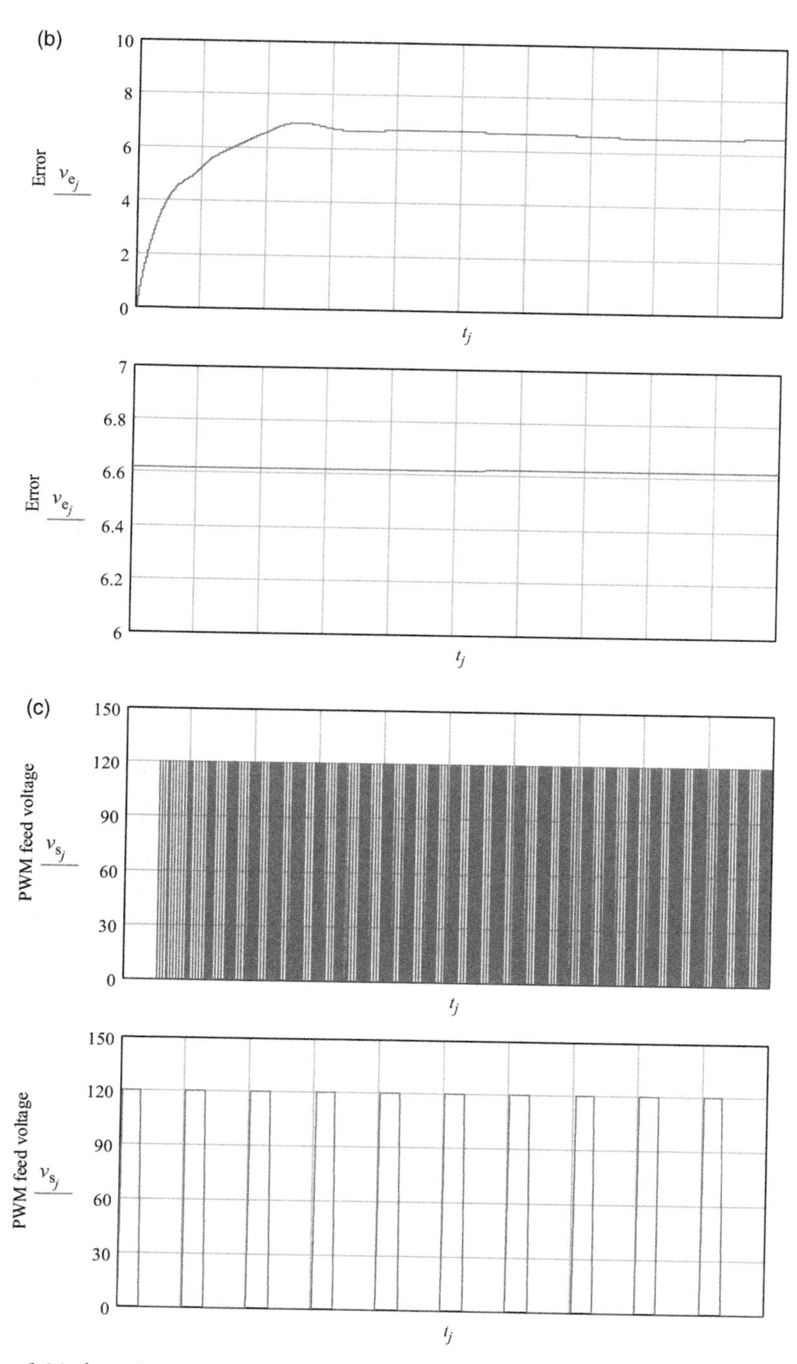

Figure 6.11 ***(cont.)***

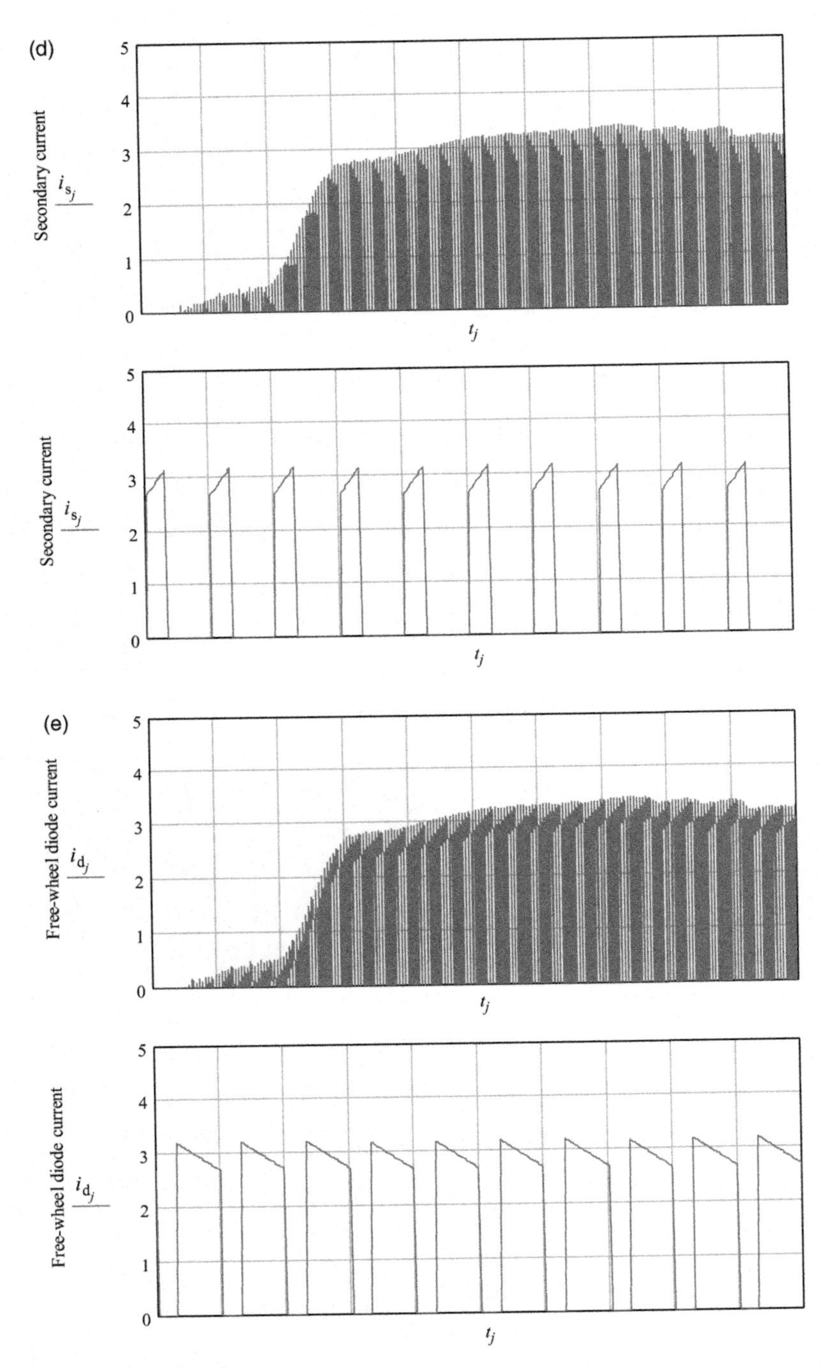

Figure 6.11 ***(cont.)***

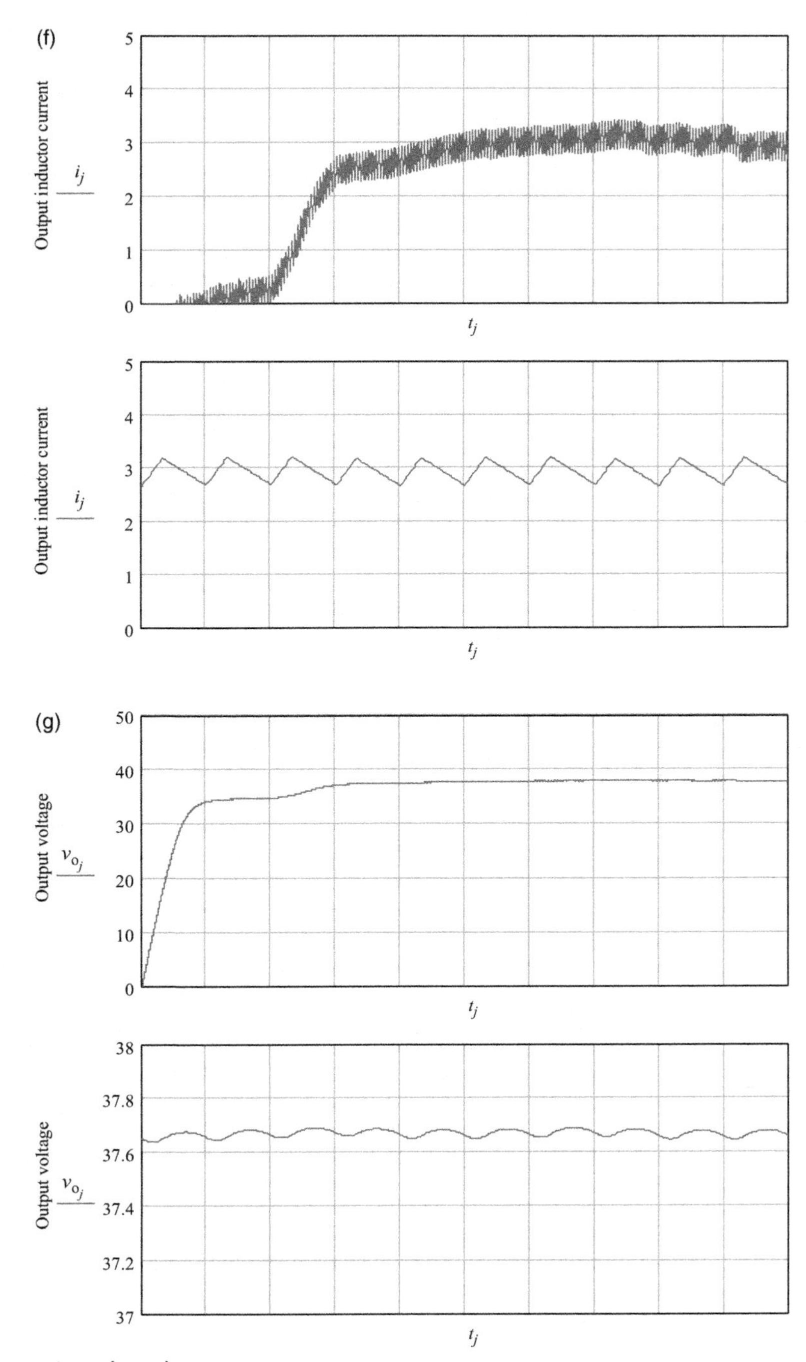

Figure 6.11 ***(cont.)***

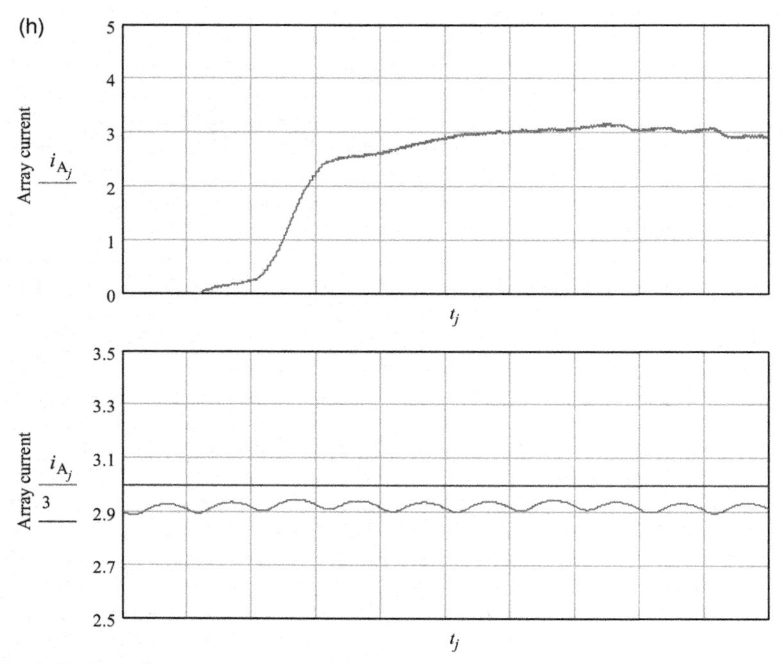

Figure 6.11 ***(cont.)***

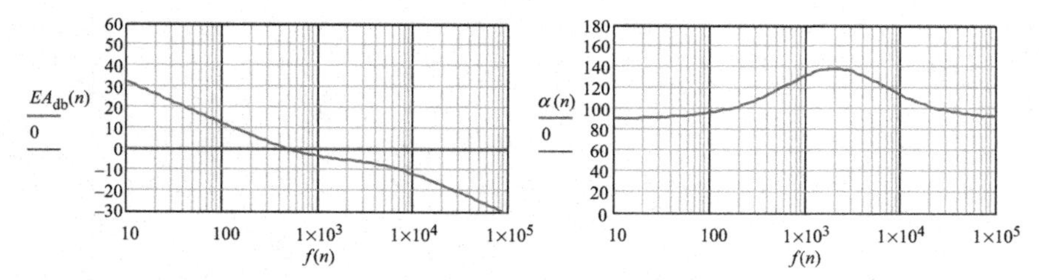

Figure 6.12 ***Frequency Responses of Type-II, Analog Error Amplifier.***

It results in an excellent match of $H_{II}(z)$ against EA(s), Figure 6.13, in terms of magnitude and phase.

The converter schematic with digital controller is given in Figure 6.14.

It turns out that with the new controller, the converter exhibits a slightly different quiescent operating point such that the optocoupler's emitter drive voltage is less. In order to compensate the condition, R_d is lowered to 685 Ω. With that modification in mind, the converter with digital controller gives a new set of difference equation (6.20):

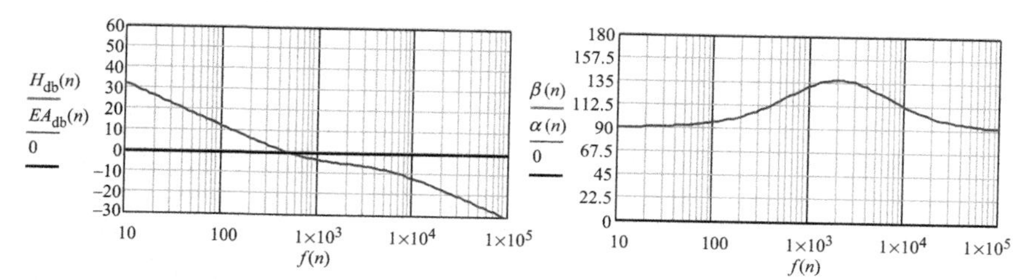

Figure 6.13 ***A Good Match of Digital Filter Against Analog Error Amplifier.***

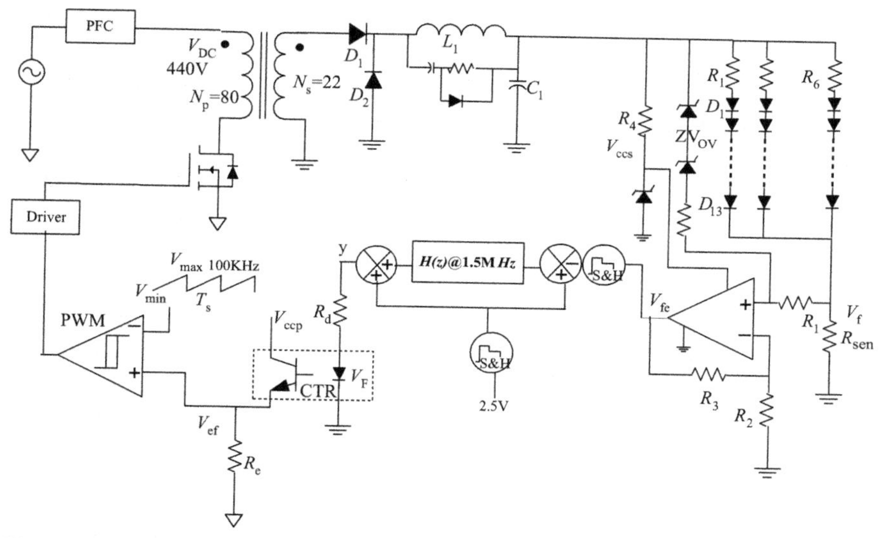

Figure 6.14 ***LED Array Driver With Digital Filter.***

$$\begin{pmatrix} i_{j+1}\\ v_{o_{j+1}}\\ y_{j+1}\\ w_{j+1}\end{pmatrix} = \begin{bmatrix} \text{if}\left[i_j<0,0,i_j+\dfrac{\delta t}{L_1}\left[\text{if}\left[\dfrac{y_j-1.2}{R_d}\text{CTR}R_e > s_{w_j}, \dfrac{N_s}{N_P}\left[V_{DC}-R_{on}\left(\dfrac{N_s}{N_P}i_j\right)\right] -V_d, -V_d-\left(r_L i_j+v_{o_j}\right)\right]\right]\right] \\ v_{o_j}+\dfrac{\delta t}{C_1}\left[i_j-n_p\left(ae^{b(v_{o_j}/m_s)}+c\right)\right] \\ \text{if}\left[y_j<0,0,\text{if}\left[y_j>12,12,v_{r_j}+a_0\left[\left[v_{r_j}-n_p\left(ae^{b(v_{o_j}/m_s)}+c\right)R_{sen}\left(1+\dfrac{R_3}{R_2}\right)\right] -b_1w_j-b_2w_{j-1}+a_1w_j+a_2w_{j-1}\right]\right]\right] \\ \left[v_{r_j}-n_p\left(ae^{b(v_{o_j}/m_s)}+c\right)R_{sen}\left(1+\dfrac{R_3}{R_2}\right)\right]-b_1w_j-b_2w_{j-1}\end{bmatrix} \tag{6.20}$$

Readers are reminded that variable y replaces v_e in (6.19), and variable w is an intermediate parameter. (6.20) gives the following performance Figure 6.15a–h.

As shown, the converter under digital control seems to take longer to settle. It may also reflect the digital, finite quantization effects or frequency beating between converter switching and digital sampling. We shall see if employing the type-III digital controller also ends up showing the similar nature.

In order to make a balanced comparison, the same requirement, 2 kHz crossover frequency and 45° phase margin, is imposed for type-III implementation. It is also important to recognize that the modulator gain, Figure 6.9, is independent of the error amplifier selection. This leads us to type-III components, (1.15), as $C_{1a} = 0.1$ μF, $C_{2a} = 0.074$ μF, $C_{3a} = 0.035$ μF, $R_{1a} = 2$K, $R_{2a} = 1.21$K, $R_{3a} = 1.46$K, and schematic as Figure 6.16.

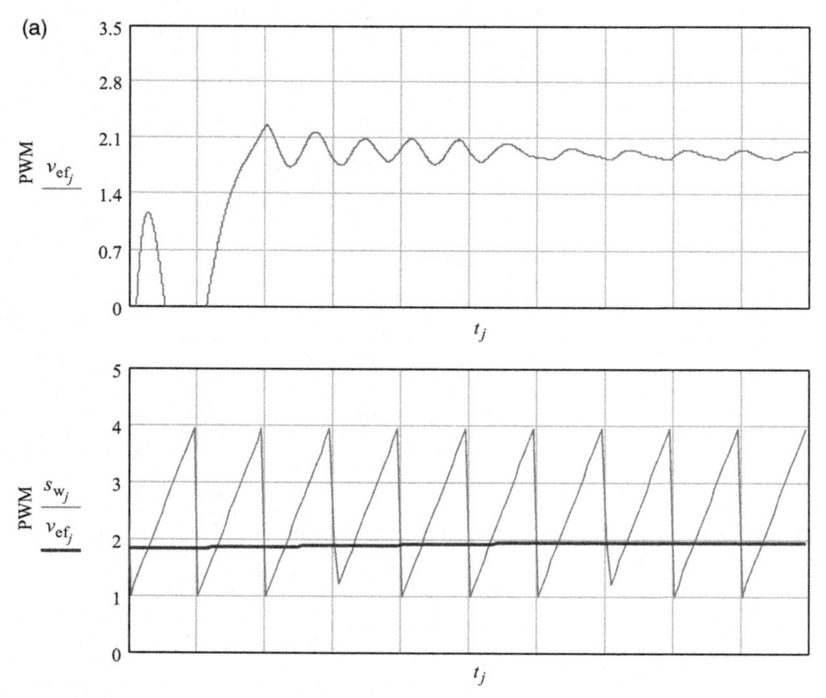

Figure 6.15 Equation set (6.20) performance (a) Error feedback at PWM block input, (b) digital error voltage, (c) D_1/D_2 common cathode voltage, (d) D_1 current, (e) D_2 current, (f) inductor current, (g) output voltage, and (h) LED array current.

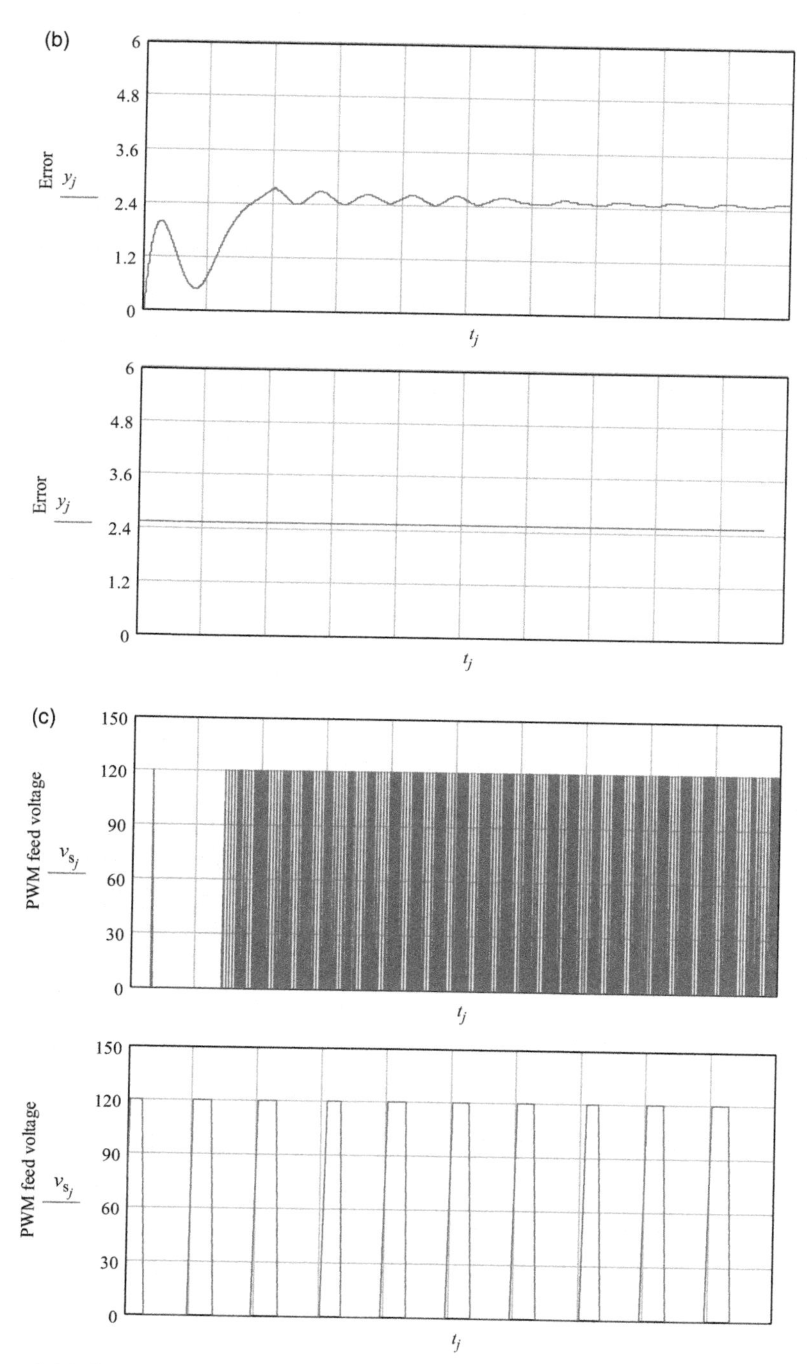

Figure 6.15 ***(cont.)***

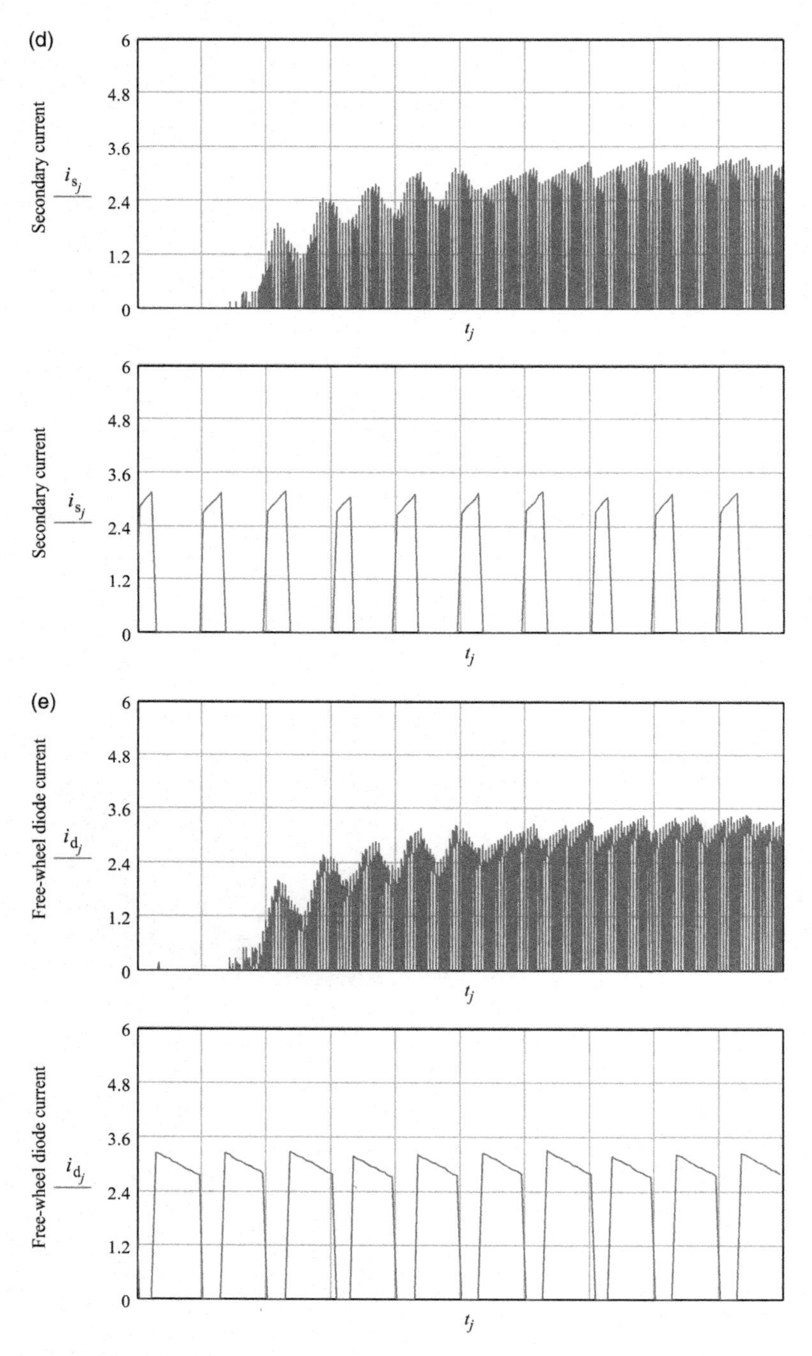

Figure 6.15 ***(cont.)***

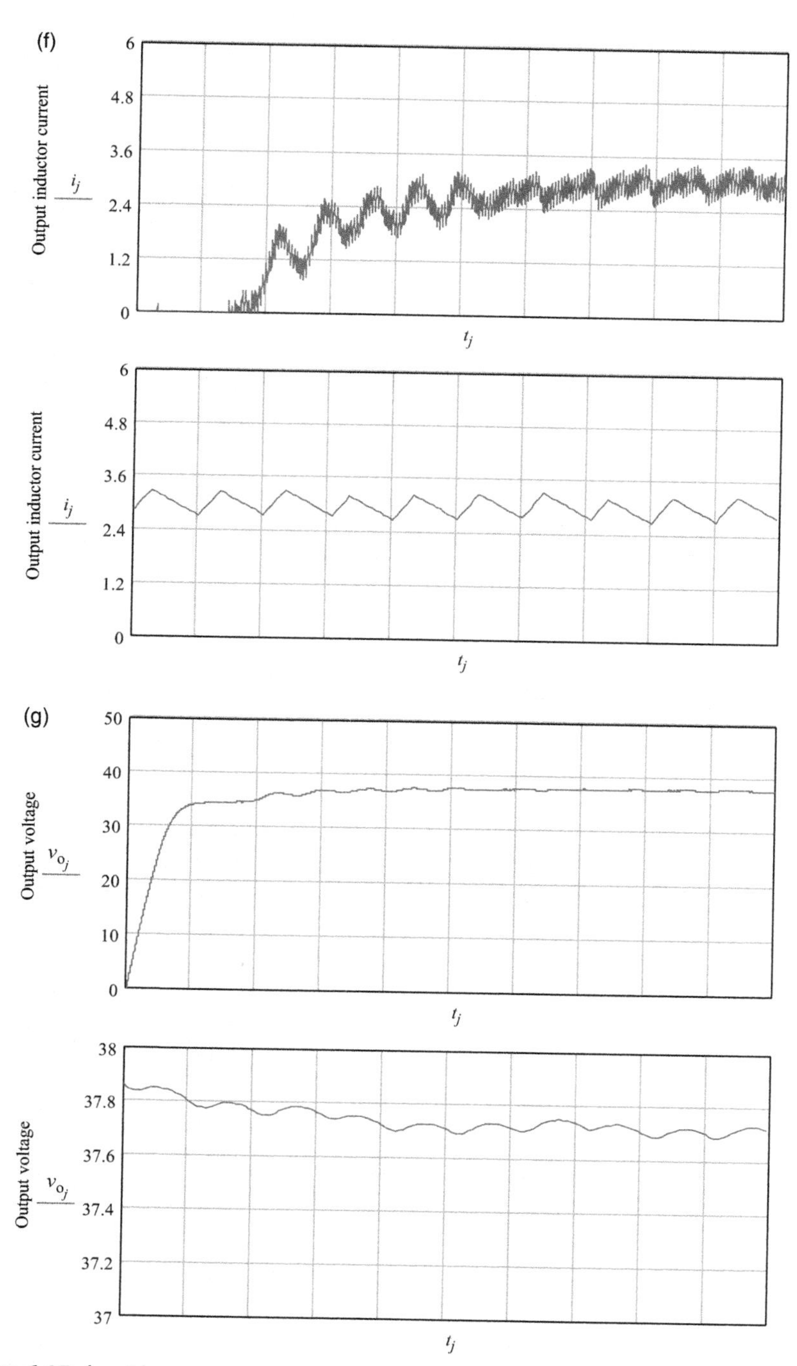

Figure 6.15 ***(cont.)***

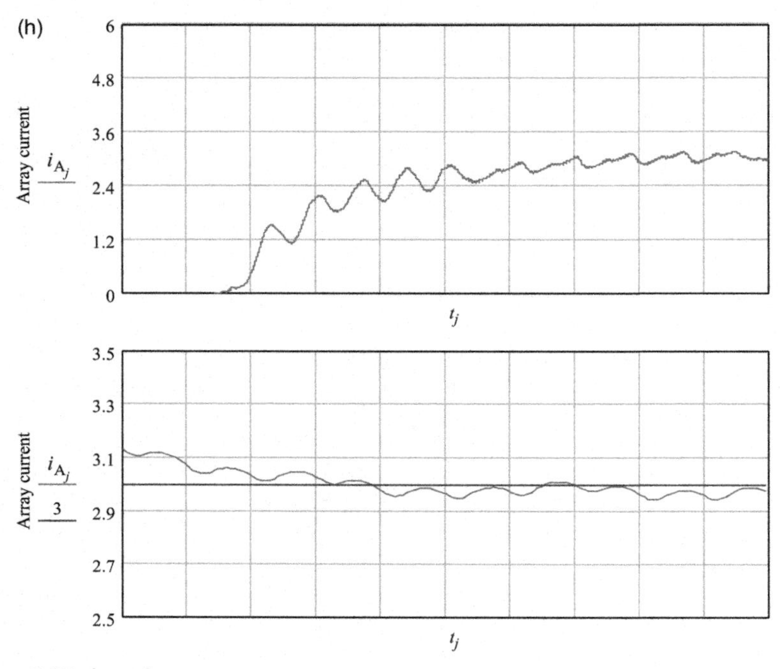

Figure 6.15 ***(cont.)***

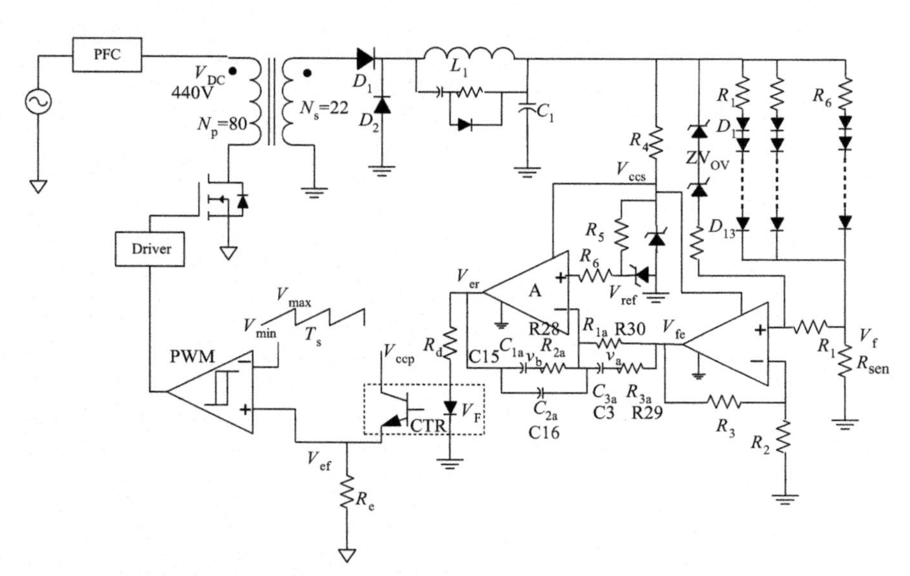

Figure 6.16 ***LED Driver With Type-III Error Amplifier.***

The corresponding difference equation set (6.21) has one additional node, v_a. It was given in the order of output filter inductor current, output voltage, error amplifier input $C_{3a}-R_{3a}$ node voltage, error amplifier feedback $C_{1a}-R_{2a}$ node, and error amplifier output.

$$\begin{pmatrix} i_{j+1} \\ v_{o_{j+1}} \\ v_{a_{j+1}} \\ v_{b_{j+1}} \\ v_{e_{j+1}} \end{pmatrix} = \begin{bmatrix} \text{if}\left[i_j<0,0,i_j+\dfrac{\delta t}{L_1}\left[\text{if}\left[\dfrac{v_{e_j}-1.2}{R_d}\text{CTR}R_e > s_{w_j}, \dfrac{N_s}{N_P}\left[V_{DC}-R_{on}\left(\dfrac{N_s}{N_P} i_j \right) \right] - V_d, -V_d \right] - \left(r_L i_j + v_{o_j} \right) \right] \right] \\ v_{o_j}+\dfrac{\delta t}{C_1}\left[i_j - n_P\left(ae^{b(v_{o_j}/m_s)}+c \right) \right] \\ v_{a_j}+\dfrac{\delta t}{C_3}\dfrac{n_P\left(ae^{b(v_{o_j}/m_s)}+c \right)R_{sen}\left(1+\dfrac{R_3}{R_2} \right) - v_{a_j}}{R29} \\ v_{b_j}+\delta t\left[\begin{array}{l} \dfrac{v_{r_j}-v_{b_j}}{R28C16} - \dfrac{n_P\left(ae^{b(v_{o_j}/m_s)}+c \right)R_{sen}\left(1+\dfrac{R_3}{R_2} \right) - v_{r_j}}{R30C16} \\ -\dfrac{n_P\left(ae^{b(v_{o_j}/m_s)}+c \right)R_{sen}\left(1+\dfrac{R_3}{R_2} \right) - v_{a_j}}{R30C16} + \dfrac{v_{r_j}-v_{b_j}}{R28C15} \end{array} \right] \\ \text{if}\left[v_{e_j}<0,0,\text{if}\left[v_{e_j}>15,15,v_{e_j}+\dfrac{\delta t}{C16}\left[\begin{array}{l} \dfrac{v_{r_j}-v_{b_j}}{R28} - \dfrac{n_P\left(ae^{b(v_{o_j}/m_s)}+c \right)R_{sen}\left(1+\dfrac{R_3}{R_2} \right) - v_{r_j}}{R30} \\ -\dfrac{n_P\left(ae^{b(v_{o_j}/m_s)}+c \right)R_{sen}\left(1+\dfrac{R_3}{R_2} \right) - v_{a_j}}{R29} \end{array} \right] \right] \right] \end{bmatrix} \tag{6.21}$$

The equation set running for 200 switching cycles gives Figure 6.17a–h. It is a surprise to see a higher array current type-III implementation seems to offer, compared with type-II results, Figure 6.11h. At the moment, no attempt is made to find out the cause, but it has significance for the end user.

We proceed with the conversion of type-III analog to digital. The following $H_{III}(z)$ polynomial coefficients, referring to (1.22) and (1.23), are obtained with sampling frequency 1.5 MHz, 15 times of the switching frequency, and bilinear transform constant $C = 3$ MHz.

$$a_0 = 5.325\times10^{-3} \quad a_1 = -5.267\times10^{-3} \quad a_2 = -5.325\times10^{-3} \quad a_3 = 5.267\times10^{-3}$$

$$b_1 = -2.974 \quad b_2 = 2.949 \quad b_3 = -0.975$$

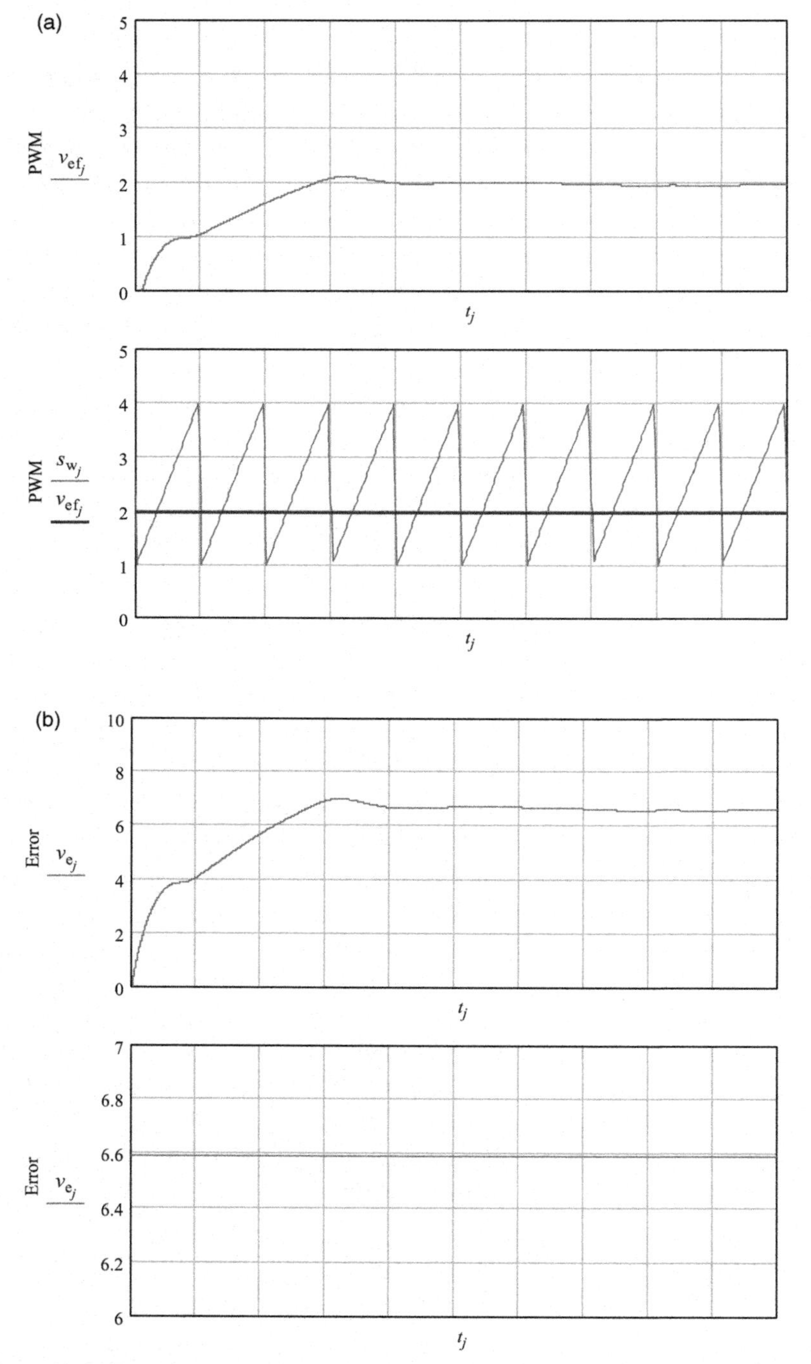

Figure 6.17 Equation set (6.21) performance (a) Error feedback at PWM block input, (b) error amplifier output voltage, (c) D_1/D_2 common cathode voltage, (d) D_1 current, (e) D_2 current, (f) inductor current, (g) output voltage, and (h) LED array current.

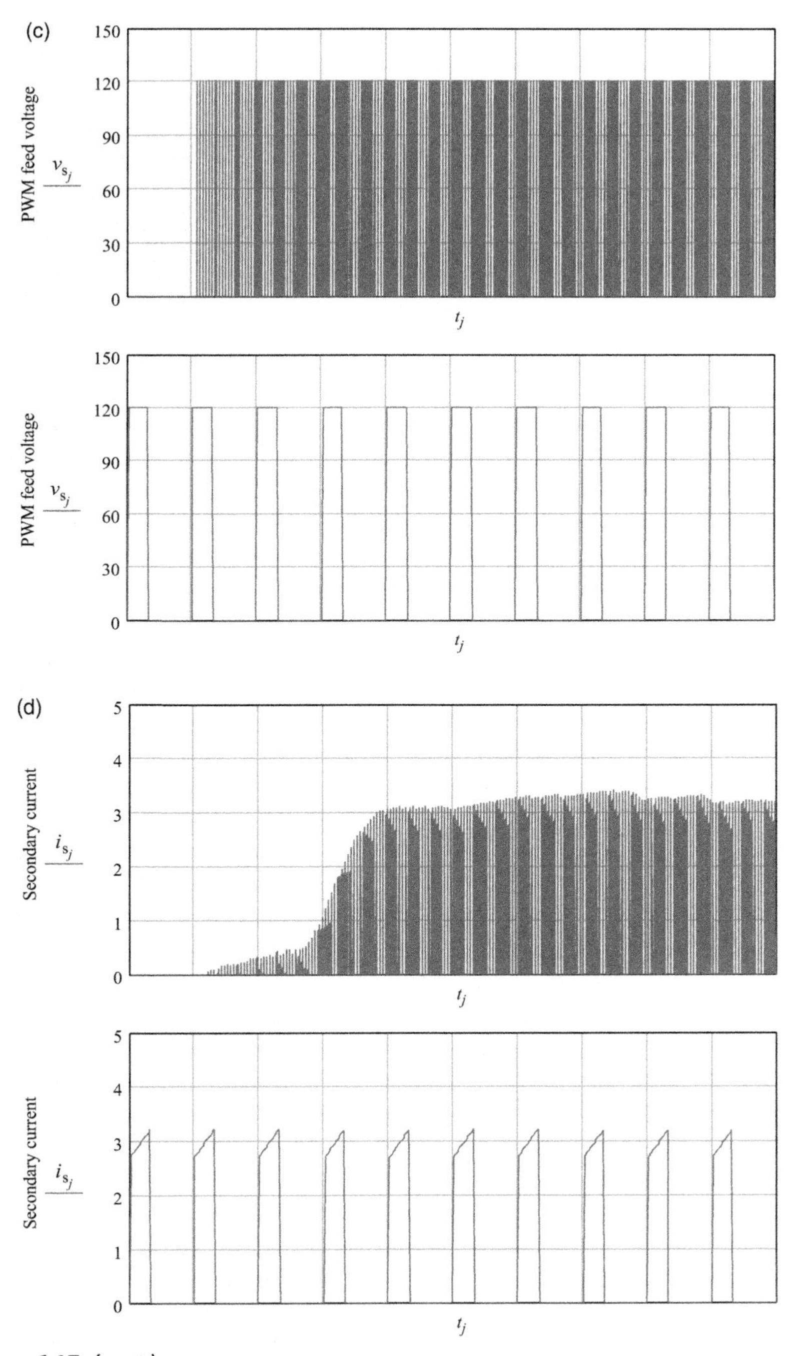

Figure 6.17 ***(cont.)***

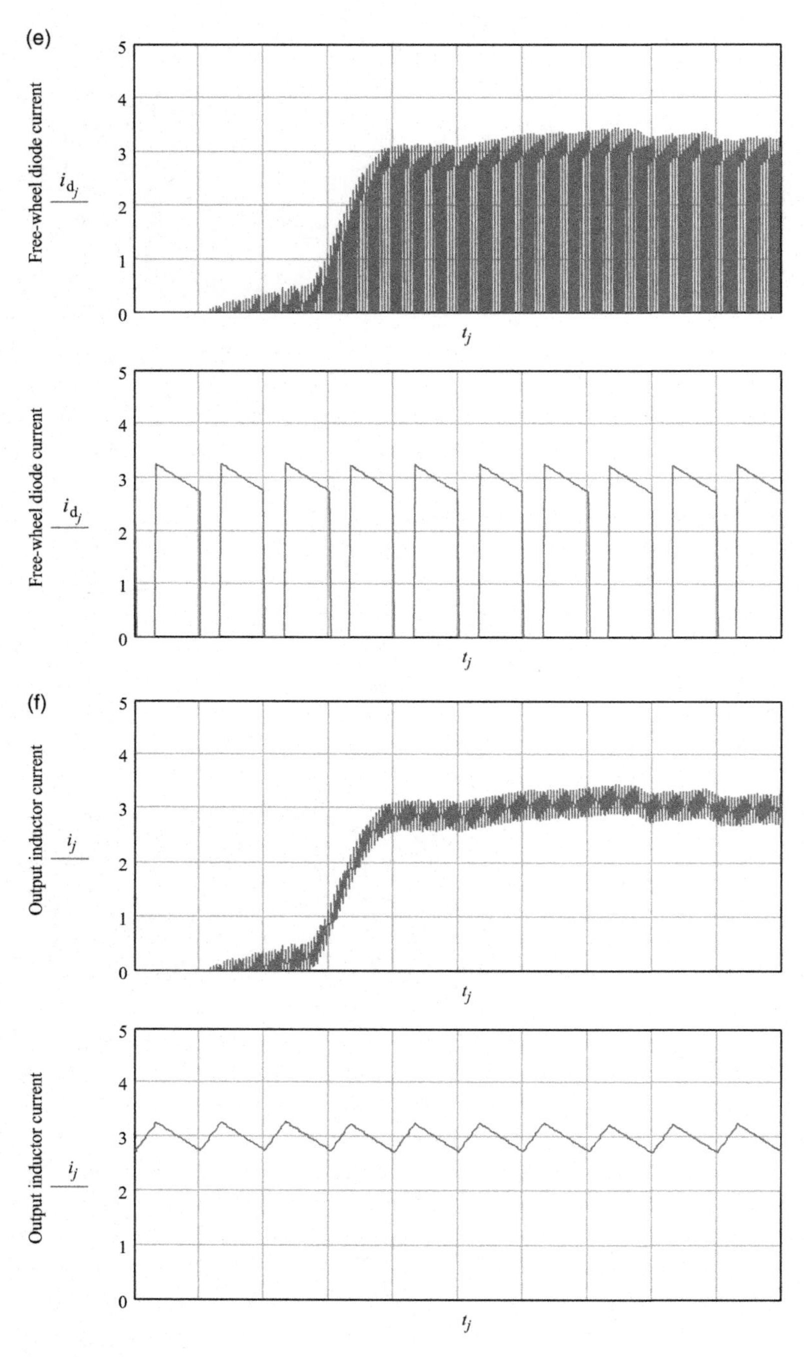

Figure 6.17 ***(cont.)***

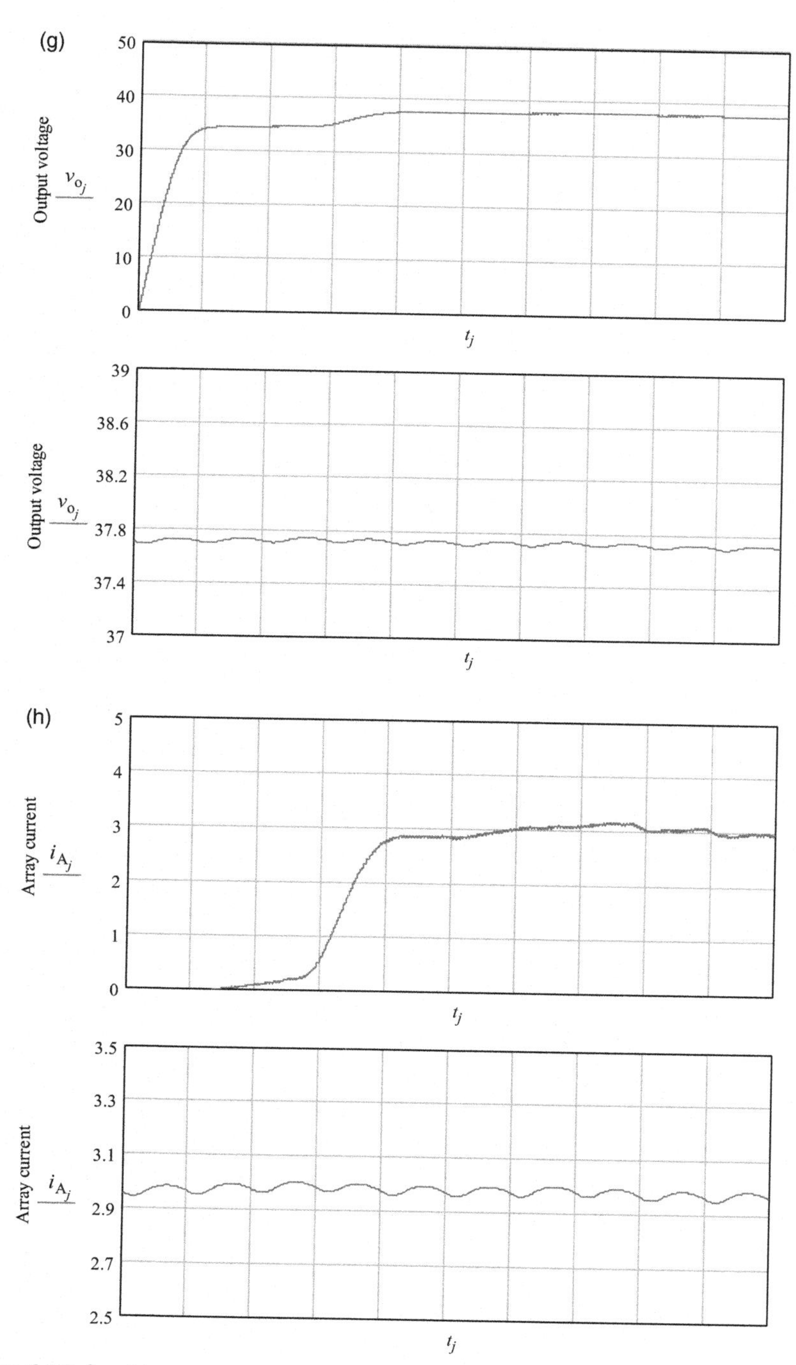

Figure 6.17 ***(cont.)***

And a new equation set (6.22) comes about

$$\begin{pmatrix} i_{j+1} \\ v_{o_{j+1}} \\ y_{j+1} \\ w_{j+1} \end{pmatrix} := \begin{bmatrix} \text{if}\left[i_j<0,0,i_j+\frac{\delta t}{L_1}\left[\text{if}\left[\frac{y_j-1.2}{R_d}\text{CTR}R_e>s_{w_j},\frac{N_s}{N_P}\left[V_{DC}-R_{on}\left(\frac{N_s}{N_P}i_j\right)\right]-V_d,-V_d\right]-\left(r_L i_j+v_{o_j}\right)\right]\right] \\ v_{o_j}+\frac{\delta t}{C_1}\left[i_j-n_P\left(ae^{b(v_{o_j}/m_s)}+c\right)\right] \\ \text{if}\left[y_j<0,0,\text{if}\left[y_j>12,12,v_{r_j}+a_0\left[\left[v_{r_j}-n_P\left(ae^{b(v_{o_j}/m_s)}+c\right)R_{sen}\left(1+\frac{R_3}{R_2}\right)\right]\right]-b_1w_j-b_2w_{j-1}-b_3w_{j-2}+a_1w_j+a_2w_{j-1}+a_3w_{j-2}\right]\right] \\ \left[v_{r_j}-n_P\left(ae^{b(v_{o_j}/m_s)}+c\right)R_{sen}\left(1+\frac{R_3}{R_2}\right)\right]-b_1w_j-b_2w_{j-1}-b_3w_{j-2} \end{bmatrix} \quad (6.22)$$

The same 200 cycles simulation gives Figure 6.18a–h.

Comparison between Figure 6.17h, LED array current with analog controller and Figure 6.18h digital version seems to indicate some peculiar behavior for the latter. It may be again attributed to digital sampling effects and deserves more studies at an appropriate time.

6.5 MATLAB SIMULINK EVALUATION

Following the same spirit of Chapters 1 and 2, the LED driver presented in Figure 6.8 is brought to MATLAB SIMULINK environment, Figure 6.19, and its performance evaluated, Figure 6.19a–h. Figure 6.20 replaces the physical error amplifier with its Laplace transfer function, while Figure 6.21 uses a digital filter equivalent. In terms of simulation time, both Figures 6.20 and 6.21 seem to run faster.

Readers may notice a subtle difference in turn-on transient profile among all three approaches. This is believed to be the effect of initial condition and how it primes the system during startup. It may have an implication for the input line current if the fuse is installed. In that case, fuse surge current rating and blow time may need to be considered.

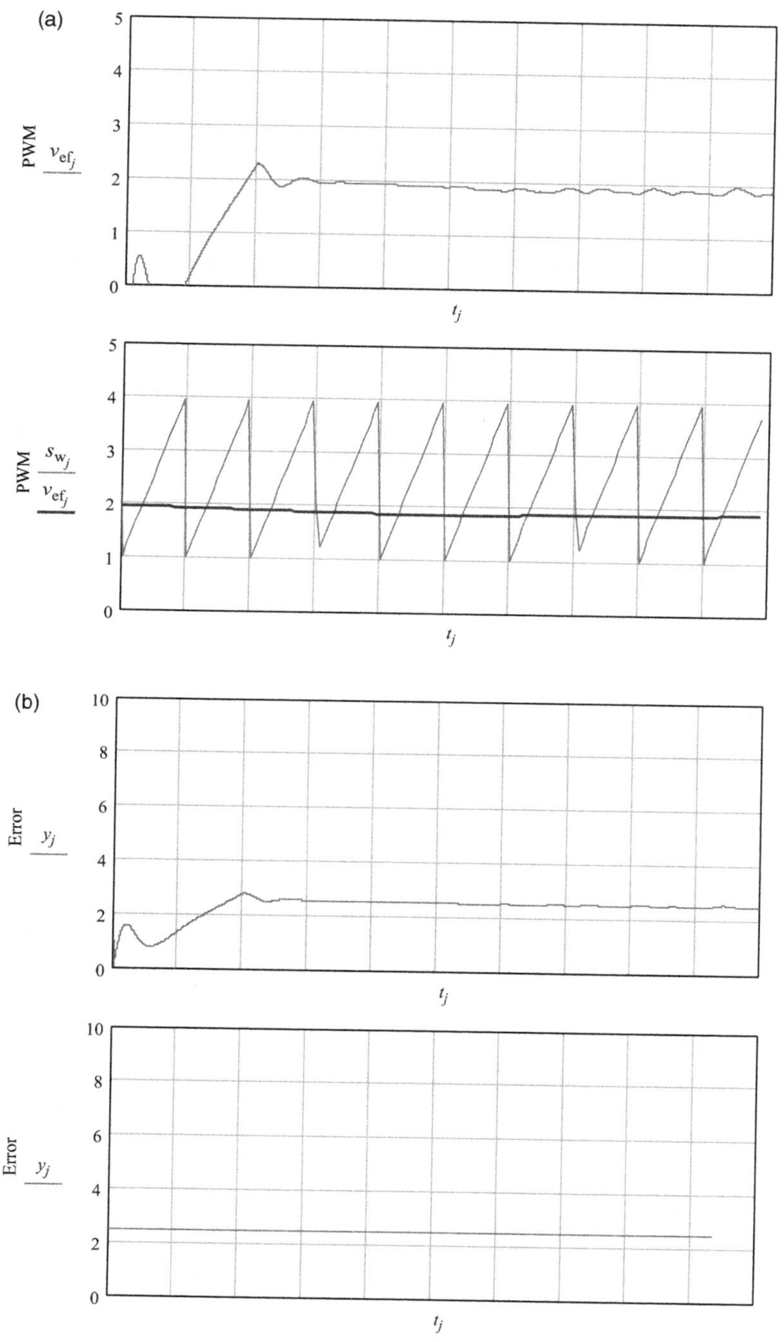

Figure 6.18 Equation set (6.22) performance (a) Error feedback at PWM block input, (b) digital error voltage, (c) D_1/D_2 common cathode voltage, (d) D_1 current, (e) D_2 current, (f) inductor current, (g) output voltage, and (h) LED array current.

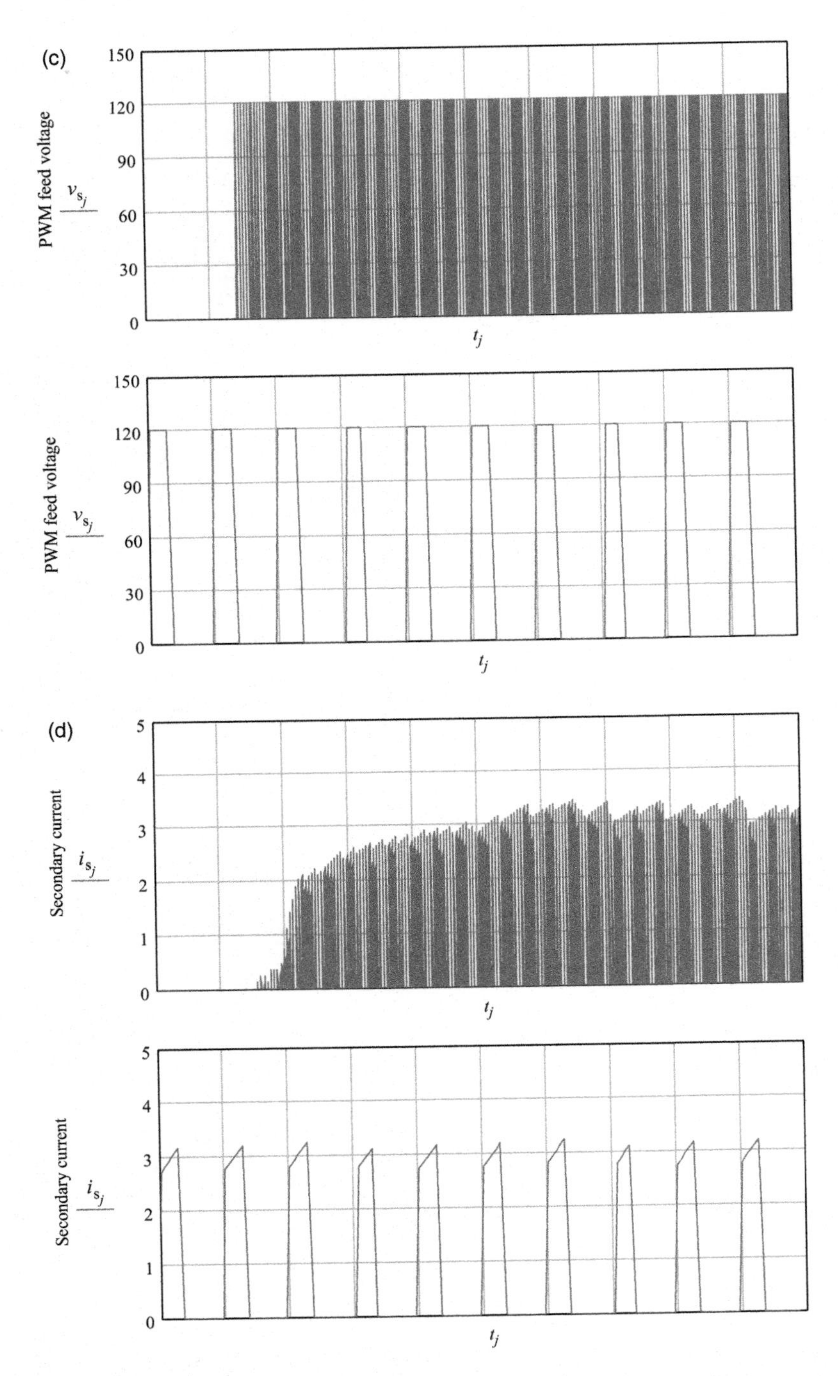

Figure 6.18 ***(cont.)***

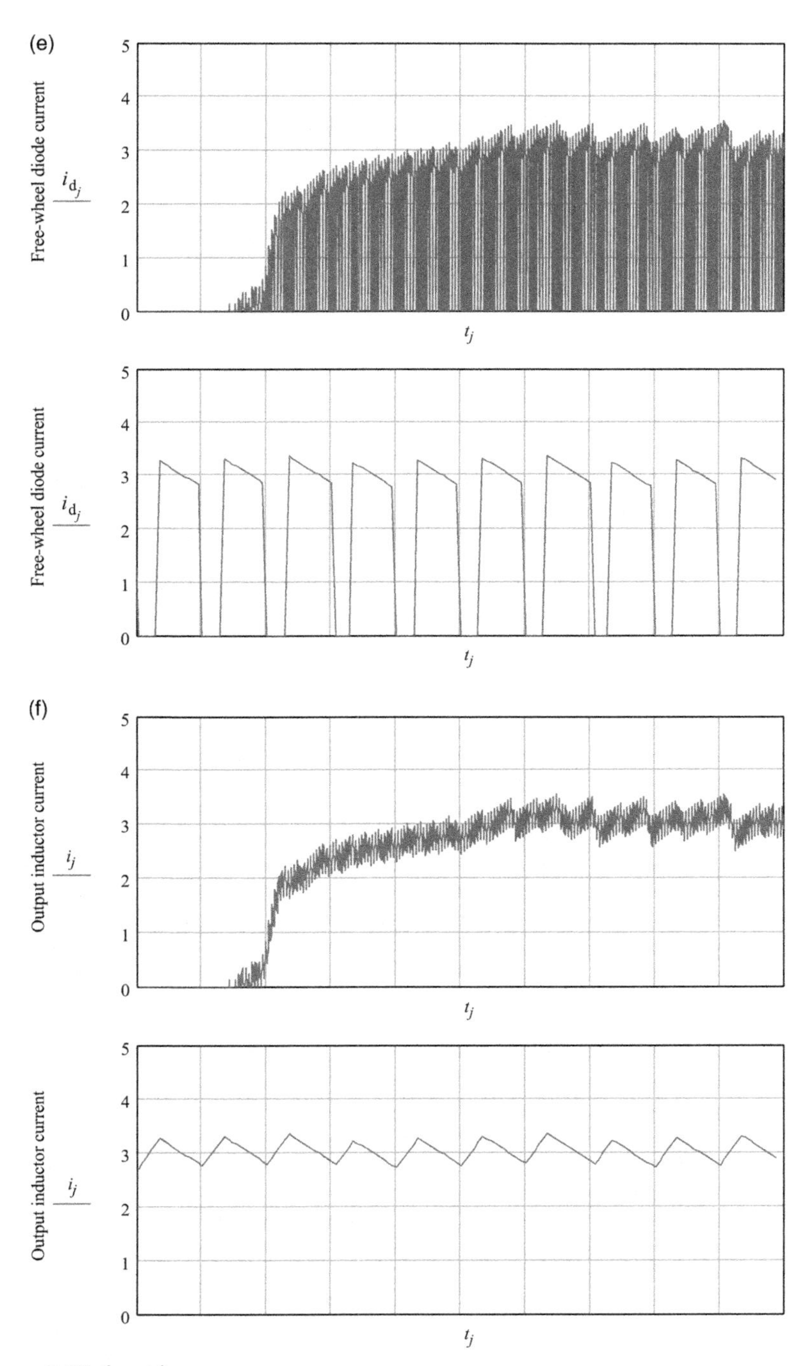

Figure 6.18 ***(cont.)***

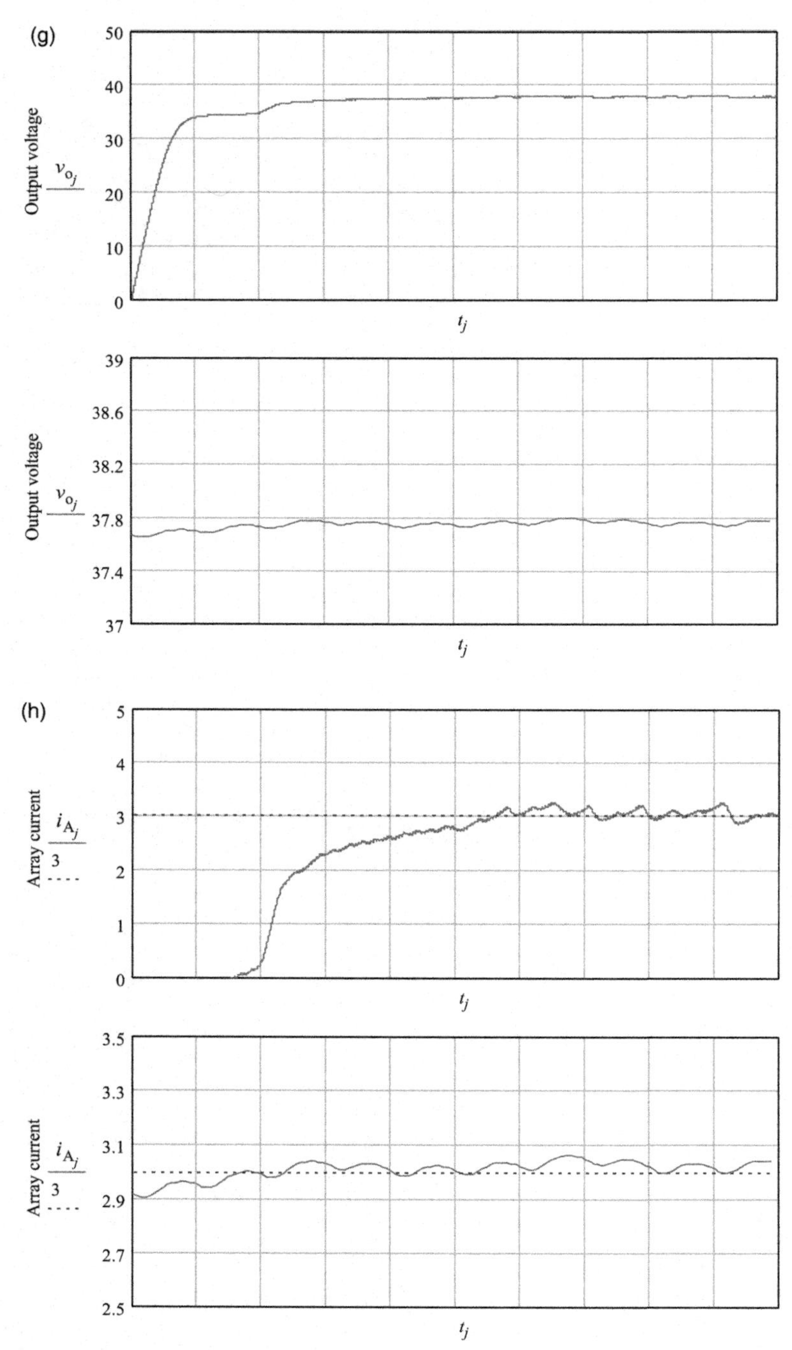

Figure 6.18 ***(cont.)***

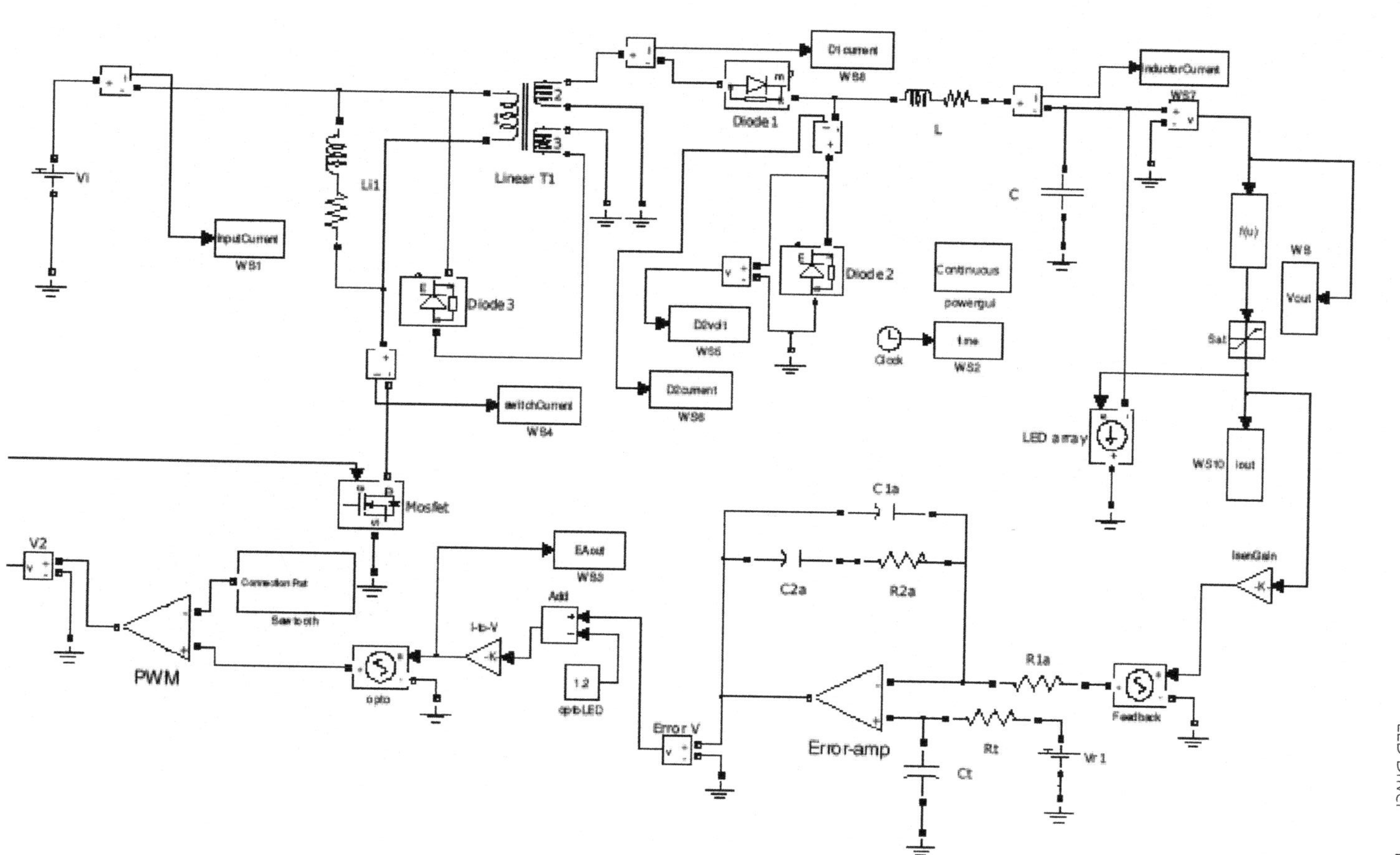

Figure 6.19 ***SIMULINK Model for LED Driver With Type-II Error Amp Represented by Physical Device.*** (a) Output voltage, (b) error voltage, (c) D_1 and D_2 cathode, (d) input current, (e) switch current, (f) inductor current, (g) D_1 current, (h) D_2 current, and (i) LED array current.

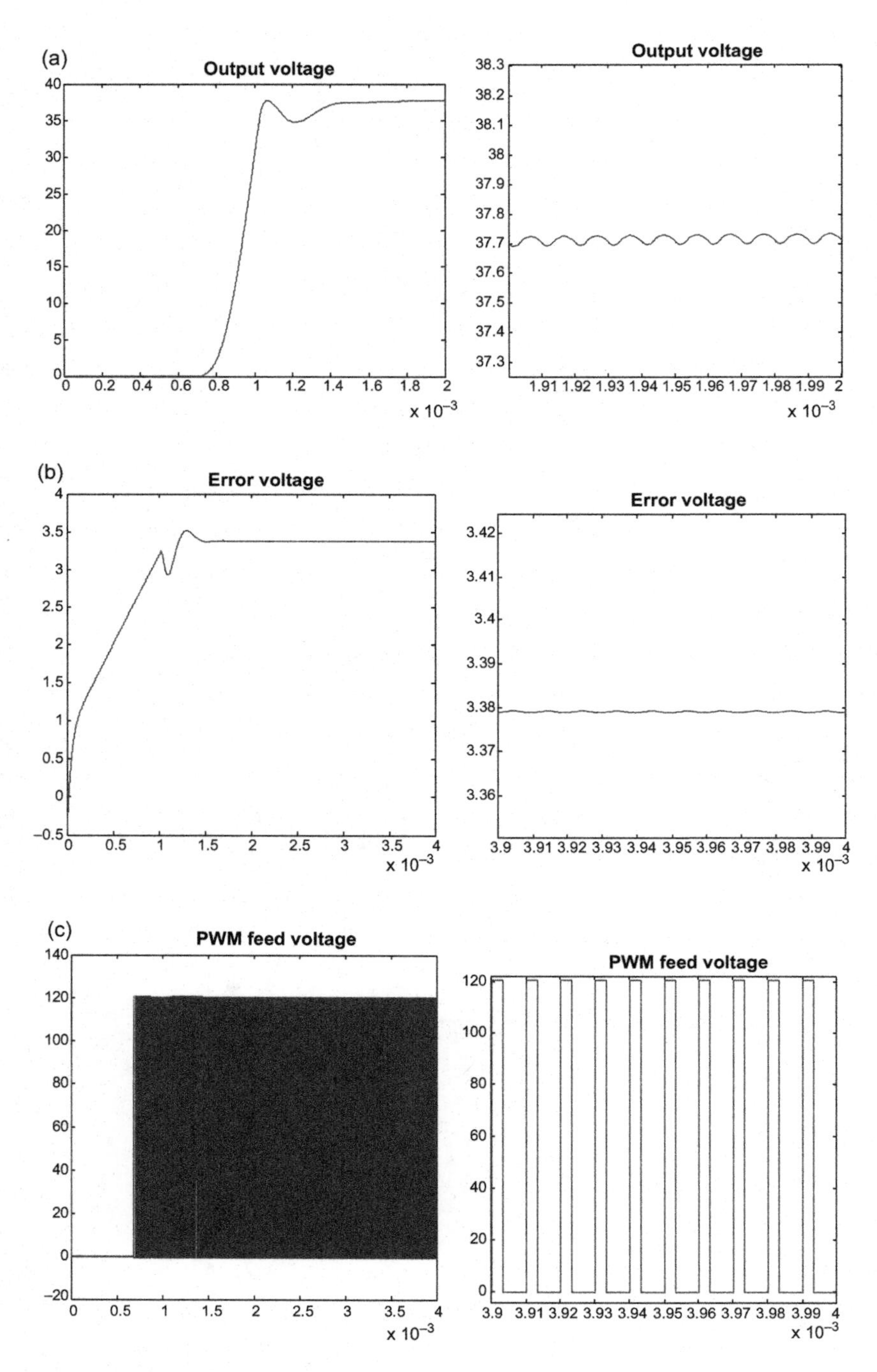

Figure 6.19 ***(cont.)***

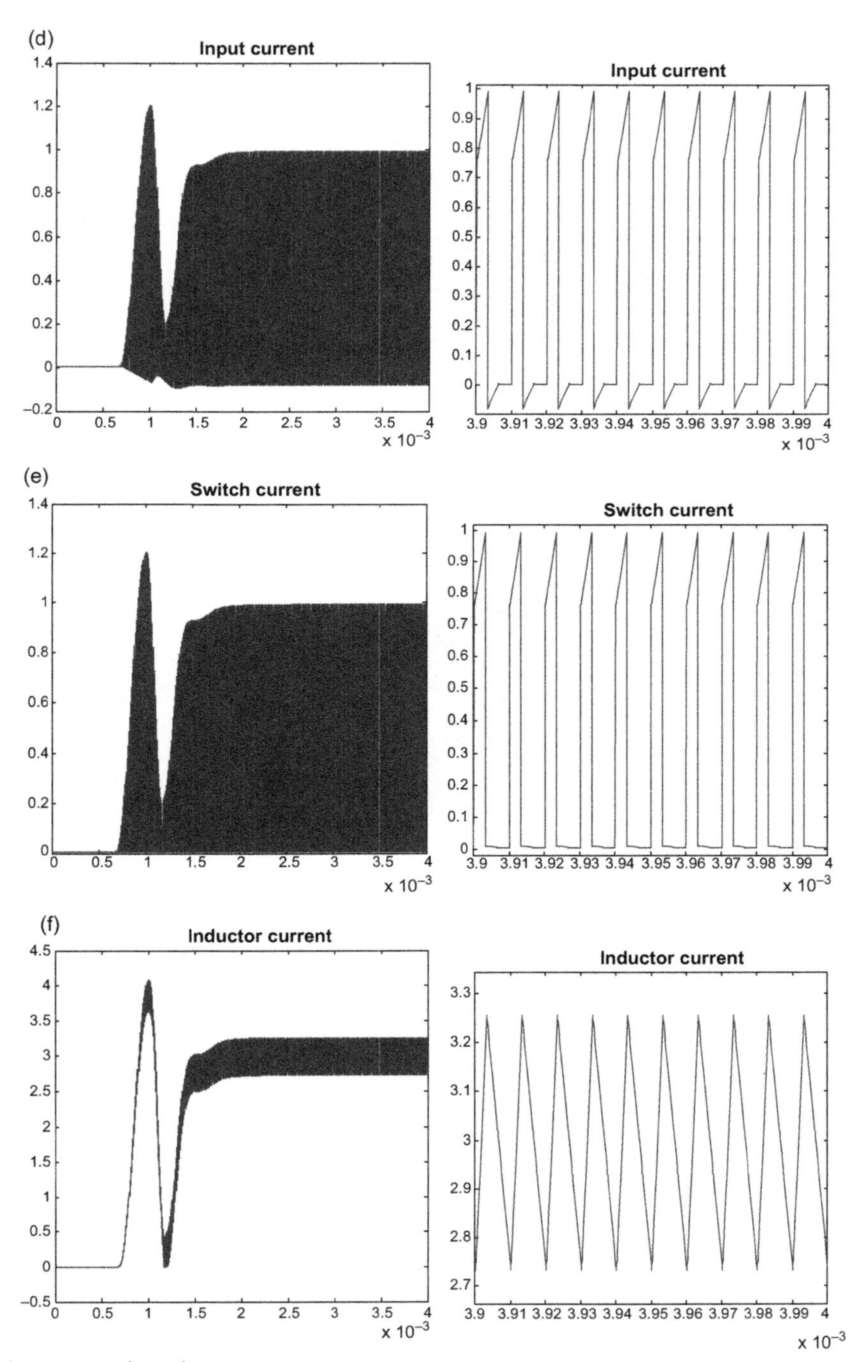

Figure 6.19 *(cont.)*

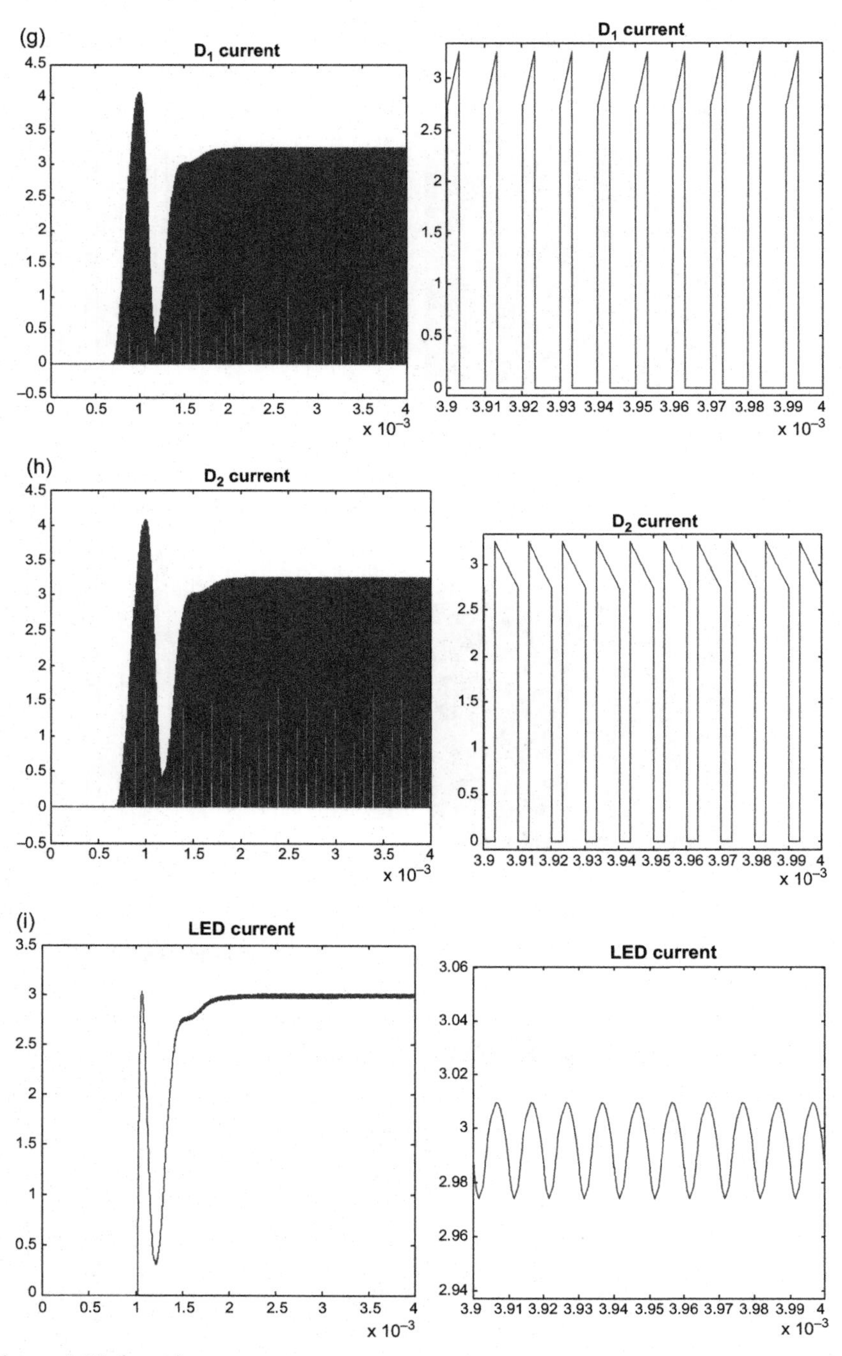

Figure 6.19 ***(cont.)***

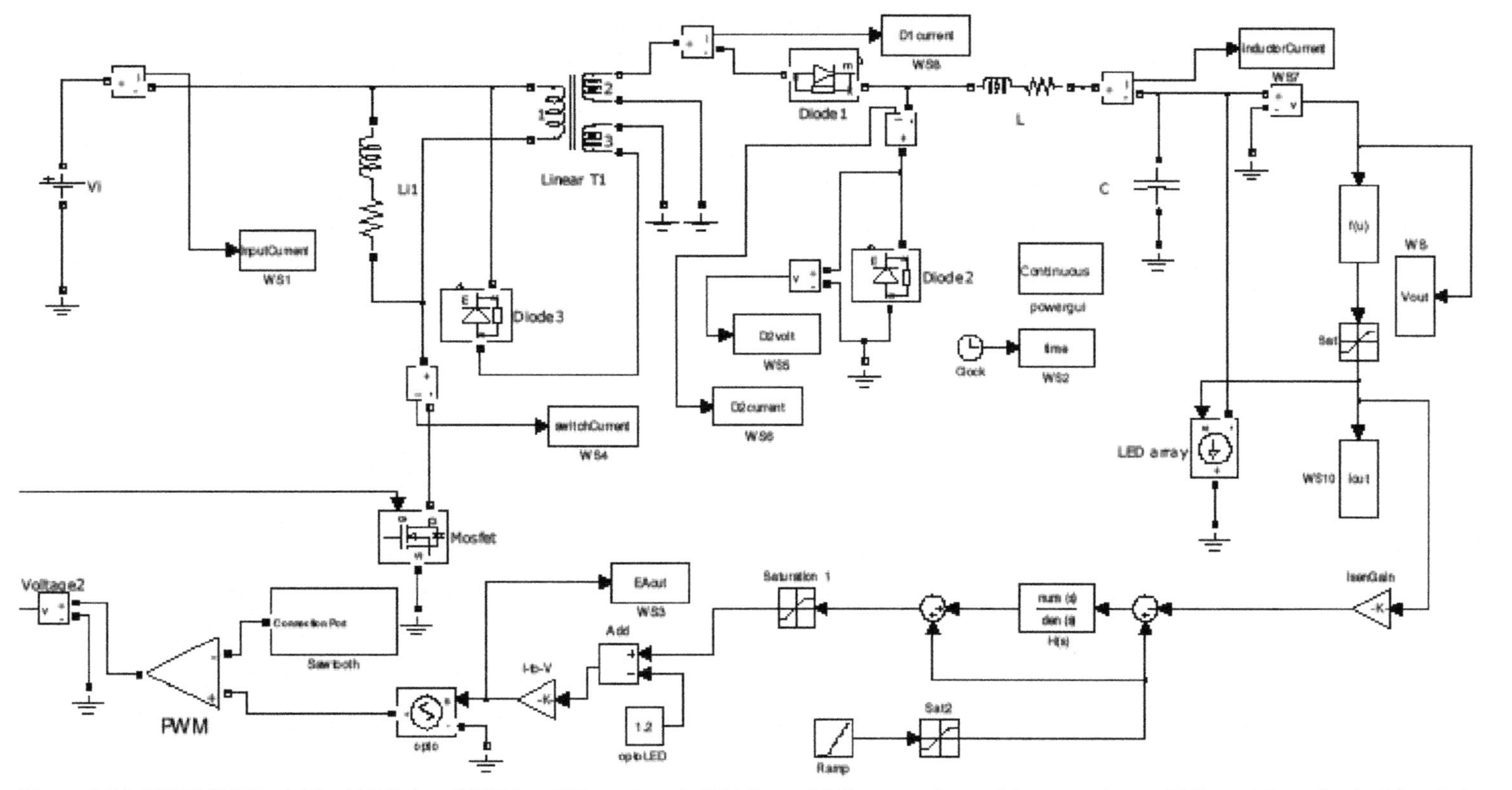

Figure 6.20 ***SIMULINK Model for LED Driver With Type-II Error Amp in*** **H(s)** ***Form.*** (a) Output voltage, (b) error voltage, (c) D_1 and D_2 cathode, (d) switch current, (e) inductor current, (f) D_1 current, (g) D_2 current, and (h) LED array current.

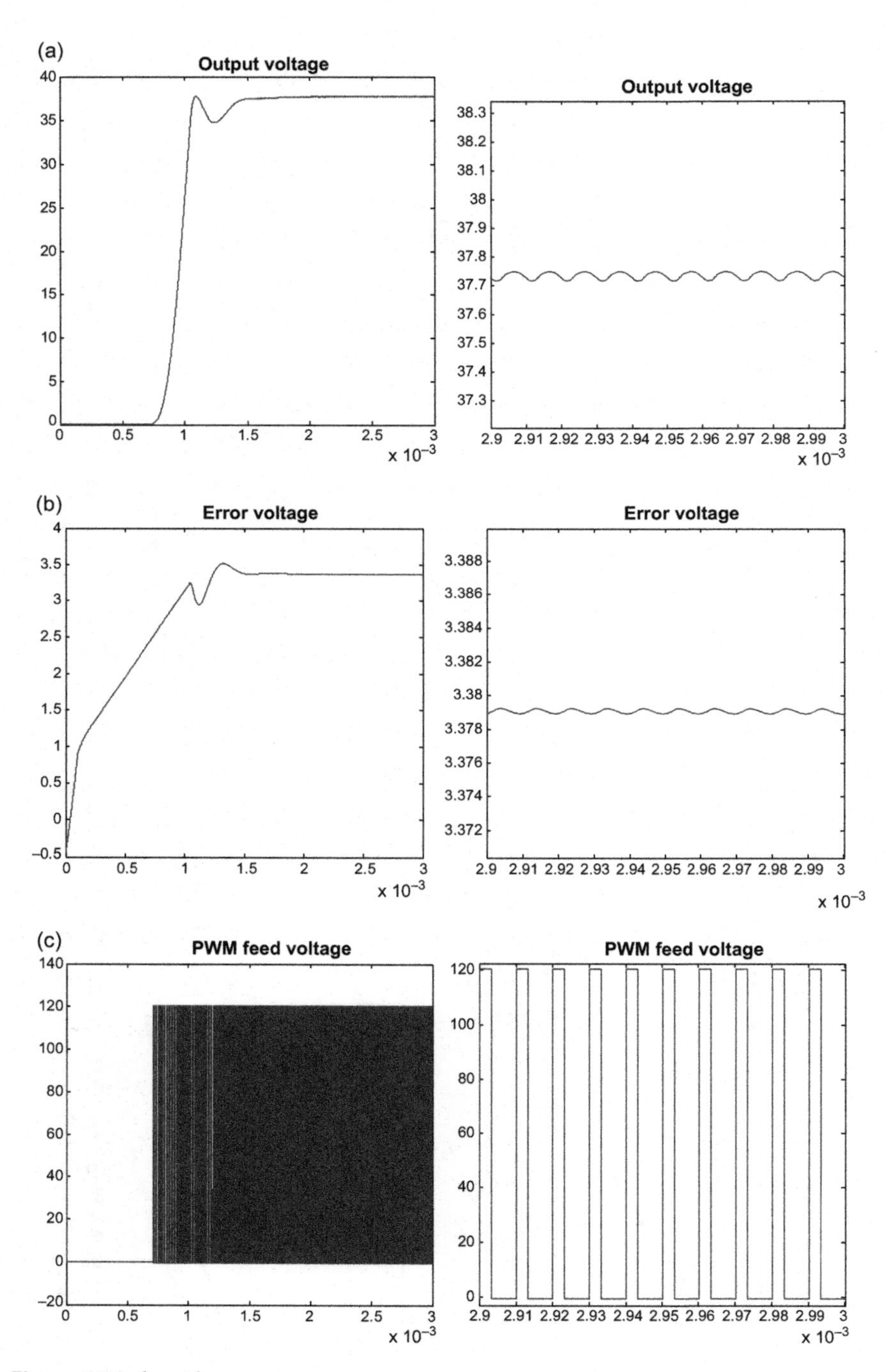

Figure 6.20 ***(cont.)***

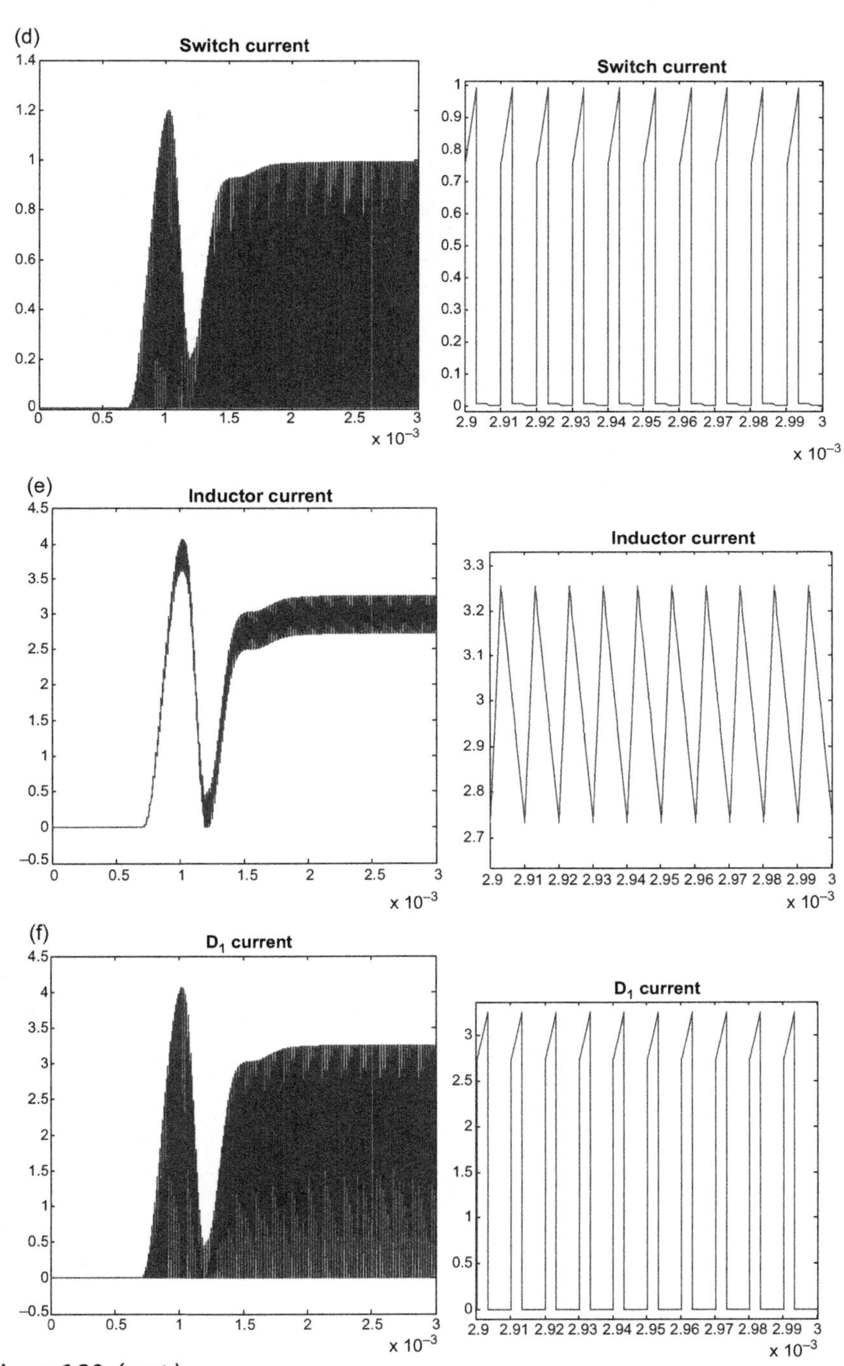

Figure 6.20 *(cont.)*

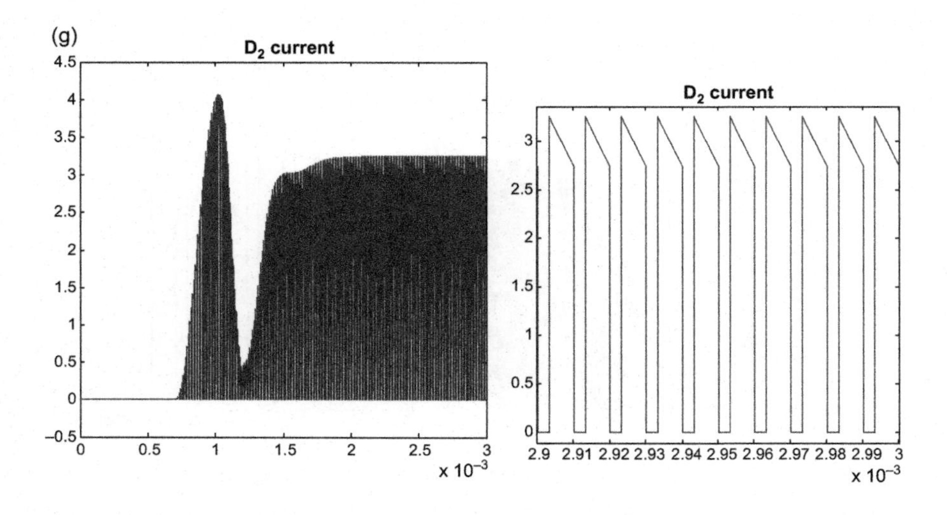

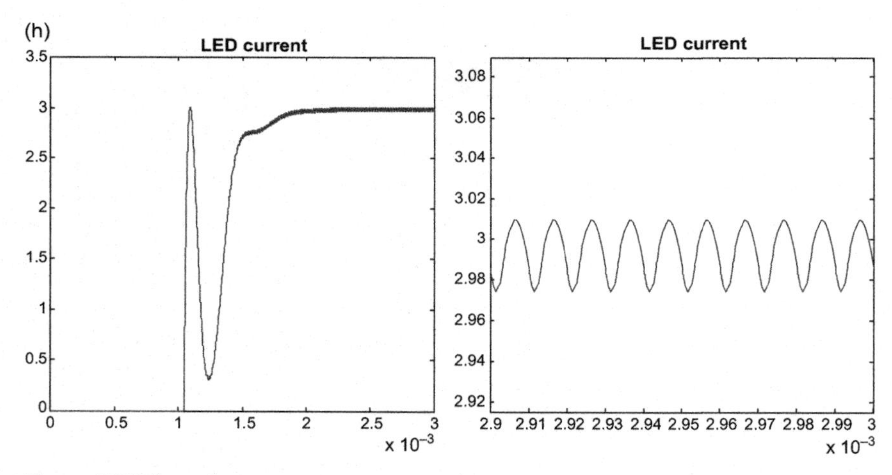

Figure 6.20 *(cont.)*

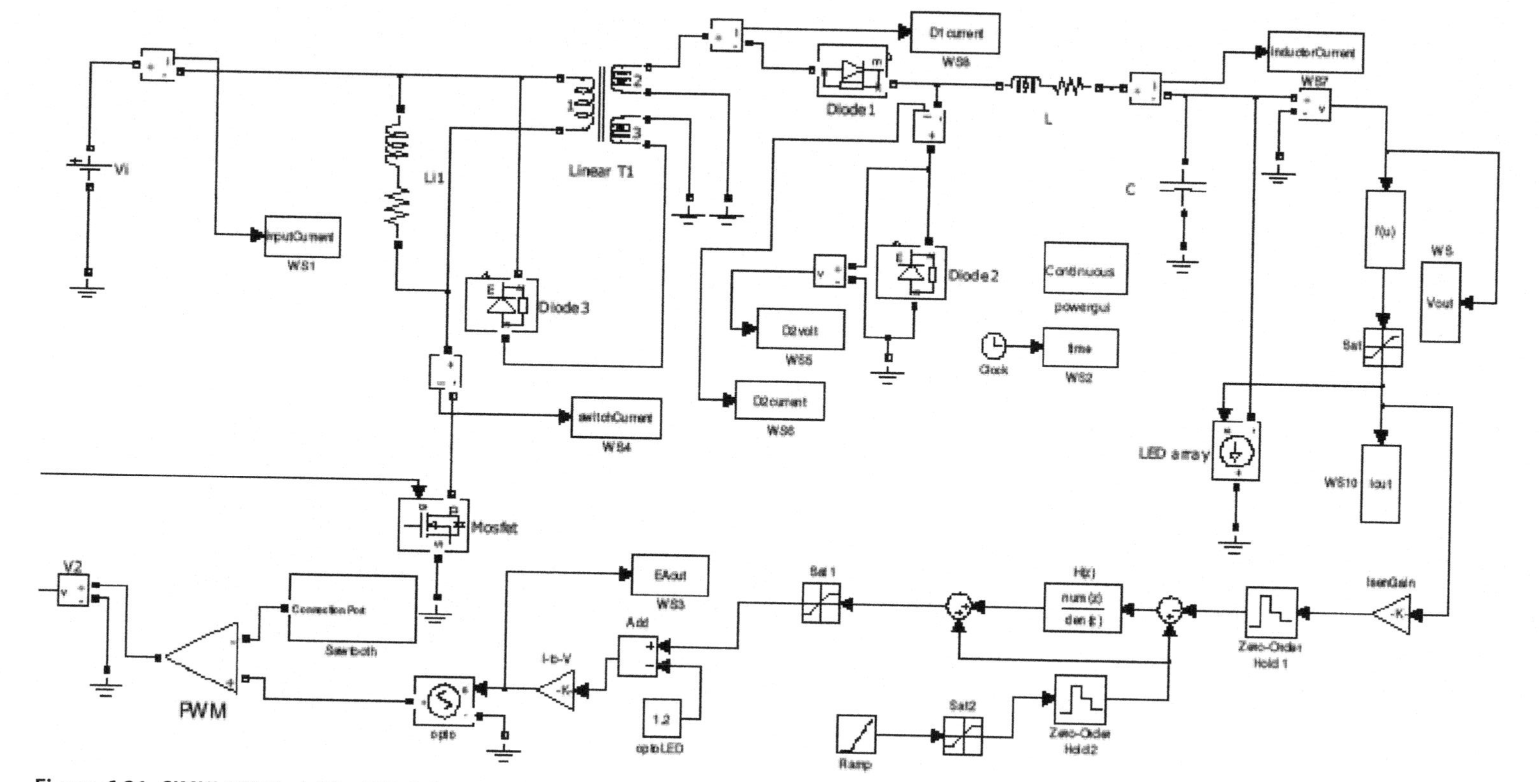

Figure 6.21 ***SIMULINK Model for LED Driver With Type-II Error Amp in Digital Filter Form.*** (a) Output voltage, (b) error voltage, (c) D_1 and D_2 cathode, (d) switch current, (e) inductor current, (f) D_1 current, (g) D_2 current, and (h) LED array current.

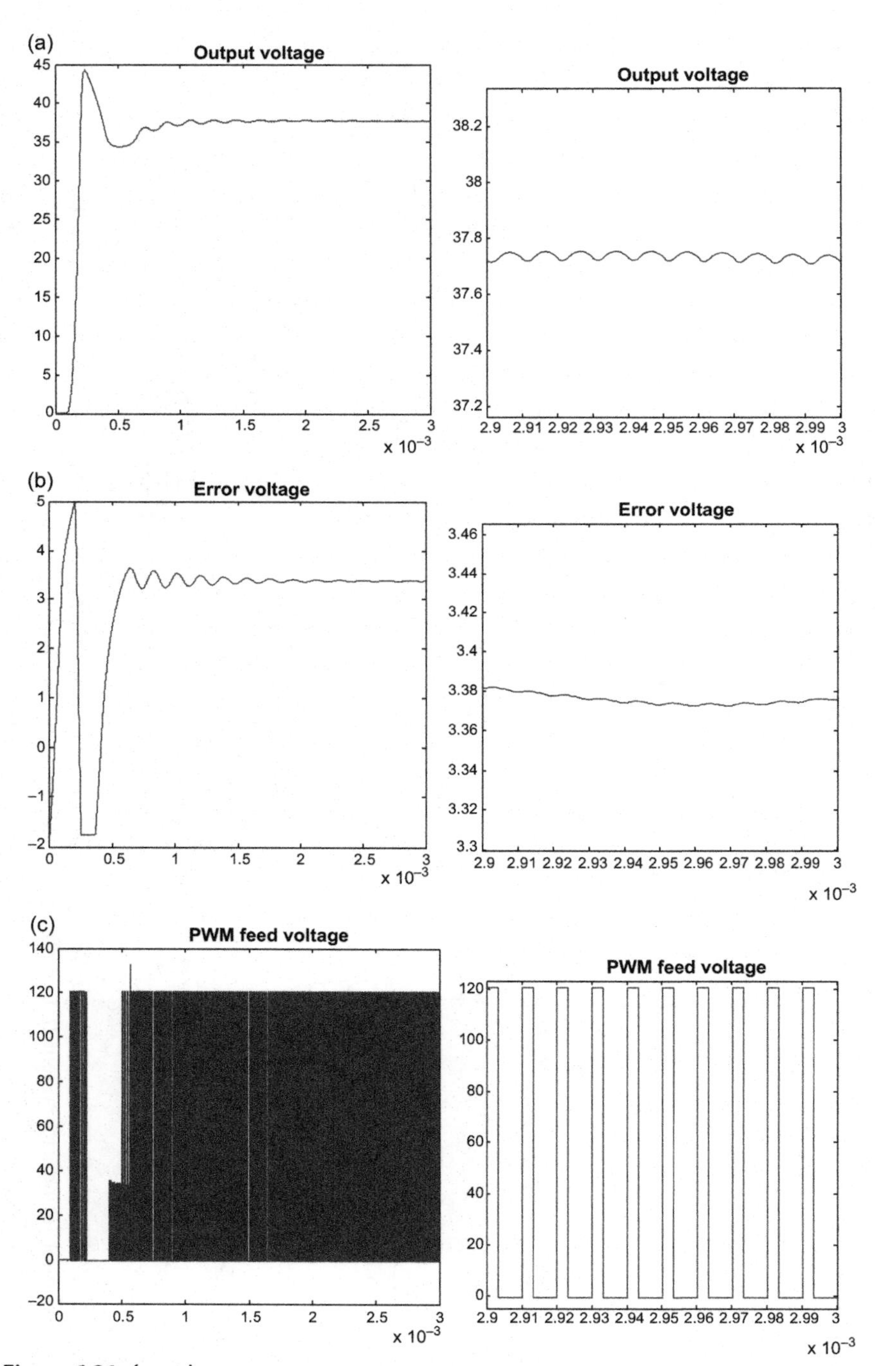

Figure 6.21 *(cont.)*

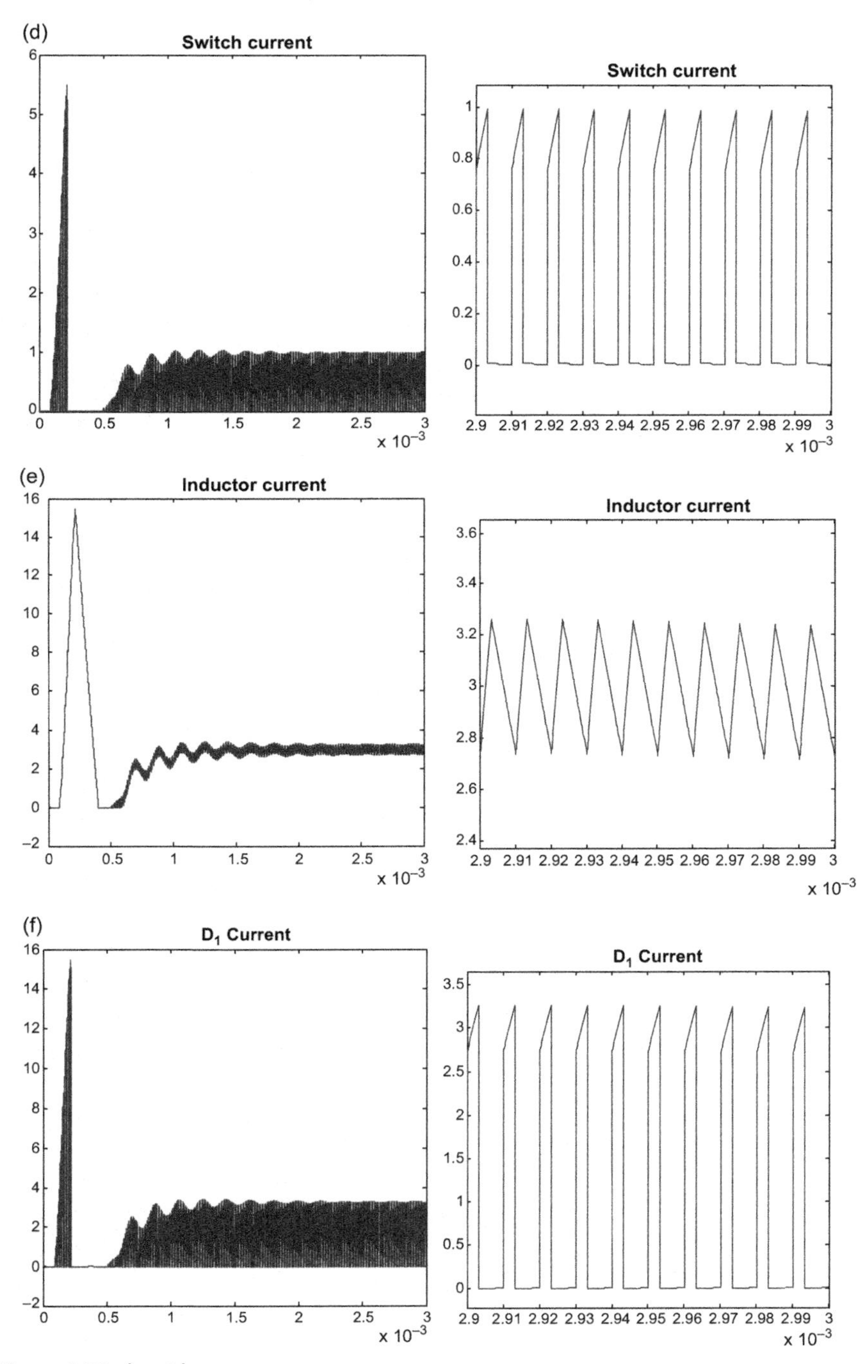

Figure 6.21 *(cont.)*

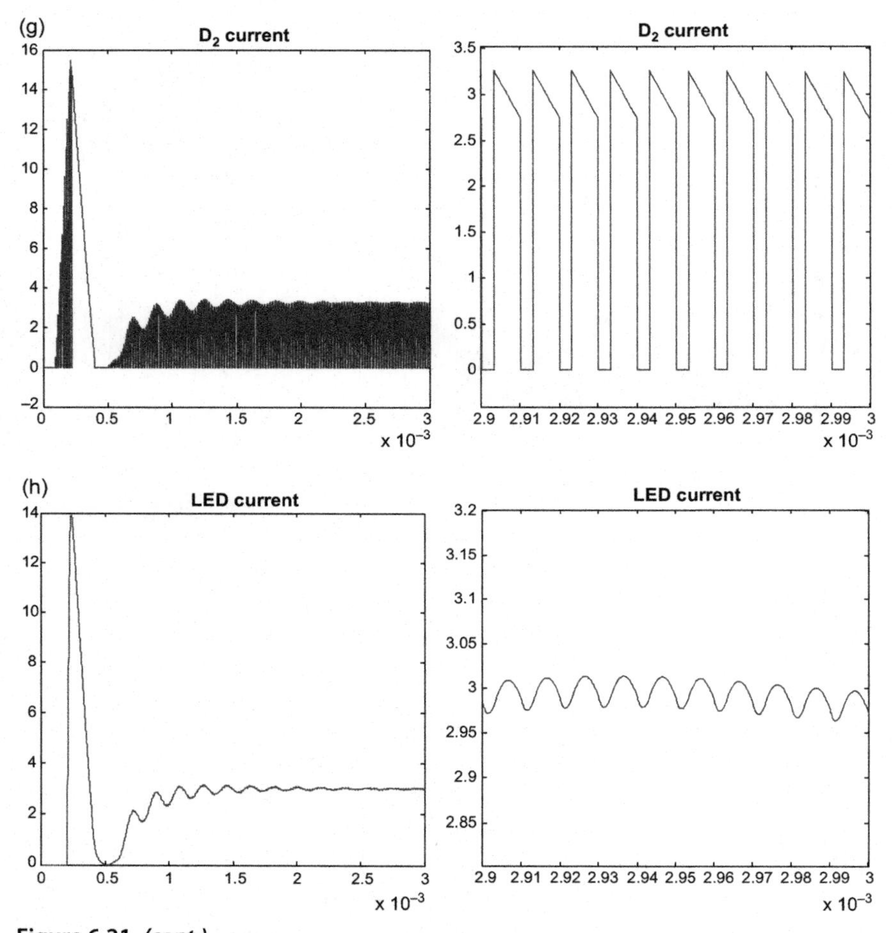

Figure 6.21 ***(cont.)***

PART IV

Boost Converters

All power processors presented so far carry a common trait; the output voltage is less than the input, that is, step–down.

Occasions arise from time to time that the input is low voltage battery sources, but outputs higher than the source are needed. In that case, step–up, that is, boost processors are called for. In theory, flyback converters given in Part II can perform both step–down and step–up. But issues of control mechanism tend to constrain its applications in step–down.

In this part, topology specifically configured to act in step–up operation is covered. Voltage-mode control is first presented, then current-mode follows.

Since the scope of this writing is not intended for basics, readers who need to refresh the procedure for selecting power train components such as, input bus capacitor, boost inductor, output capacitors, etc., can refer [3] and other materials.

CHAPTER 7

DCM Boost Converter with Voltage-Mode Control

7.1 SELECTING DISCONTINUOUS CONDUCTION MODE

Flyback converters presented in Chapters 3 and 4 operate in DCM for reasons that are not discussed here. It turns out that boost converters operating in DCM invite less trouble in designing control strategy for the same reason. Therefore, at this point, we will examine the rationale opting for DCM.

First, a fundamental difference regarding inductor operation exists between the forward converter group and the flyback/boost category. Inductor L in Figure 1.1 does not work as an energy storage device, in contrast to the coupled inductor in Figure 3.1 which stores energy when turned on and releases it when turned off. The alternation, if operating in CCM mode, presents a subtle and unintended consequence.

Here is how the undesirable effect fliesback (Figure 3.3), or boosts (Figure 7.1) if CCM exhibits in time domain.

In steady state, the inductor and output switch (diode) currents of both converters look like Figure 7.2.

In either case, the output current I_o equals to the average of the pulsating output diode current; $I_o = I_A(1-D)$, I_A = inductor average current. However, when, for example, load demand increases, I_A and I_o shall increase accordingly. I_A does increase by extending the duty cycle, D. But the action actually decreases output I_o, since the factor $(1-D)$. This peculiar behavior, unique to only flyback and boost converter operating in CCM, is also given a name, right-half-plane zero (RHP zero), in the study of control mechanism.

We attempt here to prove the existence of the RHP zero in the power train of the boost converter. Referring to Figure 7.1, the following state space matrices are obtained; (7.1a) for ON-time, while (7.1b) for OFF-time.

$$A_1 = \begin{bmatrix} -\frac{r_L}{L} & 0 \\ 0 & -\frac{1}{R_L C} \end{bmatrix}, B_1 = \begin{bmatrix} \frac{1}{L} \\ 0 \end{bmatrix}, C_1 = \begin{bmatrix} 0 & 1 \end{bmatrix} \quad (7.1a)$$

Power Converters with Digital Filter Feedback Control
http://dx.doi.org/10.1016/B978-0-12-804298-4.00007-1

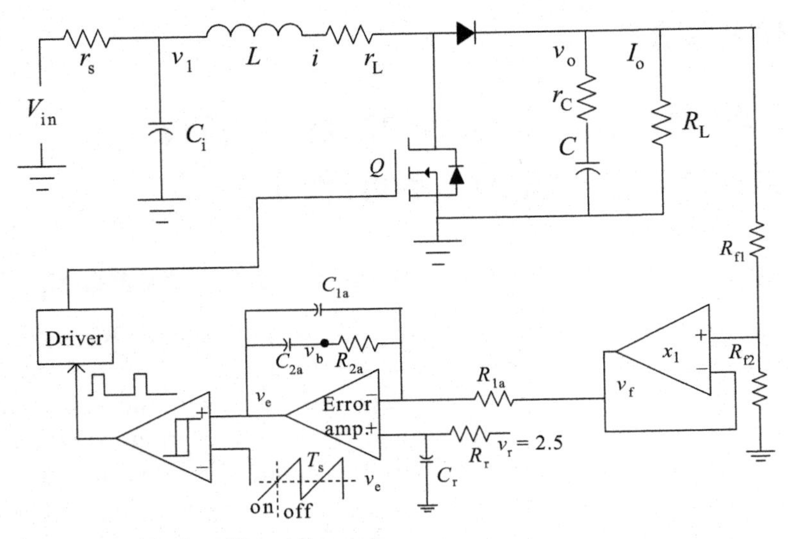

Figure 7.1 ***A Non-Isolated Boost Converter.***

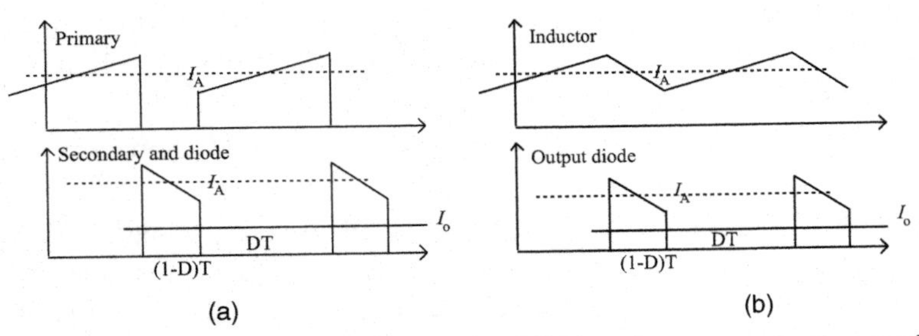

Figure 7.2 (a) Isolated flyback, and (b) boost in CCM (continuous conduction mode).

$$A_2 = \begin{bmatrix} -\frac{r_L}{L} & -\frac{1}{L} \\ \frac{1}{C} & -\frac{1}{R_L C} \end{bmatrix}, B_2 = \begin{bmatrix} \frac{1}{L} \\ 0 \end{bmatrix}, C_2 = \begin{bmatrix} 0 & 1 \end{bmatrix} \tag{7.1b}$$

With symbolic processing, (A.16) gives

$$G_{vd}(s) = \frac{-R_L L V_{in}}{\left[r_L + (1-D)^2 R_L\right]} \frac{\left[\left(s - \left[(1-D)^2 R_L - r_L\right]/L\right)\right]}{\left[s^2 LCR_L + \left(L + r_L CR_L\right)s + (1-D)^2 R_L + r_L\right]} \tag{7.2}$$

The duty cycle-to-output transfer function clearly shows a RHP zero; a numerator with a positive zero, $s = [(1-D)^2R_L - r_L]/L$. Such a factor yields a higher gain with diminishing phase, a potential response leading to instability that shall be avoided. Therefore DCM, which has three states and will not cause to squeeze OFF-time, is selected for this chapter's presentation.

7.2 A DESIGN EXAMPLE

In DCM operation, the boost inductor and the output diode currents' profile is dictated by Figure 7.3.

In addition, there are several guiding equations. Continuity of flux and balance of volt–second across the boost inductor demands,

$$V_{in}DT + \left(V_{in} - V_D - V_o\right)D_2T = 0 \tag{7.3}$$

The load demand, I_o, is related to the inductor peak current, I_p, by,

$$\frac{1}{2}I_pD_2 = I_o = \frac{V_o}{R_L} \tag{7.4}$$

While the peak current is given by,

$$I_p = \frac{V_{in}D}{Lf} \tag{7.5}$$

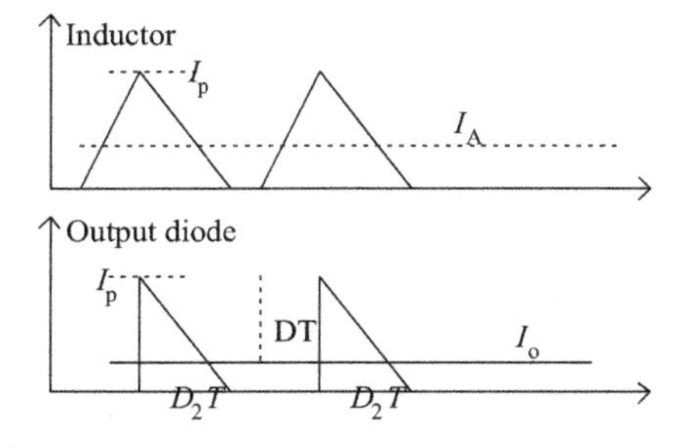

Figure 7.3 ***Currents for Boost Converter in DCM.***

Consolidation of (7.3), (7.4), and (7.5) enables determination of the boost inductor value required.

$$L = \frac{R_L V_{in}^2 D^2}{2fV_o(V_o + V_D - V_{in})} \tag{7.6}$$

in which f stands for the selected switching frequency, D is the desired switch ON-time duty cycle, and the output diode drop, V_D, is also accounted for. D and D_2 in (7.3) are constrained by $(D + D_2) < 1$.

Once the boost inductor is determined, the output diode ramp-down current profile can be expressed by,

$$i_{CR}(t) = I_p - \frac{V_D + V_o - V_{in}}{L} t \tag{7.7}$$

The ramp-down current crosses the DC load current at time T_c.

$$I_p - \frac{V_D + V_o - V_{in}}{L} T_C = I_o, \qquad T_C = \frac{L(I_p - I_o)}{V_D + V_o - V_{in}} \tag{7.8}$$

Therefore, the amount of charge entering the output capacitor is,

$$\delta Q = \frac{1}{2}(I_p - I_o)T_C = \frac{L(I_p - I_o)^2}{2(V_D + V_o - V_{in})} \tag{7.9}$$

Given a specified output voltage ripple requirement δv_o, the output capacitor value is determined. In general, the ripple requirement is specified as a certain percentage of the steady state output voltage at said full load.

$$C = \frac{\delta Q}{\delta v_o} = \frac{L(I_p - I_o)^2}{2\delta v_o(V_D + V_o - V_{in})} \tag{7.10}$$

A numerical example is in order here to give us a good sense of magnitude to be expected.

We shall select $V_{in} = 12\text{V}$, $V_o = 22\text{V}$, $\delta v_o = 1\% V_o$, $P_o = 25\text{ W}$, $f = 100\text{ kHz}$, $V_D = 0.5$, and duty cycle $D = 0.4$.

With operating parameters selected, (7.6) gives $L = 9.65\ \mu\text{H}$, (7.5) $I_p = 4.972\text{ A}$, $I_o = 1.135\text{ A}$, (7.10) $C = 30\ \mu\text{F}$, and $D_2 = 0.457$. Constraint

$(D + D_2) = 0.857 < 1$ is satisfied. This allows a non-zero D_3 $(=1 - D - D_2)$, slightly less than 15%, for transient response cushion without concern of instability.

7.3 DERIVATION OF MODULATOR GAIN

Following the same procedure, as we did for flyback converter in Chapter 3, the small-signal model invoking state space averaging is called out. In the case of Figures 3.4 and 3.5, isolation was involved, and it led to a more complex source and impedance transformation. In the current case for non-isolated boost in DCM, such transformation is not needed. The simplification gives Figure 7.4. Besides, based on the same NASA report cited in Section 3.2, all power stage model parameters are given as the follows,

$$\begin{aligned} K &= \frac{2Lf}{R} \quad M = \frac{1}{2}\left(1+\sqrt{1+4\frac{D^2}{K}}\right) \quad j_1 = \frac{2V_o}{R}\sqrt{\frac{M}{K(M-1)}} \\ r_1 &= \frac{M-1}{M^3}R \quad g_1 = \frac{M}{(M-1)R} \quad j_2 = \frac{2V_o}{R\sqrt{KM(M-1)}} \\ r_2 &= \frac{(M-1)R}{M} \quad g_2 = \frac{M(2M-1)}{(M-1)R} \end{aligned} \tag{7.11}$$

with $R = R_L$.

Since no isolation is involved, the output-to-duty cycle transfer function can be easily obtained by setting $n = 1$ in (3.9), and it leads to

$$G_{vd}(s) = \frac{j_1 g_2 - j_2\left[\left(Z_s(s)\right)^{-1} + r_1^{-1}\right]}{g_1 g_2 - \left[\left(Z_s(s)\right)^{-1} + r_1^{-1}\right]\left((1/r_2) + Cs + (1/R)\right)} \tag{7.12}$$

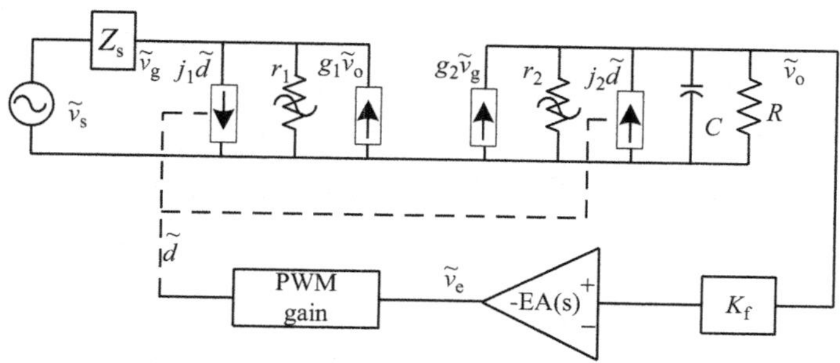

Figure 7.4 ***Small-Signal Model for Boost Converter in DCM.***

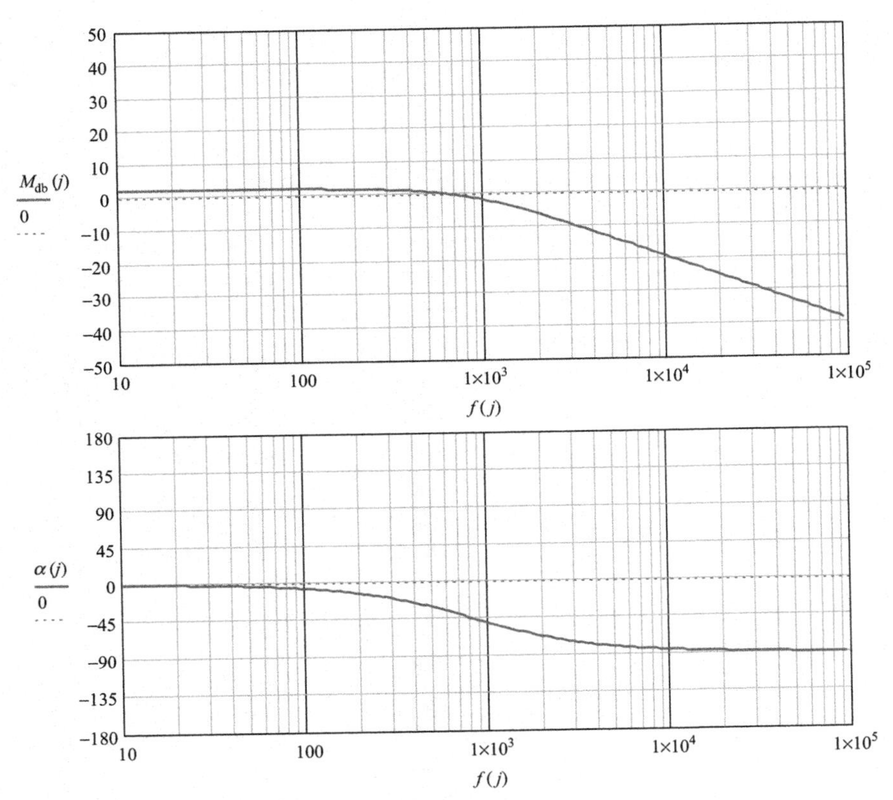

Figure 7.5 ***Modulator Gain of Boost Converter in DCM.***

The PWM gain, F_m, for voltage mode control is a straightforward number equal to 0.98/(sawtooth swing), while the feedback factor, K_f, a simple resistive voltage division ratio.

For the current example given in Section 7.2, for a 3 V sawtooth swing and a 2.5 V reference voltage, the resulting modulator gain, $M(s)$, equals $K_f F_m G_{vd}(s)$ and shows a frequency profile Figure 7.5.

7.4 DESIGNING ANALOG ERROR AMPLIFIER

The modulator gain and phase plot gives a −2 db and −50° at 1 kHz. If that frequency is selected as the close–loop crossover frequency, and if a 45°-phase margin is desired, (1.12) indicates that only a 5°-phase boost

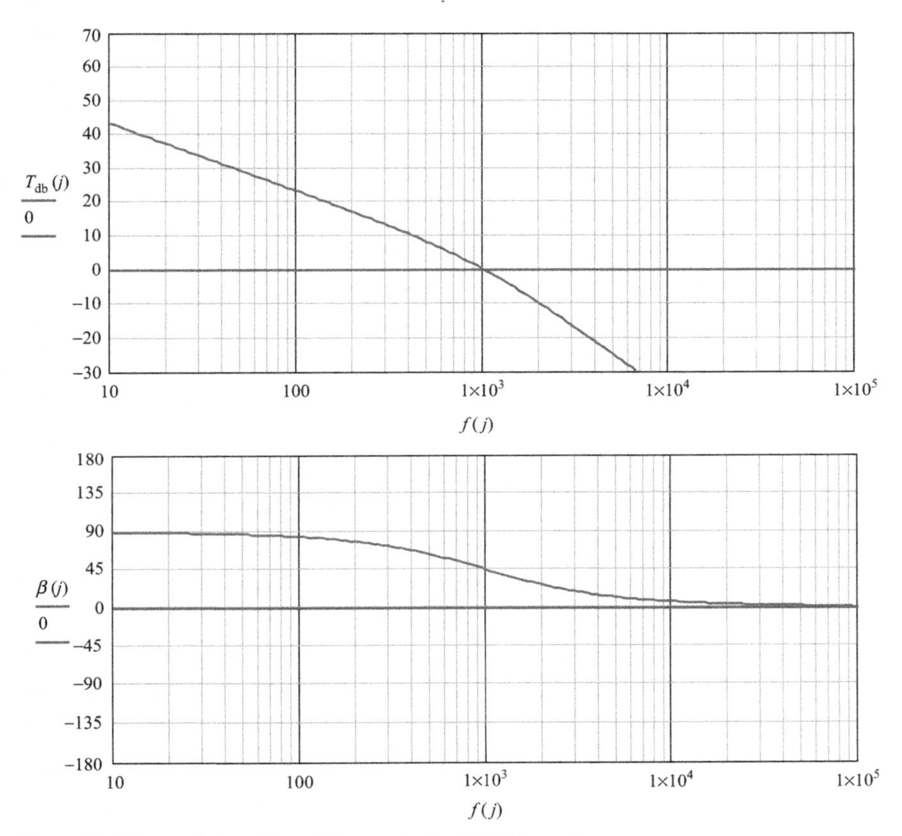

Figure 7.6 ***Loop Gain of Boost Converter in DCM Operation.***

is required. This low-boost requirement can be easily met by a type-II amplifier, (1.13) and (1.14). Numerical computation leads to R_{1a} = 2K, C_{1a} = 0.058 μF, R_{2a} = 15.7 K, and C_{2a} = 0.011 μF. And the overall loop gain, $T(s) = -E_A(s)M(s)$, proves to be as shown in Figure 7.6.

As shown in Figure 7.6, it does cross at 1 kHz.

7.5 PERFORMANCE OF CONVERTER WITH ANALOG CONTROL

We are ready here to study the converter performance in time domain. There are five state variables, as given in Figure 7.1; input capacitor node v_1, boost inductor current i, output capacitor node v_c, error amplifier output

v_e, and error amplifier local feedback v_b. All five variables give a set of discrete difference equations suitable for iterative computation starting with a zero state.

$$\begin{pmatrix} v_{1_{j+1}} \\ i_{j+1} \\ v_{c_{j+1}} \\ v_{e_{j+1}} \\ v_{b_{j+1}} \end{pmatrix} = \begin{bmatrix} v_{1_j} + \delta t\left[\left(V_{in} - v_{1_j}\right)\frac{1}{r_s C_i} - \frac{1}{C_i} i_j\right] \\ \text{if}\left[i_j < 0, 0, i_j - \delta t\left(\frac{r_L}{L} i_j + \frac{1}{L}\text{if}\left(v_{e_j} > s_{w_j}, 0, R_p \text{if}\left(v_{e_j} > s_{w_j}, 0, i_j\right) + k_R v_{c_j}\right)\right) + \frac{\delta t v_{1_j}}{L}\right] \\ v_{c_j} + \frac{\delta t}{r_c C}\left(R_p \text{if}\left(v_{e_j} > s_{w_j}, 0, i_j\right) + k_R v_{c_j} - v_{c_j}\right) \\ \text{if}\left[v_{e_j} < 0, 0, \text{if}\left[v_{e_j} > 12, 12, v_{e_j} + \left[\begin{array}{l} \frac{\delta t}{C_{1a}}\left(\frac{1}{R_{1a}} + \frac{1}{R_{2a}}\right)V_{R_j} - \frac{\delta t}{\left(R_{2a} C_{1a}\right)} v_{b_j} \\ -\frac{\delta t}{\left(R_{1a} C_{1a}\right)} K_f\left(R_p \text{if}\left(v_{e_j} > s_{w_j}, 0, i_j\right) + k_R v_{c_j}\right) \end{array}\right]\right]\right] \\ \begin{array}{l} \left[1 - \left[\frac{1}{\left(R_{2a} C_{2a}\right)} + \frac{1}{\left(R_{2a} C_{1a}\right)}\right]\delta t\right] v_{b_j} + \left[\frac{1}{\left(R_{2a} C_{2a}\right)} + \frac{1}{C_{1a}}\left(\frac{1}{R_{1a}} + \frac{1}{R_{2a}}\right)\right]\delta t V_{R_j} \\ \qquad -\frac{\delta t}{\left(R_{1a} C_{1a}\right)} K_f\left(R_p \text{if}\left(v_{e_j} > s_{w_j}, 0, i_j\right) + k_R v_{c_j}\right) \end{array} \end{bmatrix} \tag{7.13}$$

where,

$$R_p = \frac{r_c R_L}{r_c + R_L} \quad k_R = \frac{R_L}{r_c + R_L} \tag{7.14}$$

With computation resolution at 100 point/cycle, that is, $\delta t = 0.01$ T, and computing 150 cycles, Figure 7.7a–d present the turn on transient and the steady state.

7.6 CONVERSION TO DIGITAL CONTROL

We now want to convert the analog error amplifier, $E_A(s)$, identified in Section 7.4, to its digital equivalent $H(z)$. It turns out that some property not well understood, at least to this author, exists. From the $E_A(\omega)$ plot in

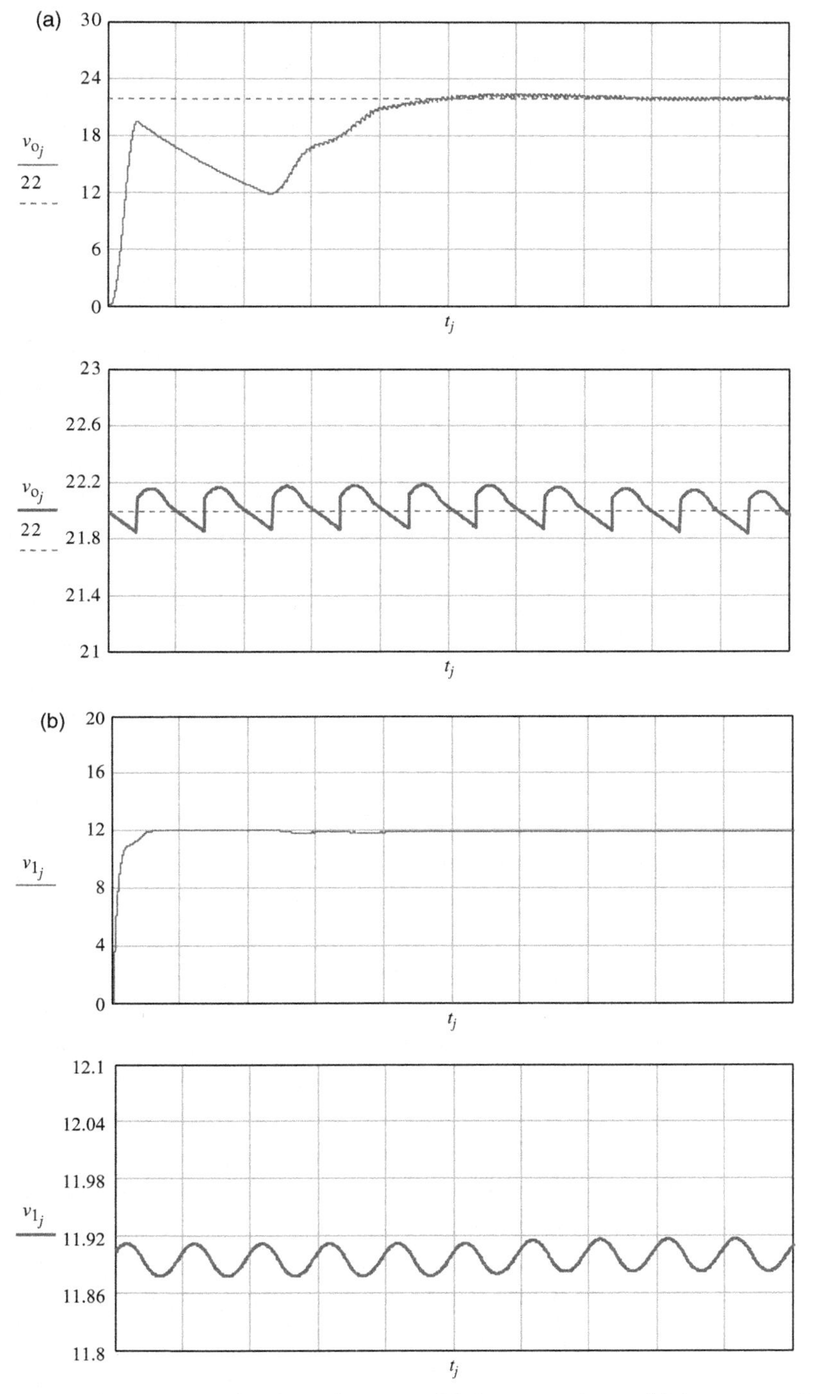

Figure 7.7 Equation set (7.13) performance (a) Output voltage, (b) input capacitor voltage, (c) inductor current, and (d) error voltage and sawtooth.

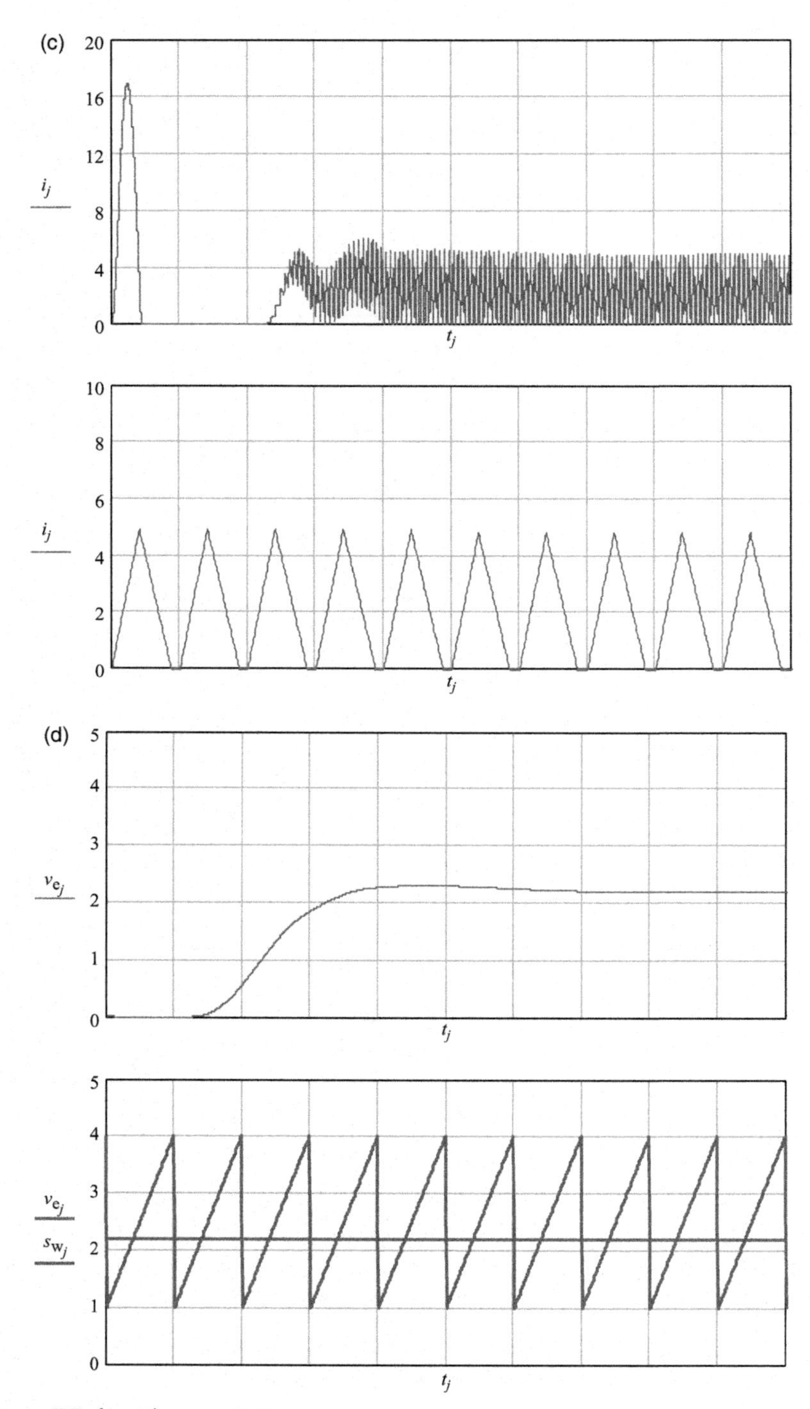

Figure 7.7 ***(cont.)***

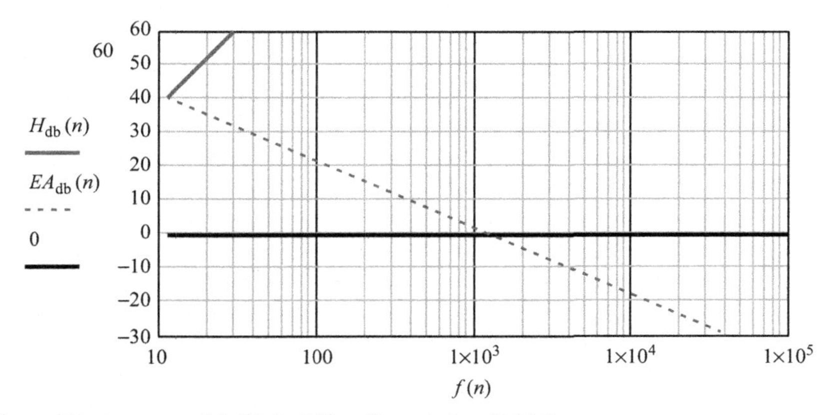

Figure 7.8 ***Incompatible Digital Filter Sampled at 100 kHz.***

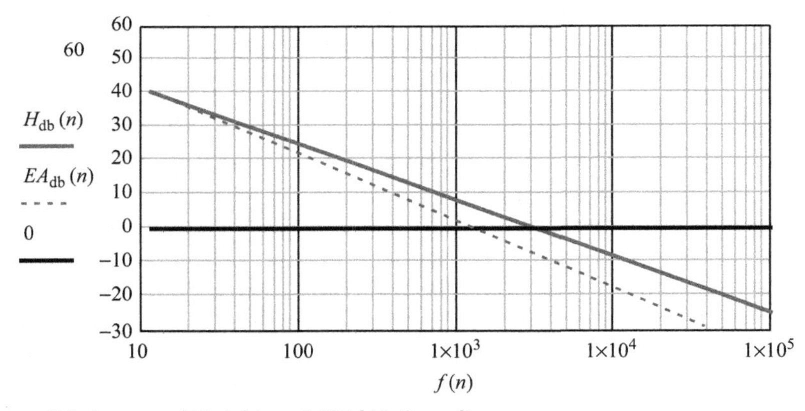

Figure 7.9 ***Improved Matching at 300 kHz Sampling.***

frequency domain, 100 kHz sampling was thought to be a good choice, but it ends up having a frequency response, $H(\omega)$, that is utterly incompatible as shown in Figure 7.8, with the analog version.

Increasing the sampling frequency to 300 kHz improves matching, Figure 7.9, with losses in high frequency.

Increasing the sampling frequency further to 800 kHz improves more, Figure 7.10, but further increase to 900 kHz makes it worse. We therefore settle for sampling at 800 kHz.

At 800 kHz sampling, the polynomial coefficients for the corresponding $H_{II}(z)$, (1.19), is given as,

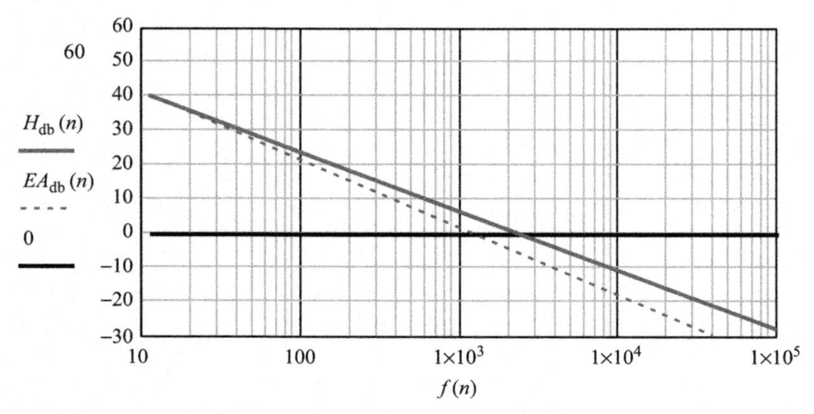

Figure 7.10 ***Better Matching at High Frequency with 800 kHz Sampling.***

$$a_0 = 5.392 \times 10^{-3} \quad a_1 = 38.663 \times 10^{-6} \quad a_2 = -5.353 \times 10^{-3}$$
$$b_1 = -1.991 \times 10^{-0} \quad b_2 = 991.465 \times 10^{-3}$$

With the digital filter identified, Figure 7.1 is transformed to its digital control version, Figure 7.11. The digital filter is also proved to be stable, Figure 7.12, with poles within the unit circle of z-plane.

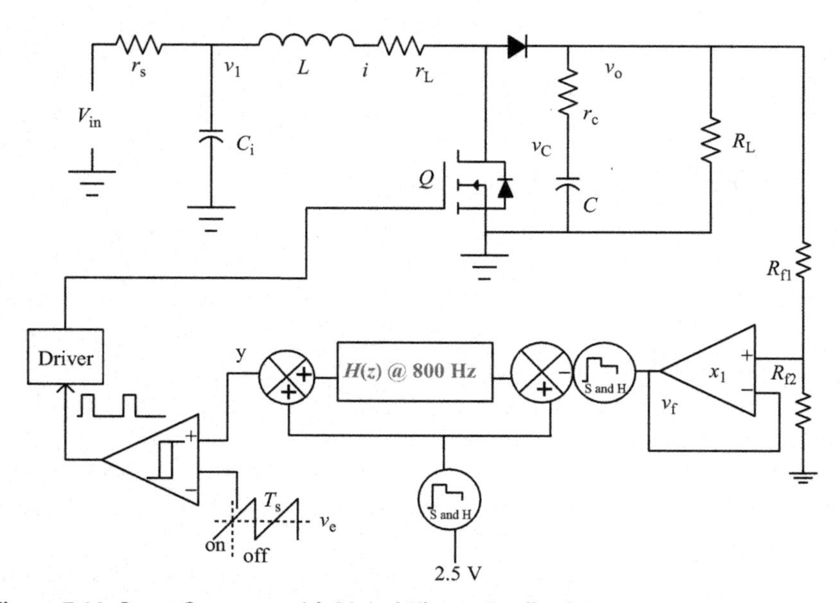

Figure 7.11 ***Boost Converter with Digital Filter in Feedback Loop.***

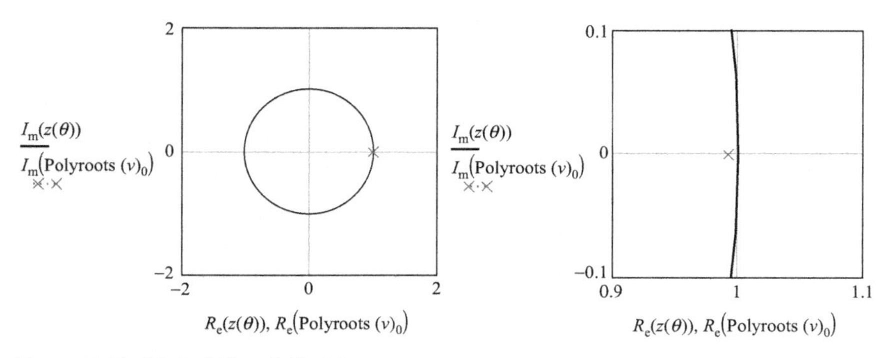

Figure 7.12 ***Digital Filter is Stable.***

7.7 PERFORMANCE OF CONVERTER WITH DIGITAL CONTROL

Direct form implementation, Figure 1.6 and (1.27), for the digital filter worked out above at this step replaces two analog state variables, v_e and v_b, in the analog equation set (7.13). The new set with digital feedback results is as follows:

$$\begin{pmatrix} v_{1_{j+1}} \\ i_{j+1} \\ v_{c_{j+1}} \\ y_{j+1} \\ w_{j+1} \end{pmatrix} = \begin{bmatrix} v_{1_j} + \delta t\left[\left(V_{\text{in}} - v_{1_j}\right)\frac{1}{r_s C_i} - \frac{1}{C_i} i_j\right] \\ \text{if}\left[i_j < 0, 0, i_j - \delta t\left(\frac{r_L}{L} i_j + \frac{1}{L}\text{if}\left(K_a y_j > s_{w_j}, 0, R_p \text{if}\left(K_a y_j > s_{w_j}, 0, i_j\right) + k_R v_{c_j}\right)\right) + \frac{\delta t v_{1_j}}{L}\right] \\ v_{c_j} + \frac{\delta t}{r_c C}\left(R_p \text{if}\left(K_a y_j > s_{w_j}, 0, i_j\right) + k_R v_{c_j} - v_{c_j}\right) \\ \text{if}\left[y_j < 0, 0, \text{if}\begin{bmatrix} K_a y_j > 12, 12, v_{r_j} \\ +a_0\left[\left[v_{r_j} - K_f\left(R_p \text{if}\left(K_a y_j > s_{w_j}, 0, i_j\right) + k_R v_{c_j}\right)\right] - b_1 w_j - b_2 w_{j-1}\right] \\ +a_1 w_j + a_2 w_{j-1} \end{bmatrix}\right] \\ \left[v_{r_j} - K_f\left(R_p \text{if}\left(K_a y_j > s_{w_j}, 0, i_j\right) + k_R v_{c_j}\right)\right] - b_1 w_j - b_2 w_{j-1} \end{bmatrix} \tag{7.15}$$

An important note shall be added here. Figure 7.10 indicates a slight discrepancy in digital gain compared to the analog error amplifier. This is treated as a small gain adjustment factor, $K_a = 0.81$, multiplying the digital filter output, symbol y, and equation set (7.15) yields performance as shown in Figure 7.13a–d.

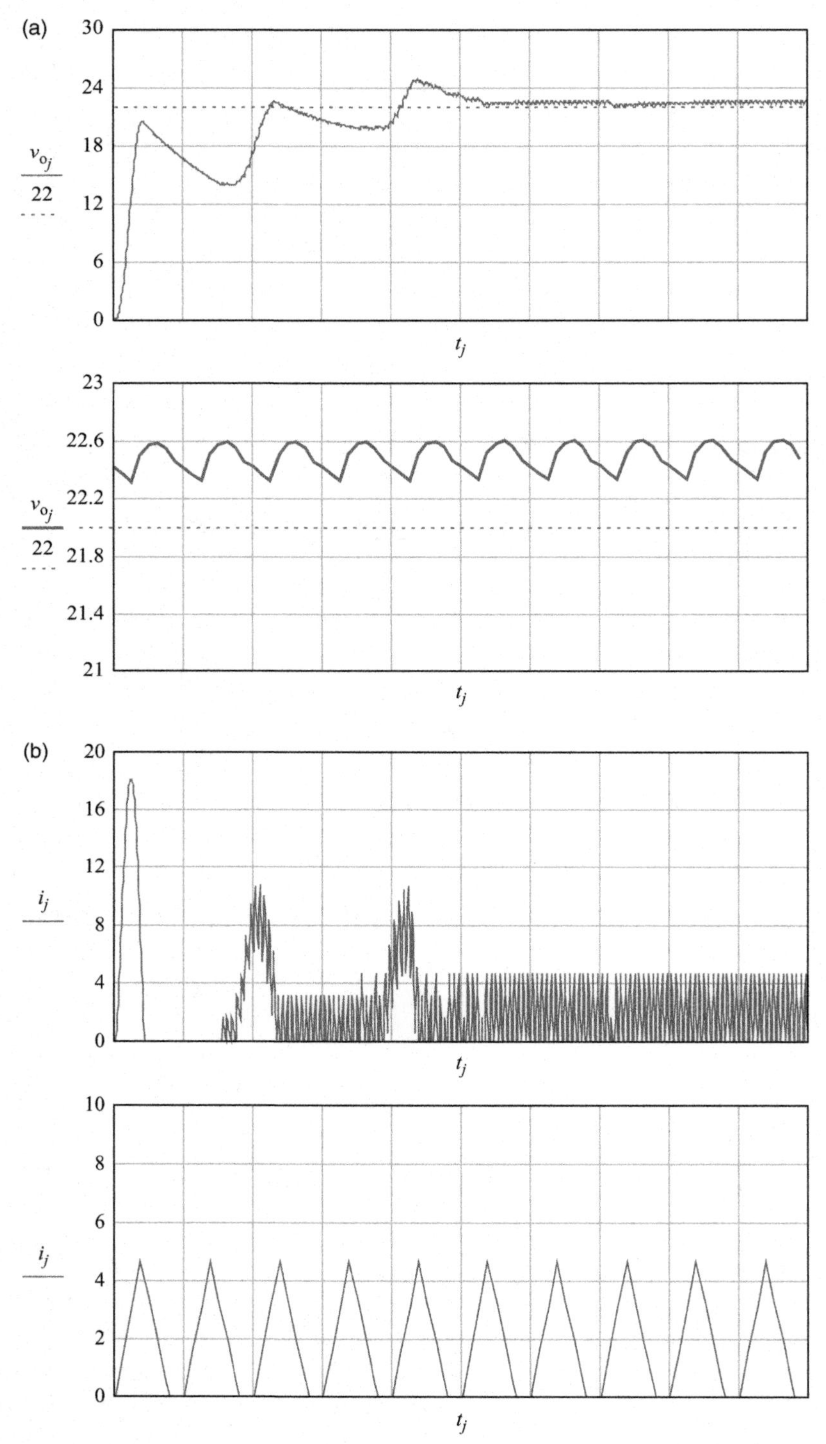

Figure 7.13 Equation set (7.15) performance (a) Output voltage, (b) inductor current, (c) diode current, and (d) error voltage and sawtooth.

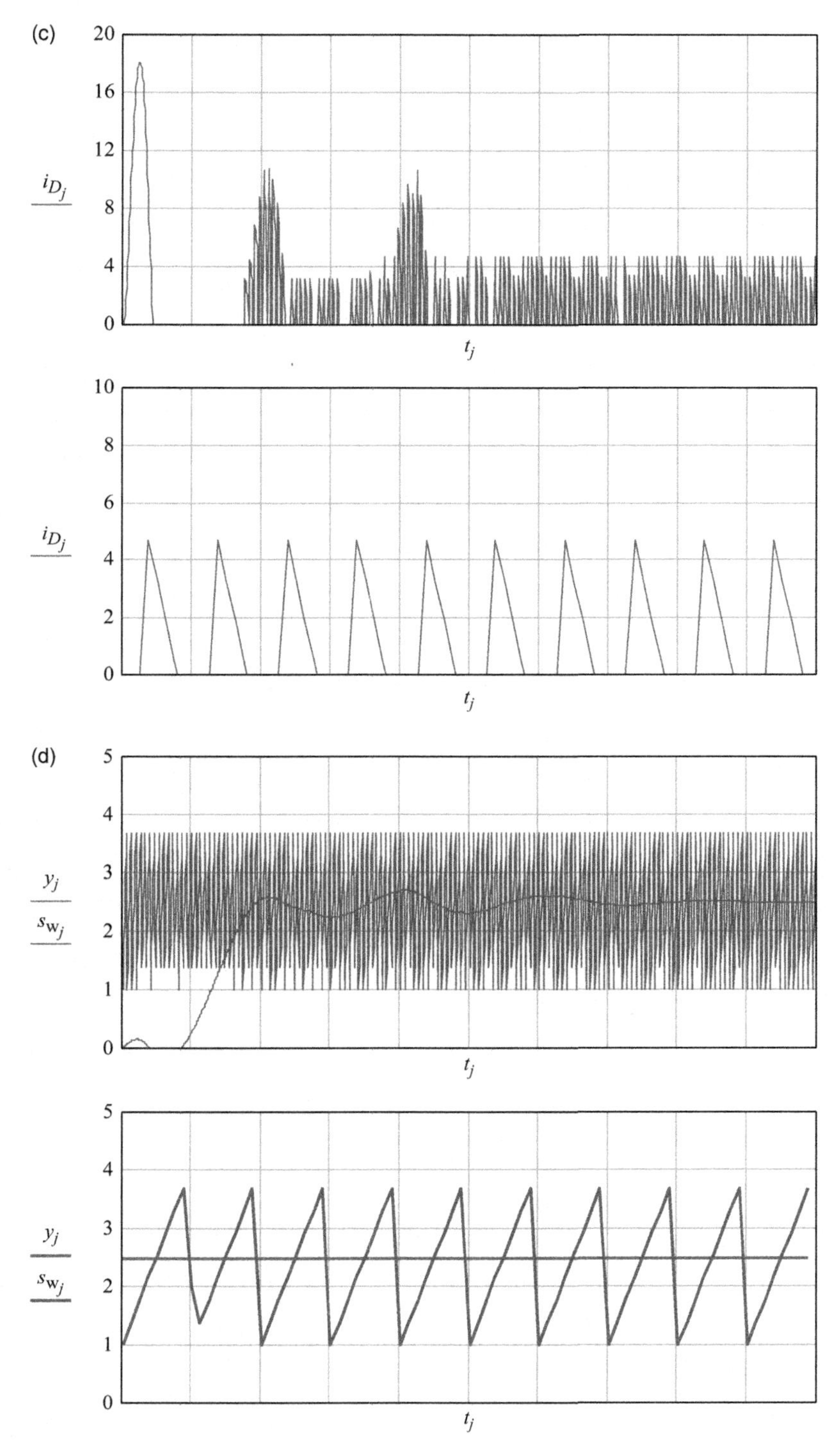

Figure 7.13 ***(cont.)***

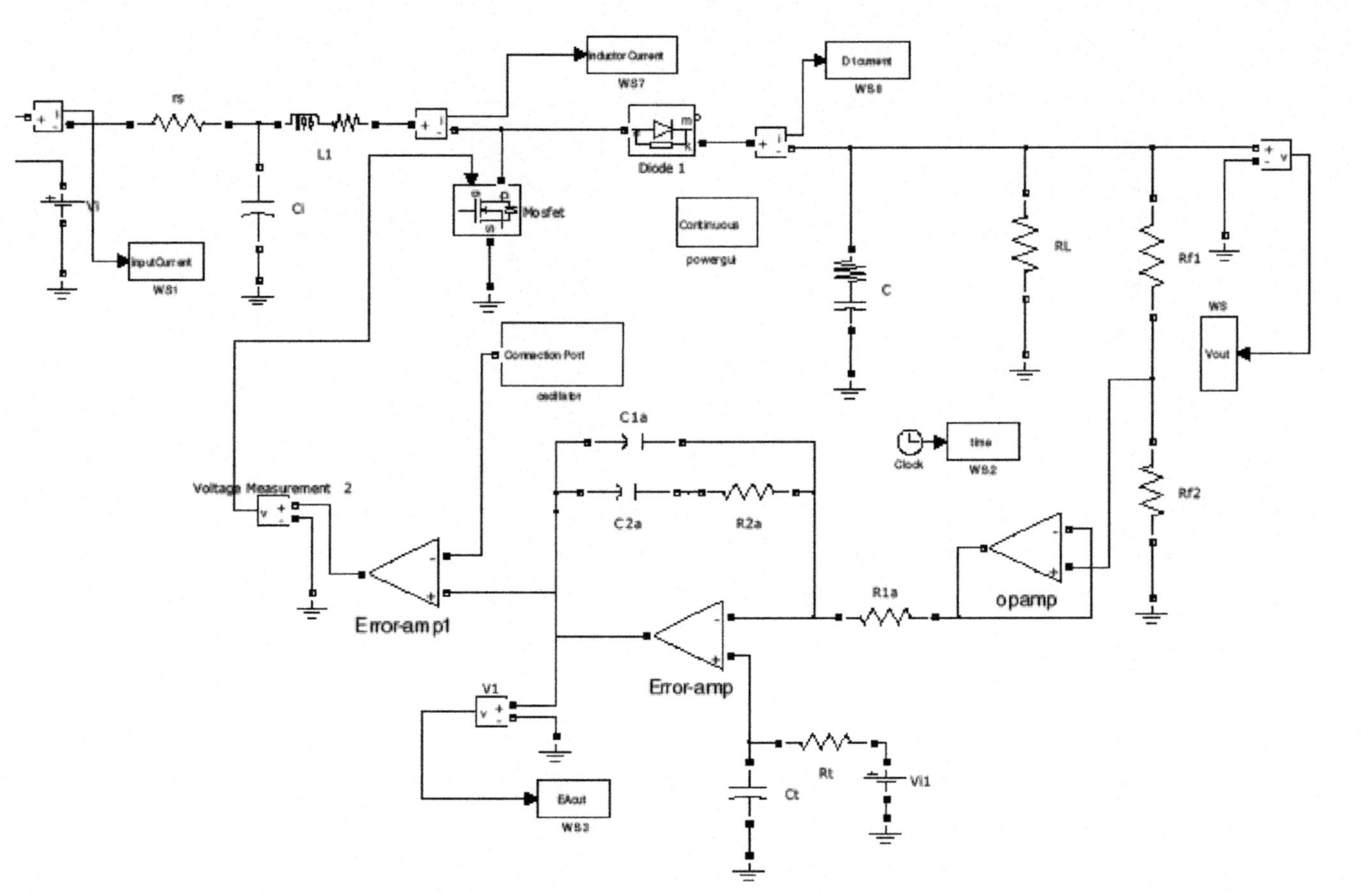

Figure 7.14 ***SIMULINK Model for Boost Converter in DCM with Error Amplifier Represented by Physical Devices.*** (a) Output voltage, (b) inductor current, (c) diode current, and (d) error voltage.

7.8 PERFORMANCE VERIFICATION WITH SIMULINK

Next, we invoke SIMULINK to conduct further verification. First, Figure 7.1 is rebuilt in the SIMULINK environment. It results in Figure 7.14. R_t and C_t perform slow starting. Figure 7.14a–d give output voltage, inductor current, diode current, and error voltage.

Then, Figure 7.12 is transported to SIMULINK with the digital filter coefficients given in Section 7.6. Figure 7.15 show the schematic, while 7.15a–d show performance in time domain.

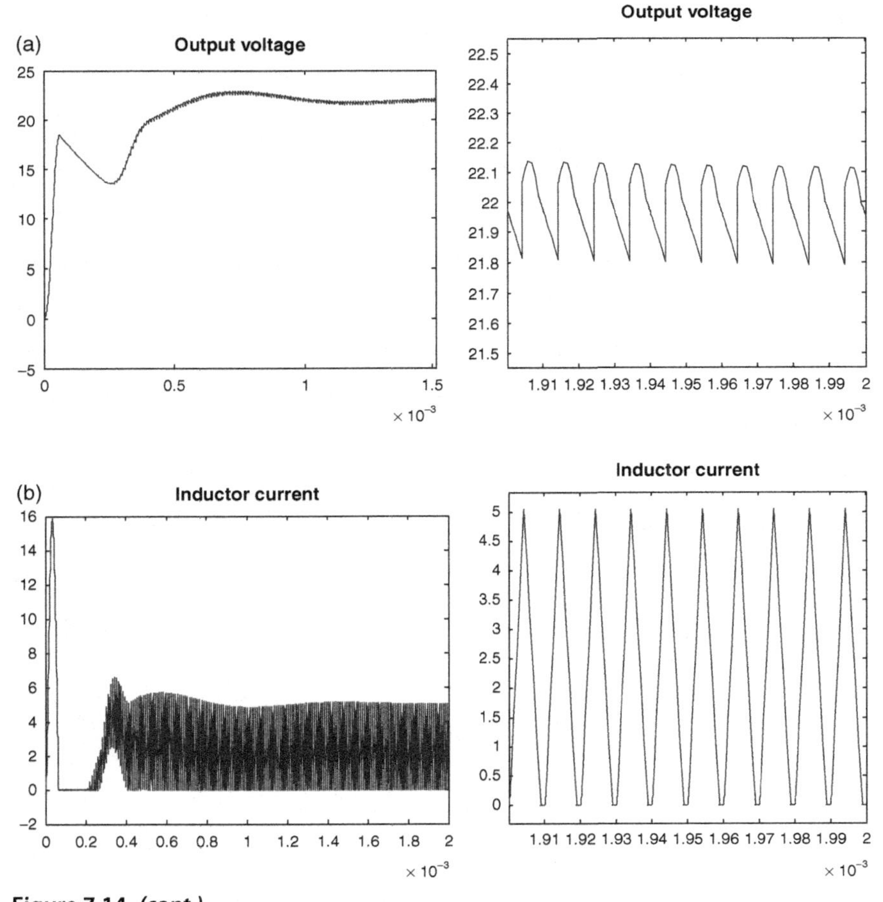

Figure 7.14 *(cont.)*

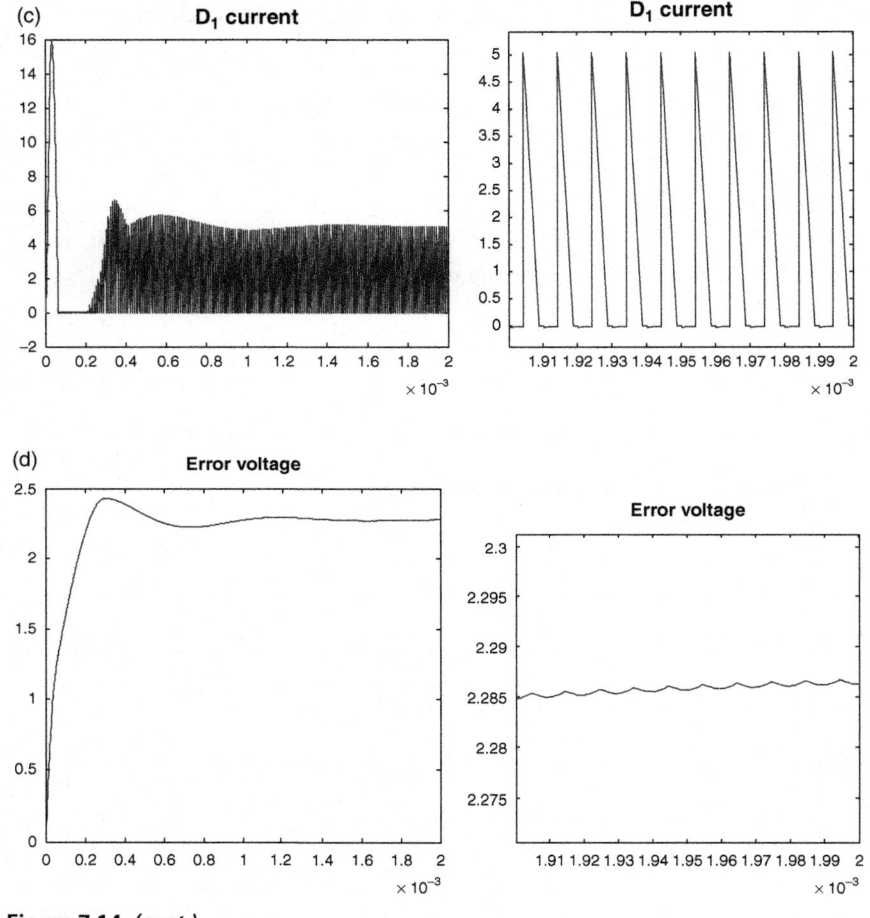

Figure 7.14 ***(cont.)***

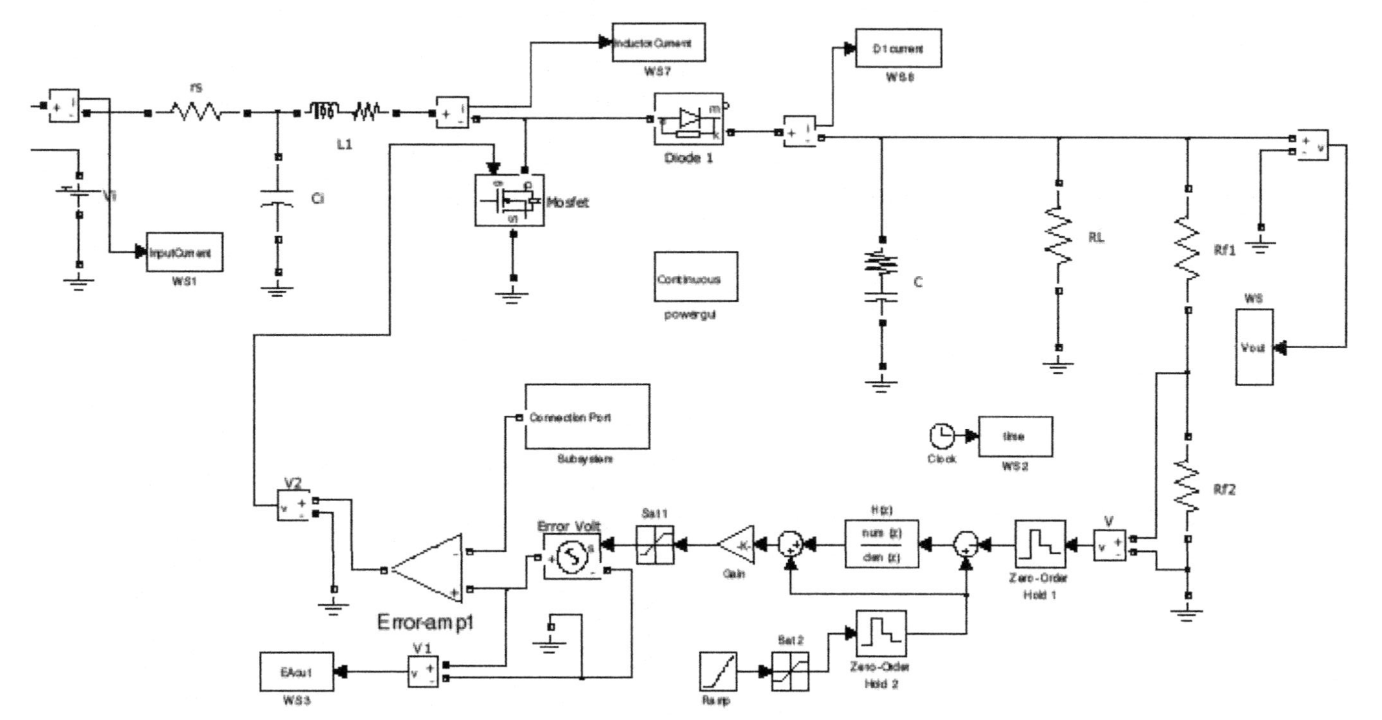

Figure 7.15 ***SIMULINK Model for Boost Converter in DCM with Error Amplifier in* H(z).** (a) Output voltage, (b) inductor current, (c) diode current, and (d) error voltage.

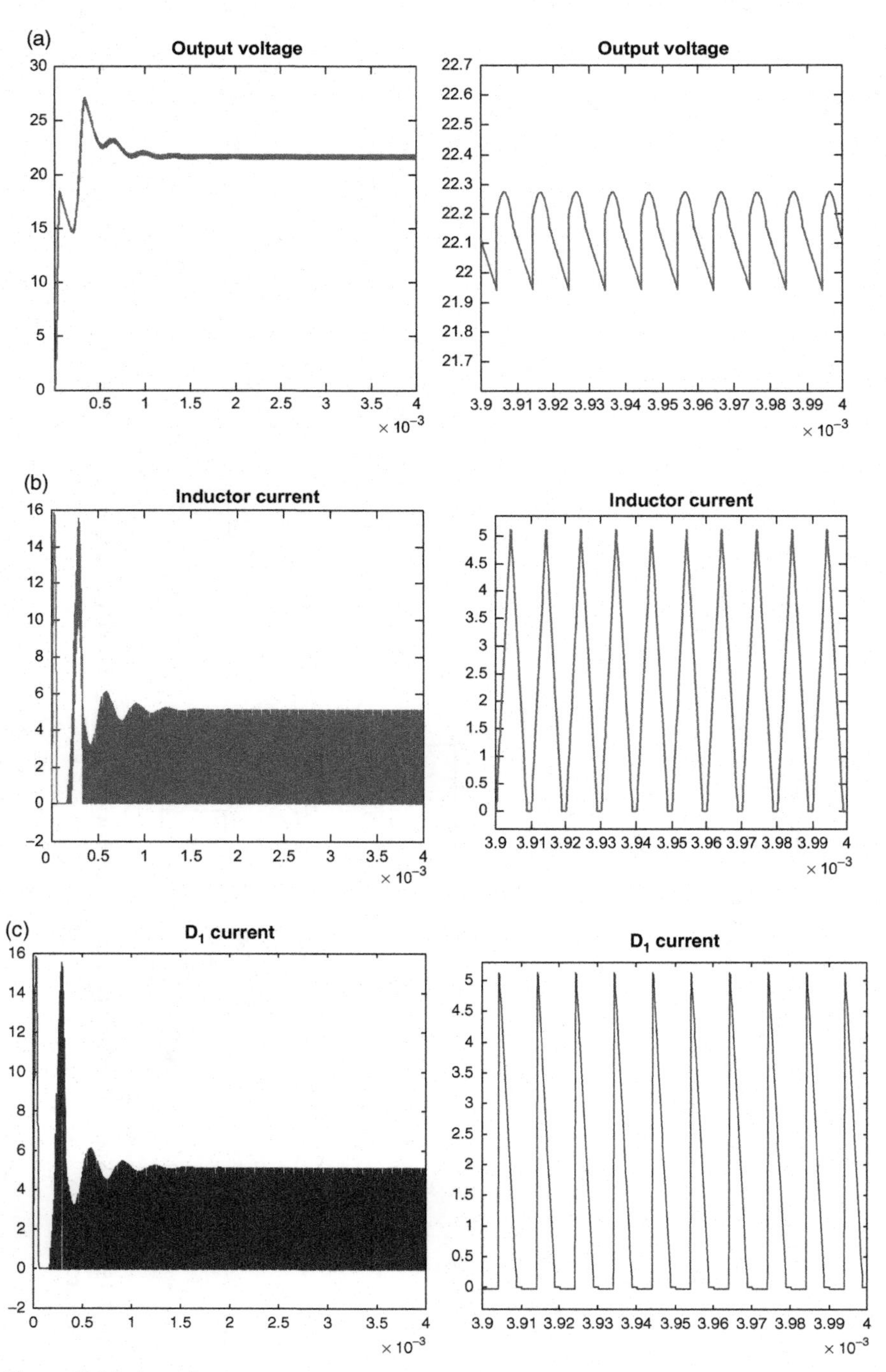

Figure 7.15 *(cont.)*

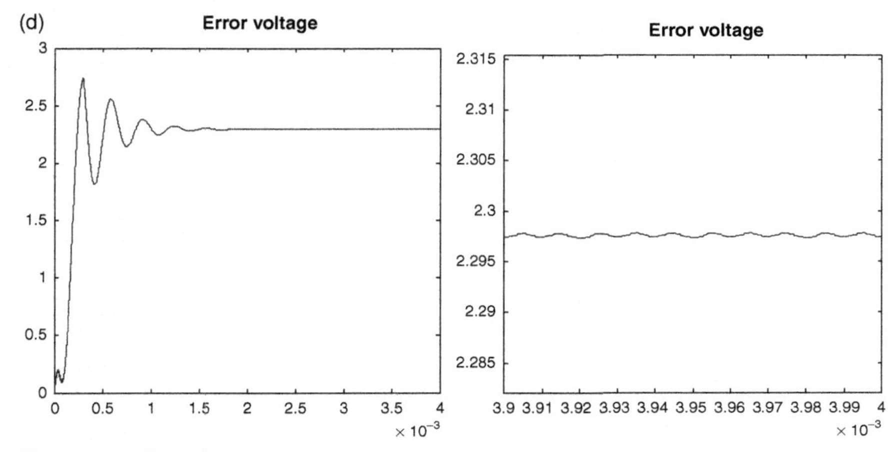

Figure 7.15 ***(cont.)***

CHAPTER 8

DCM Boost Converter with Current-Mode Control

8.1 SCHEMATIC WITH CURRENT-MODE CONTROL

In contrast to the flyback converter in DCM with current-mode control presented in Section 4.1, which uses a resistive sensor, it is of course feasible to implement current sensing with a magnetic device. Figure 8.1 shows exactly such an implementation with a current transformer of n_i turn and a simple current-to-voltage converting resistor, $R_{s_{en}}$.

8.2 PWM GAIN AND MODULATOR

To a large extent, the current sensing mechanism depicted in Figure 8.1 is basically identical to that of Figure 4.1 except for a minor difference in symbolical form. In the current case, implicit function determining the steady state duty cycle D is,

$$f(V_{in}, v_e, D) = \frac{V_{in}}{r_L + R_{on}}\left(1 - e^{-\frac{D}{[L/(r_L + R_{on})]f}}\right)\frac{R_{s_{en}}}{n_i} - v_e \qquad (8.1)$$

Similar to (4.5) and (4.6), the PWM gain, F_m, is obtained and modulator gain expressed, $M(s) = K_f F_m G_{vd}(s)$.

8.3 DESIGN EXAMPLE

We shall retain exactly the same example ($V_{in} = 12$ V, $V_o = 22$ V, $\delta v_o = 1\%$ V_o, $P_o = 25$ W, $f = 100$ KHz) presented in Chapter 7, but implement current-mode control as called out in Figure 8.1.

The current sensing transformer is selected as $n_i = 50$, while the sensing resistor $R_{s_{en}} = 25\ \Omega$. By design, v_e is nearly equal to 2.5. With $D = 0.4$, MOSFET on resistance R_{on} a few milliohm, (8.1) then yields $F_m = 0.162$ for the same boost inductor and switching frequency given in Chapter 7. There is no change in the feedback factor and the duty cycle-to-output gain. As a result, the modulator gain with current-mode control is significantly changed because of the difference in the PWM gain F_m. Figure 8.2 shows the modulator gain response.

Power Converters with Digital Filter Feedback Control
http://dx.doi.org/10.1016/B978-0-12-804298-4.00008-3

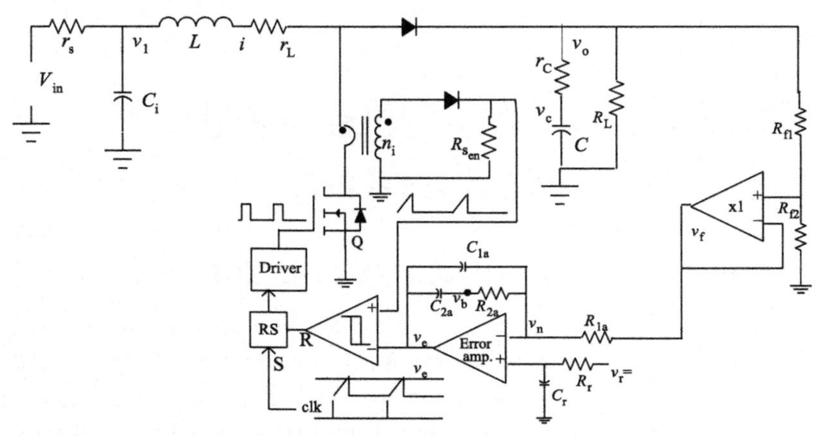

Figure 8.1 ***Boost Converter in DCM with Current-Mode Control.*** $v_r = 2.5$.

At the same desired crossover frequency, it gives a −8 db gain and −50° phase. For the same phase margin of 45° selected, again a type-II error amplifier, $E_A(s)$, can do the job. (1.13) and (1.14) give $R_{1a} = 2$ K, $C_{1a} = 0.028$ μF, $R_{2a} = 32.06$ K, and $C_{2a} = 0.0054$ μF. Together with the modulator, the overall loop gain $T(s) = E_A(s)M(s)$ is given in Figure 8.3.

It does cross at 1 kHz with the selected phase margin.

8.4 PERFORMANCE VERIFICATION WITH MATHCAD

Operation of Figure 8.1 can be expressed in a set of discrete, difference equations,

$$\begin{pmatrix} q_{j+1} \\ v_{1_{j+1}} \\ i_{j+1} \\ v_{c_{j+1}} \\ v_{e_{j+1}} \\ v_{b_{j+1}} \end{pmatrix} = \begin{bmatrix} -\left[\left(\text{if}\left(\frac{R_{s_{en}}}{n_i} i_j \ge v_{e_j}, 1, 0\right)\right) \vee -\left[q_j \vee \left(\text{if}\left(c_{lk_j} = 1, 1, 0\right)\right)\right]\right] \\ v_{1_j} + \delta t\left[\left(V_{in} - v_{1_j}\right)\frac{1}{r_s C_i} - \frac{1}{C_i} i_j\right] \\ \text{if}\left[i_j < 0, 0, i_j - \delta t\left(\frac{r_L}{L} i_j + \frac{1}{L}\,\text{if}\left(q_j = 1, 0, R_p\,\text{if}\left(q_j = 1, 0, i_j\right) + k_R v_{c_j}\right)\right) + \frac{\delta t v_{1_j}}{L}\right] \\ v_{c_j} + \frac{\delta t}{r_c C}\left(R_p\,\text{if}\left(q_j = 1, 0, i_j\right) + k_R v_{c_j} - v_{c_j}\right) \\ \text{if}\left[v_{e_j} < 0, 0, \text{if}\left[v_{e_j} > 12, 12, v_{e_j} + \left[\frac{\delta t}{C_{1a}}\left(\frac{1}{R_{1a}} + \frac{1}{R_{2a}}\right)\text{VR}_j - \frac{\delta t}{(R_{2a}C_{1a})} v_{b_j} - \frac{\delta t}{(R_{1a}C_{1a})} K_f \left(R_p\,\text{if}\left(q_j = 1, 0, i_j\right) + k_R v_{c_j}\right)\right]\right]\right] \\ \left[1 - \left[\frac{1}{(R_{2a}C_{2a})} + \frac{1}{(R_{2a}C_{1a})}\right]\delta t\right] v_{b_j} + \left[\frac{1}{(R_{2a}C_{2a})} + \frac{1}{C_{1a}}\left(\frac{1}{R_{1a}} + \frac{1}{R_{2a}}\right)\right]\delta t \text{VR}_j - \frac{\delta t}{(R_{1a}C_{1a})} K_f\left(R_p\,\text{if}\left(q_j = 1, 0, i_j\right) + k_R v_{c_j}\right) \end{bmatrix} \tag{8.2}$$

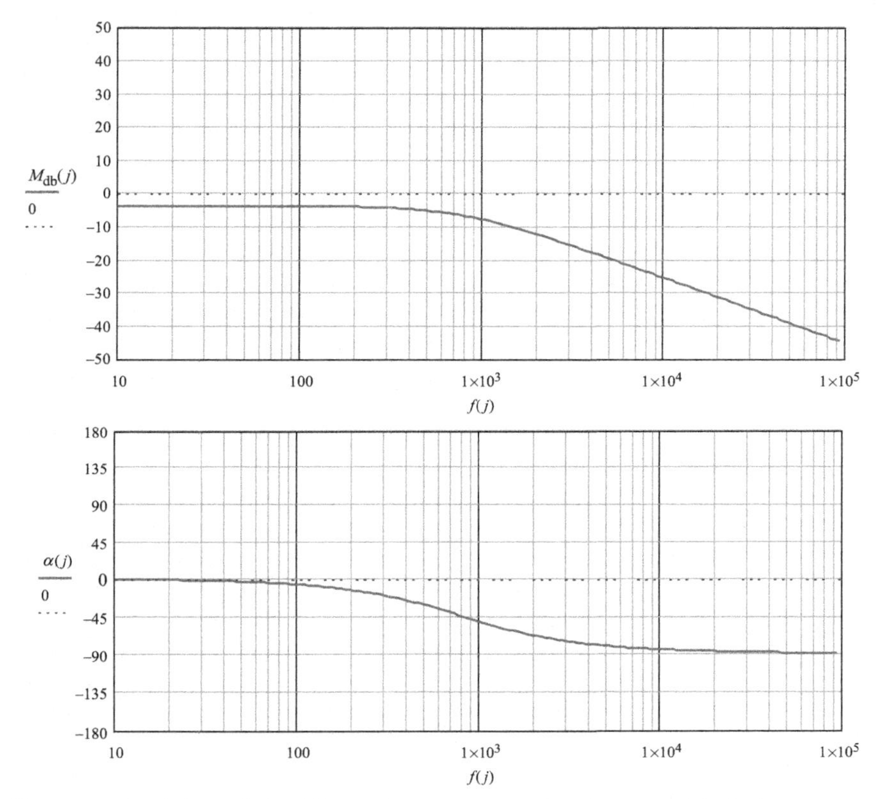

Figure 8.2 ***Modulator Gain of Boost Converter in DCM with Current-Mode Control.***

in which the first statement works as a reset-set flip flop (RSFF), incorporating current sensing, comparison, and a clock at 100 kHz. Figure 8.4a–d present turn-on transient and zoom-in on the steady state, for four key state variables.

The type-II error amplifier identified in Section 8.3 is converted to a digital filter with sampling frequency F_s = 800 kHz and bilinear transformation constant $C = 2F_s$. The process yields $H_{II}(z)$, from (1.19), polynomial coefficients for this implementation.

$$a_0 = 11.008 \times 10^{-3} \quad a_1 = 78.939 \times 10^{-6} \quad a_2 = -10.929 \times 10^{-3}$$

$$b_1 = -1.991 \times 10^0 \quad b_2 = 991.465 \times 10^{-3}$$

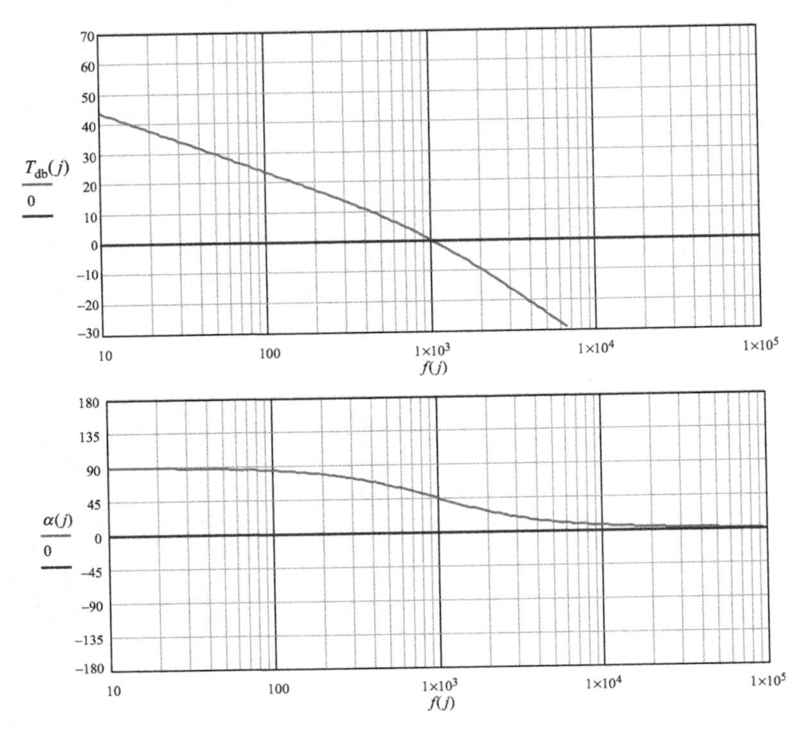

Figure 8.3 ***Loop Gain of Boost Converter in DCM with Current-Mode Control.***

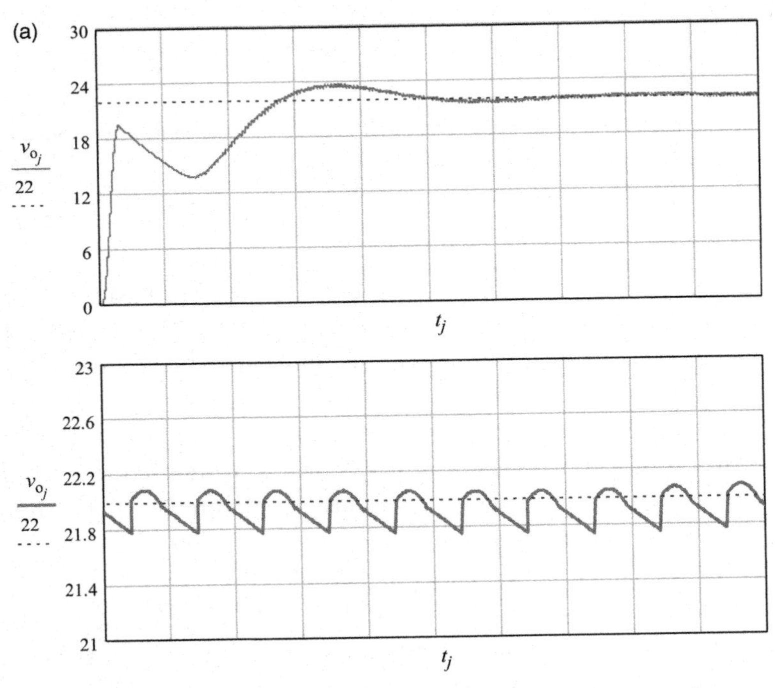

Figure 8.4 Equation set (8.2) performance (a) Output voltage, (b) error voltage, current feedback, and RSFF output, (c) inductor current, and (d) diode current.

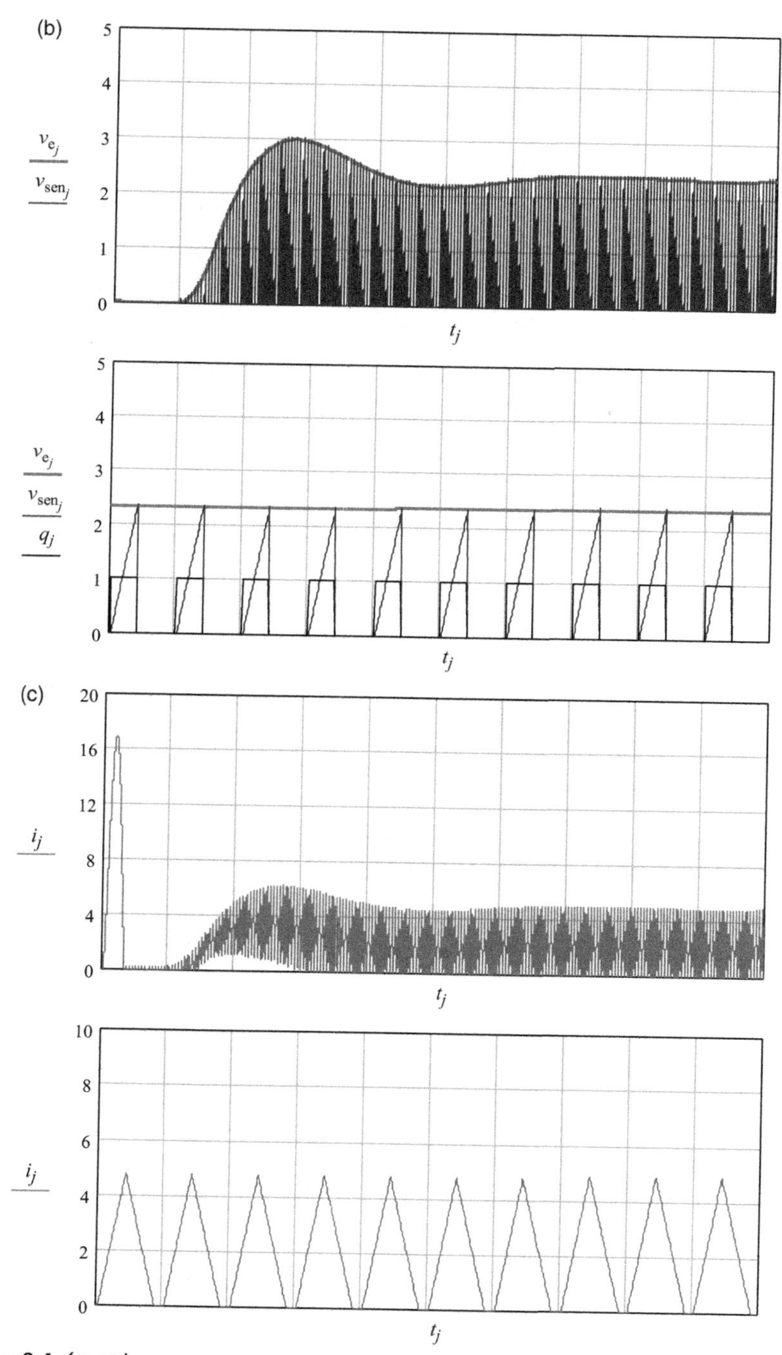

Figure 8.4 *(cont.)*

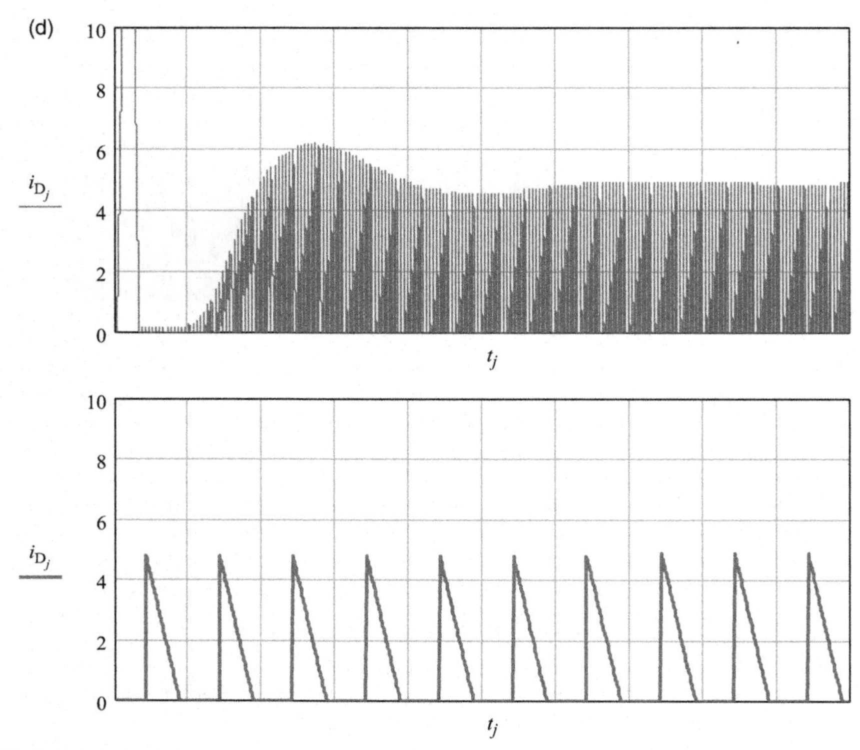

Figure 8.4 *(cont.)*

8.5 PERFORMANCE VERIFICATION WITH SIMULINK

Figure 8.1 is brought to the MATLAB SIMULINK platform, and it results in Figure 8.5. Figure 8.5a–e shows simulation results.

The analog error amplifier in Figure 8.5 is replaced with its equivalent digital filter, $H_{II}(z)$, derived in Section 8.4, and it results in Figure 8.6. Thereafter, simulation output plots follow.

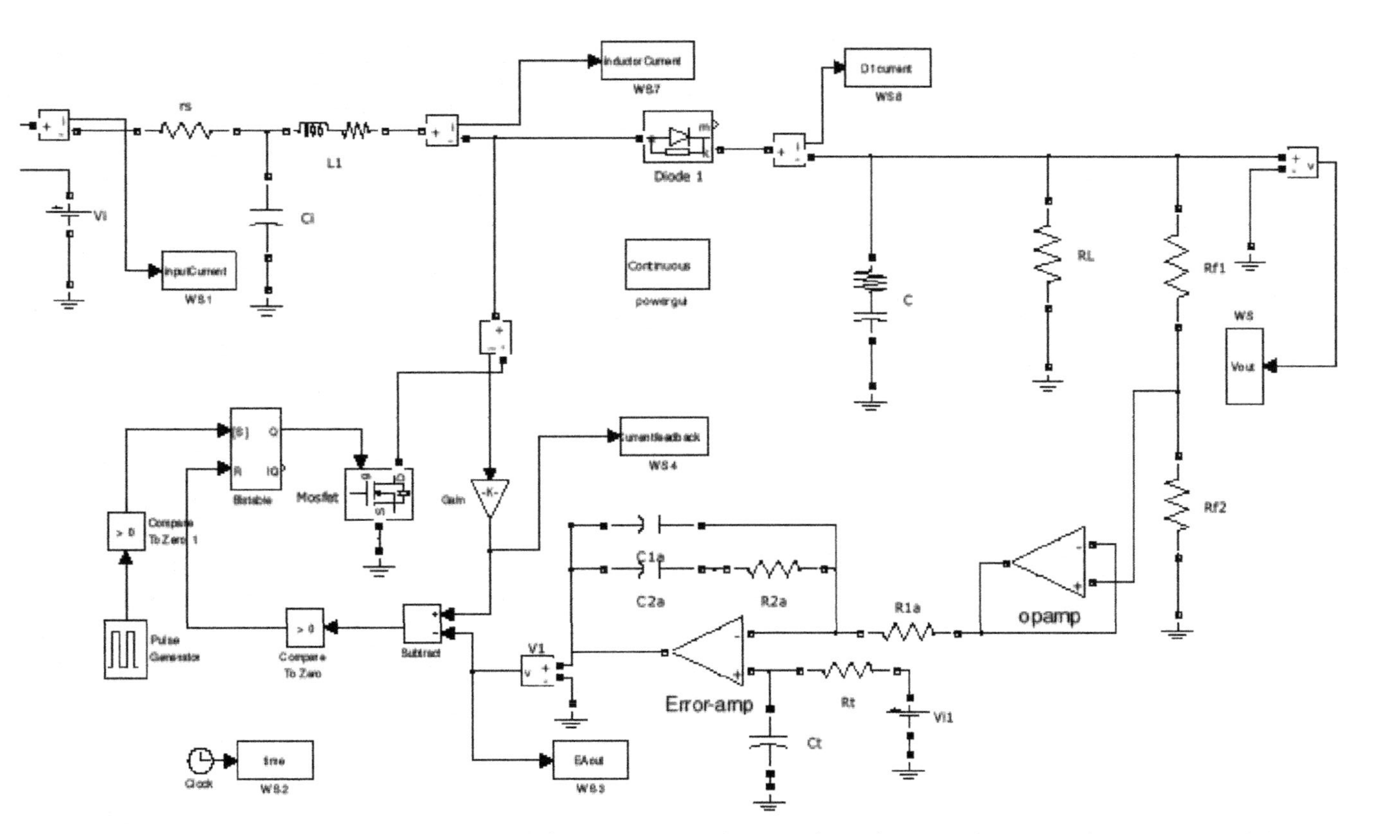

Figure 8.5 ***Boost Converter with I-Mode, Analog Control in SIMULINK.*** (a) Output voltage, (b) error voltage, (c) inductor current, (d) diode current, and (e) current feedback.

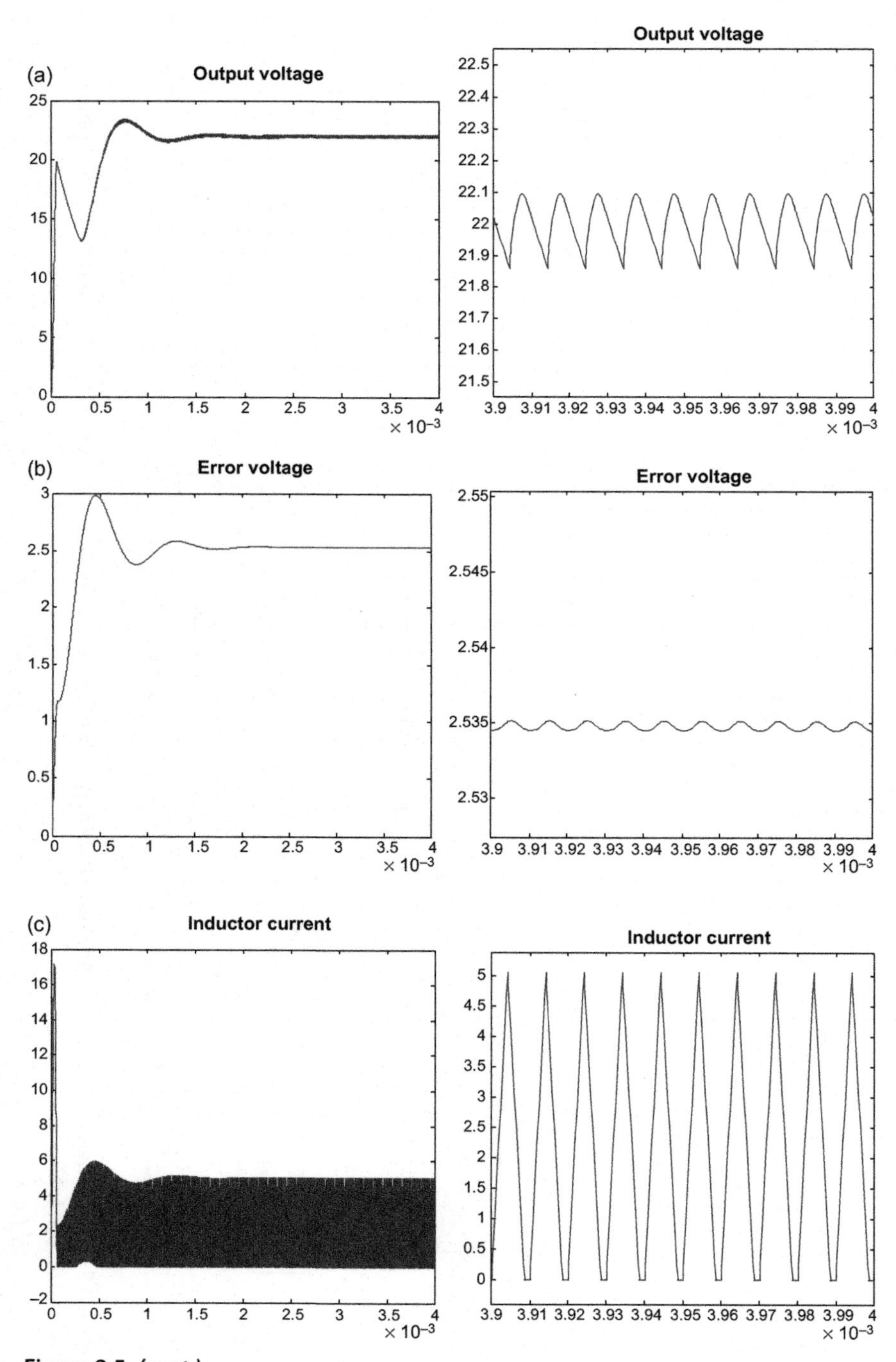

Figure 8.5 ***(cont.)***

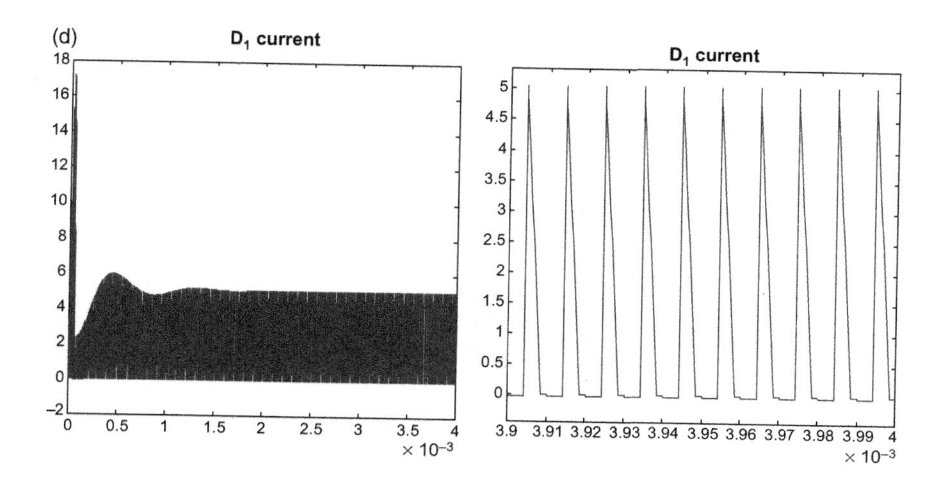

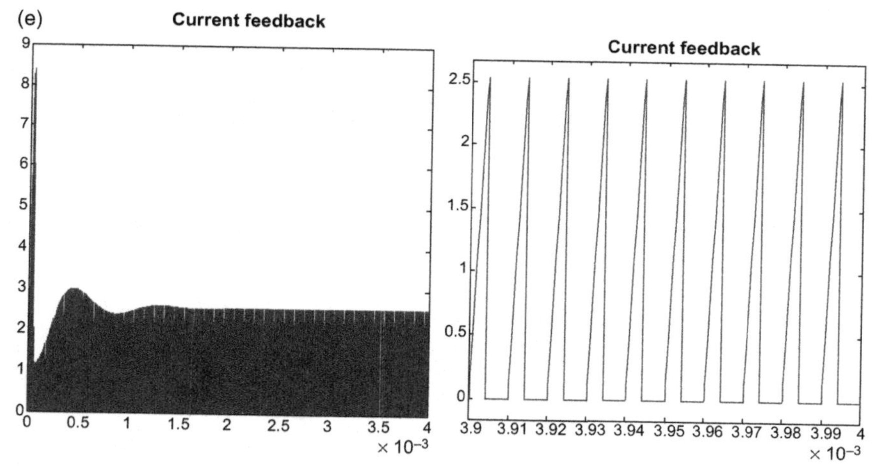

Figure 8.5 *(cont.)*

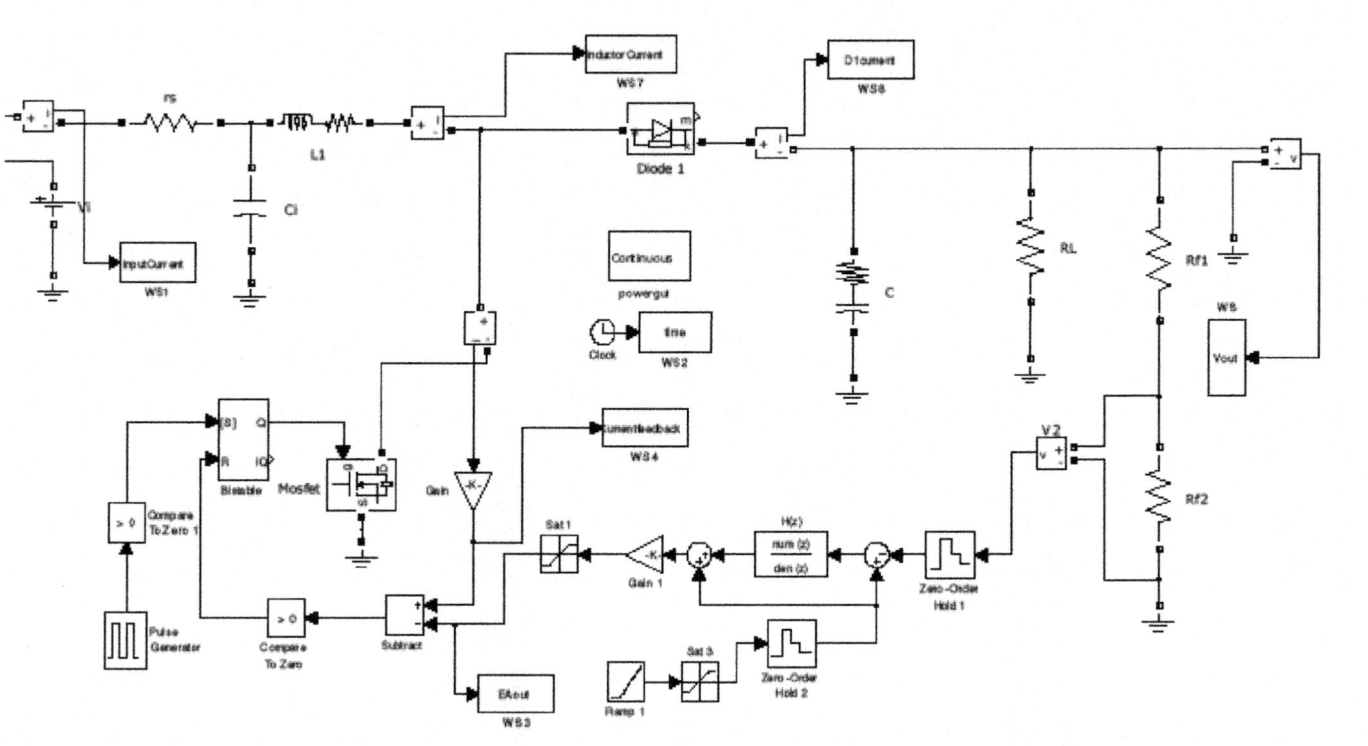

Figure 8.6 ***Boost Converter with I-Mode and Digital Filter.*** (a) Output voltage, (b) error voltage, (c) inductor current, (d) diode current, and (e) current feedback.

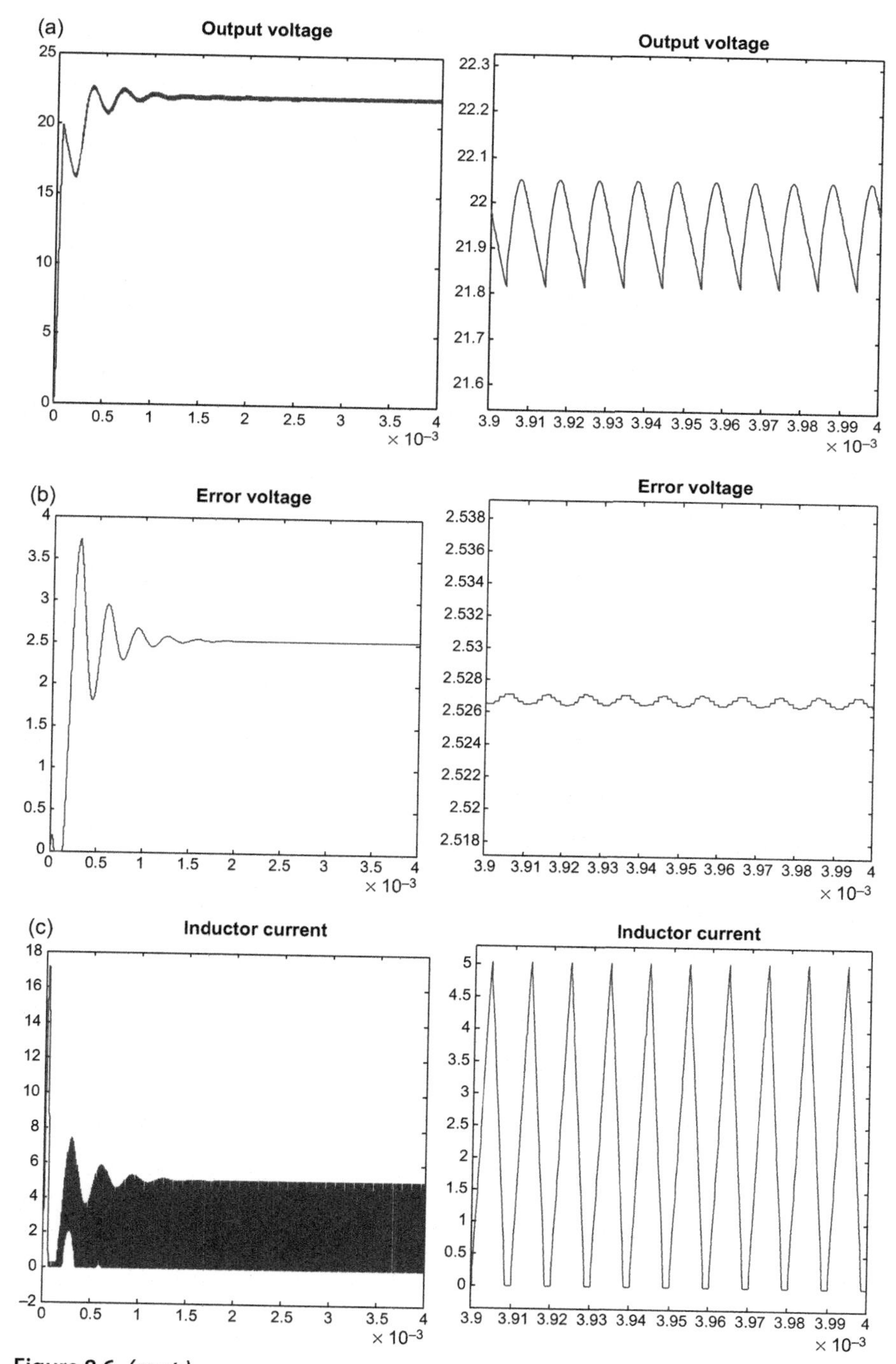

Figure 8.6 ***(cont.)***

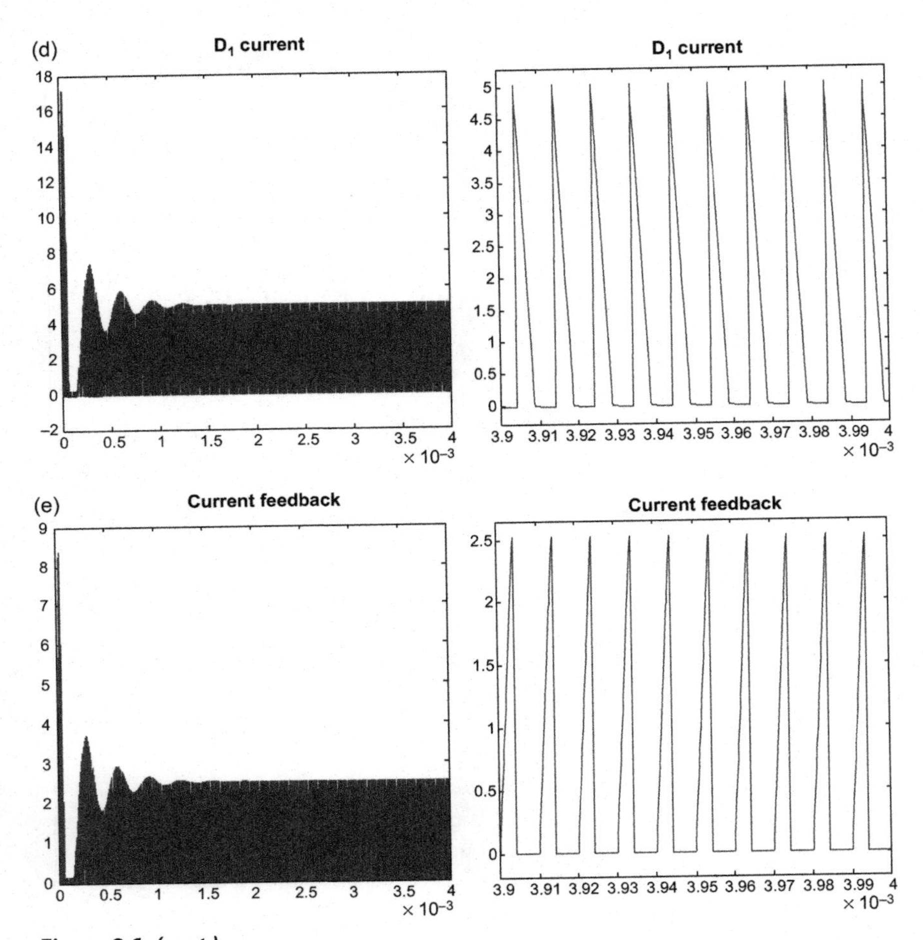

Figure 8.6 ***(cont.)***

PART V

Special Converters

A picture emerges in all time-domain simulations conducted earlier. That is the sharp transitions of either current or voltage in all switches; power transistors or rectifier diodes, in particular in the output side circuits. Such movements of currents at high time-rate generate a lot of conducted or radiated emissions that easily become sources of electromagnetic interference (EMI). A need, therefore, arises to tame the situation. And Chapter 9 presents one of the ways to meet the goal: reducing EMI at the source.

Another picture also emerges in all chapters presented so far. That is the voltage sourcing nature of all converter outputs shown earlier. Without additional measure, voltage source is more susceptible to overload (overcurrent) damage. Chapter 10 gives an approach that provides power supply in current sourcing form. With that, output over loading is curtailed to some extent.

CHAPTER 9

Resonant Converter

9.1 RIPPLE CONTENT

In Figures 1.1 and 2.1, the feed voltage at D_1 and D_2 junction is a pulsating rectangular wave. The current i_s of Figure 3.1, i_1 of Figure 4.1, and the main diode current of Figures 7.1 and 8.1 are also of pulsating nature in triangular form. Those types of hard-switching voltages and currents contain rich Fourier contents at high frequencies. As a result, it needs more filtering to meet output ripple specification and EMI requirement.

Naturally, and in order to alleviate the demand on filter design, the idea arose instead to reduce harmonic components at the feed source.

In case of Figures 1.1 and 2.1, the output filters work to extract the DC (average) content of the rectangular, driving source voltage. It is therefore plausible to conclude that, as long as the feed source contains the identical DC component, the output filter does not really care if the driving source consists of a waveform with fewer harmonics. Actually, it welcomes such a waveform.

Waveforms with lower harmonics have been well understood and documented. What easily first comes to mind is the sinusoidal wave. Figure 9.1 shows three waveforms with the same DC content of 0.5. The solid trace is a square wave at 50% duty cycle, the dotted trace is a level-shifted sine wave, and the dot-dash is a time-shifted sine wave. The square wave has the highest Fourier content, while the other two have only one frequency, 1 kHz. A filter fed by the square wave will give an output with a high ripple. In contrast, the other two feeds will generate a very low output ripple.

9.2 GENERATING SINUSOIDAL WAVEFORM

A typical second order system, for instance, an LRC series network, when subjected to a step perturbation, gives a damped oscillatory response, Figure 9.2.

Given the same system, a single pulse generates response of Figure 9.3.

Figure 9.3 shows that a certain portion of the response is negative. If some mechanism is provided to clip off the negative, the pulse response becomes Figure 9.4.

http://dx.doi.org/10.1016/B978-0-12-804298-4.00009-5

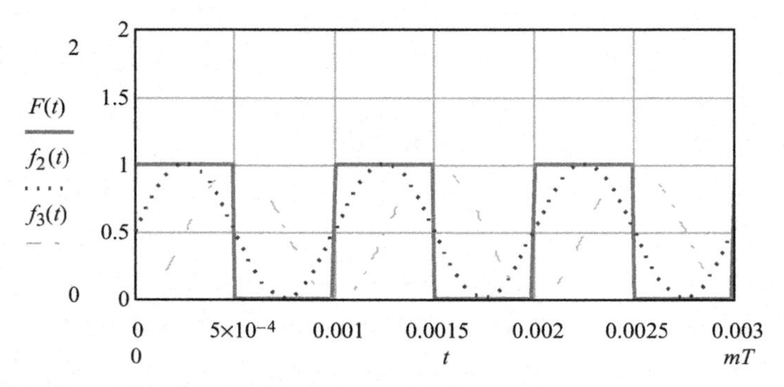

Figure 9.1 ***Three Waveforms with the Same DC (Average).***

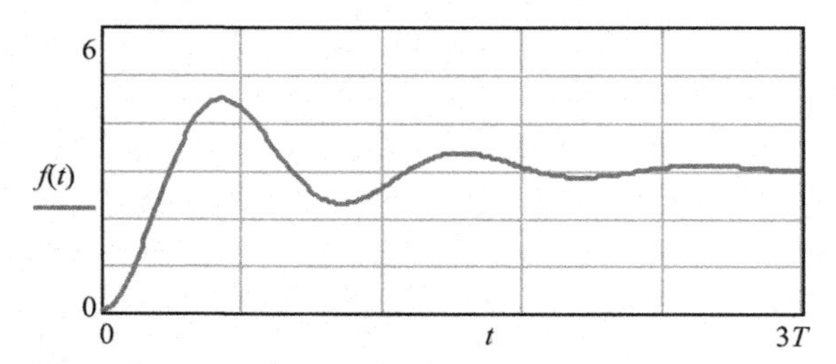

Figure 9.2 ***Step Response of a Second Order System.***

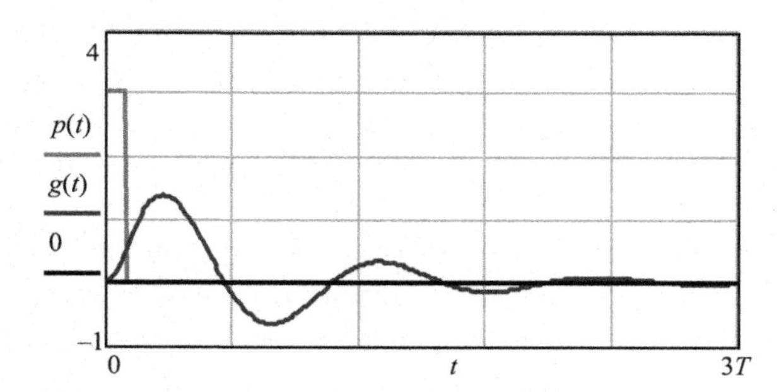

Figure 9.3 ***Pulse Response of a Second Order System.***

Figure 9.4 begins to show some sign of a single sine-like response corresponding to a single drive on a one-to-one basis. With proper timing and an additional mechanism, it is therefore totally possible to do just that.

9.3 QUASIRESONANT CONVERTER

The required mechanism outlined in the last section is implemented by the circuit given in Figure 9.5, the power stage of a quasiresonant converter.

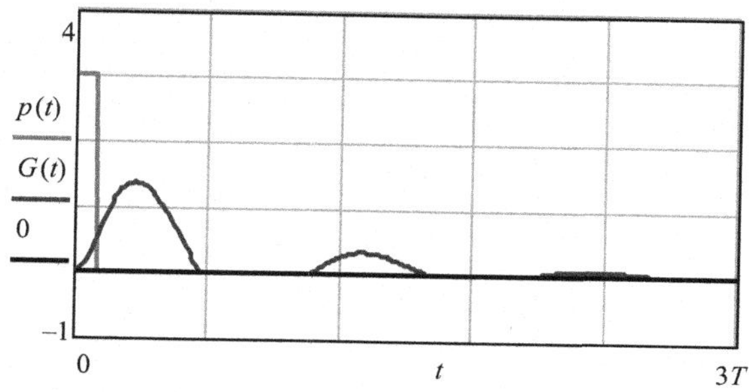

Figure 9.4 *Positive-Only Pulse Response.*

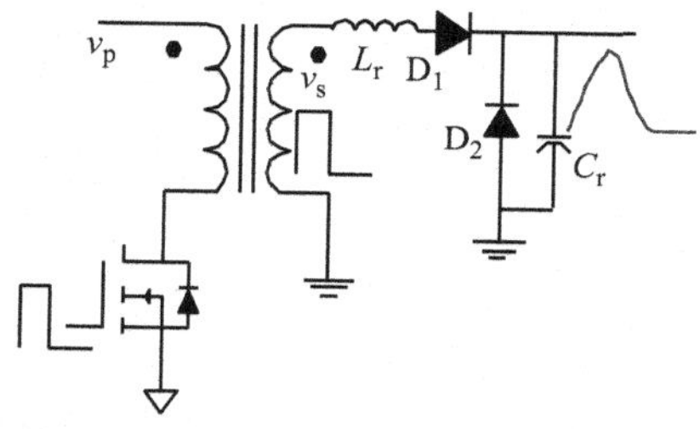

Figure 9.5 *Power Stage of Quasiresonant Converter.*

When driven by a voltage pulse, the L_r and C_r tank circuit goes through a resonance. However, diode D_1 permits only current flow in one direction, forward from drive source, v_s, to the output port, and diode D_2 prevents the output voltage from going negative. In effect, D_1 performs autonomous zero-current switching (ZCS) while D_2 acts as a zero-voltage switch (ZVS). As long as the drive source duration meets the time constraint set by the resonant tank components, the sine-like output wave is generated every time the drive source activates. Readers are referred to Chapter 8 of Ref. [3] and US Patent 4,415,959 for detailed analysis. We focus here instead on the control loop and the conversion of control amplifier to its corresponding digital counterpart. The analytical procedure given in Ref. [3] will not be duplicated. But we want to take some of its end results in developing the control loop.

9.4 FREQUENCY MODULATION VERSUS PULSE WIDTH MODULATION

So far and up to Chapter 8, pulse width modulation (PWM) is the sole mechanism employed to regulate converter output when either the loading or input source is fluctuating. The approach changes the active time duration of a rectangular drive source at a fixed repetition frequency.

In the current case of shaping the drive source waveform in order to reduce harmonic contents, output regulation by way of time duration is no longer valid, since the waveform is intentionally shaped and cannot be altered in any form or shape once the resonant tank components are selected and the minimum pulse duration is determined.

With the previous understanding, the only technique left for performing regulation is to change the repetition rate of the preshaped waveform: frequency modulation.

Figure 6.8 is a good candidate for implementing the quasiresonant approach with frequency modulation (FM). We therefore go ahead and modify the circuit. Figure 9.6 shows a different power stage and an FM feedback loop.

The error feedback, V_{ef}, is buffered and feeds a frequency-setting resistor network, R_{in} and R_{set}, for a voltage-controlled oscillator (VCO), which outputs a 50% square wave. The leading, positive edge of each oscillation cycle triggers a one-shot and set the D-flip-flop (DFF). It subsequently turns on the power switch and initiates a resonant cycle. As a result, the resonant tank consisting of L_r, D_1, C_r, and D_2 generates a sinusoidal current and voltage. The current rings up, then down, and returns to zero. The same current is reflected to the primary side and is sensed by R_{sen}. At the time the resonant current returns to zero, the zero-crossing detector and its associated one-shot issue a pulse that resets the DFF and terminates the power switch conduction. The switch cycle repeats when the next clock comes along.

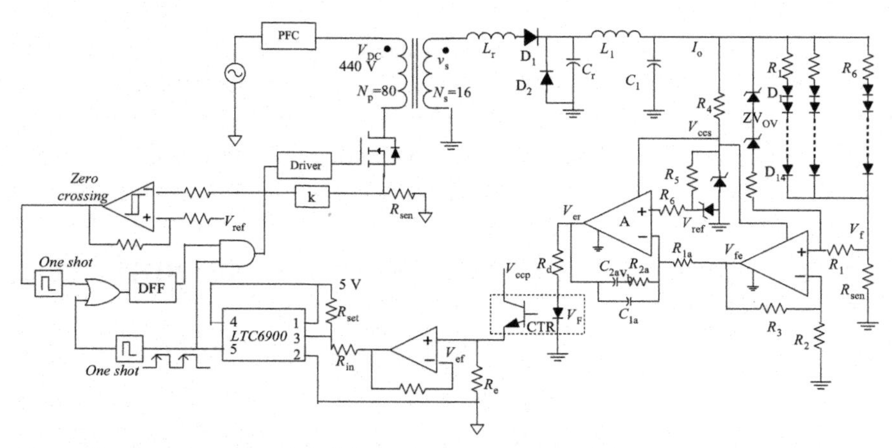

Figure 9.6 *Quasiresonant LED Driver with Frequency Modulation.*

9.5 VCO MODULATION GAIN

In Figure 9.6, a VCO circuit is configured using LTC6900, a Linear Technology IC. Based on the application note [5] for the IC, the output square wave has a frequency range as function of the control input v_{ef}, if it is powered by a 5V supply.

$$f(v_{ef}) = 2\times 10^{9}\left(\frac{R_{set}+R_{in}}{R_{set}R_{in}}\right)\left(1+\frac{v_{ef}-5}{1.1}\frac{1}{1+(R_{in}/R_{set})}\right) \tag{9.1}$$

Therefore, the VCO modulation gain in hertz/volt, is

$$G_{vco} = \frac{df(v_{ef})}{dv_{ef}} = \frac{2\times 10^{9}}{1.1R_{in}} \tag{9.2}$$

9.6 POWER STAGE GAIN

Given the resonant tank component in Figures 9.5 and 9.6, the following parameters are defined to ease analytical formulation.

$$\omega = \frac{1}{\sqrt{L_r C_r}},\ f = \frac{\omega}{2\pi},\ T = \frac{1}{f},\ Z_n = \sqrt{\frac{L_r}{C_r}},\ \alpha = \pi + \sin^{-1}\left(\frac{I_o Z_n}{v_s}\right) \tag{9.3}$$

The resonant tank when energized by a pulse v_s is understood to go through four operation phases (linear, resonant, recovery, and freewheel) for each switching cycle, [3]. The time duration of the first three phases are given by

$$T_{d0} = \tfrac{L_r I_o}{v_s},\ T_{d1} = \tfrac{\alpha}{\omega},\ T_{d2} = \frac{C_r v_s(1-\cos\alpha)}{I_o} \tag{9.4}$$

Duration T_{d3} is frequency dependent.

Furthermore, if time-zero is tentatively set at the beginning of T_{d1}, the voltage across C_r can be expressed as

$$\begin{aligned} v_{cr}(t) &= v_s(1-\cos\omega t)\left[u(t)-u(t-T_{d1})\right]+ \\ &\left[v_s(1-\cos\omega T_{d1})-\tfrac{I_o}{C_r}(t-T_{d1})\right]\left[u(t-T_{d1})-u(t-T_{d1}-T_{d2})\right] \end{aligned} \tag{9.5}$$

However, it is more desirable to observe waveform referred to the switch-on instant that is the beginning of phase one T_{d0}. In that case, both the resonant current and the capacitor voltage are expressed as

$$\begin{aligned} i_r(t) &= \tfrac{v_s}{L_r} t\left[u(t) - u(t - T_{d0})\right] + \\ &\left[I_o - \tfrac{v_s}{Z_n} \sin\omega(t - T_{d0})\right]\left[u(t - T_{d0}) - u(t - T_{d0} - T_{d1})\right] \\ v_r(t) &= v_{cr}(t - T_{d0}) \end{aligned} \tag{9.6}$$

In this form, it enables a designer to go one last step. That is

$$\tfrac{1}{T}\int_0^T v_r(t)dt = f_x \int_0^{1/f_x} v_{cr}(t - T_{d0})\,dt = V_o \tag{9.7}$$

where V_o is the desired output voltage; in this case the LED string voltage plus voltage drop across R_{sen}. Therefore, (9.7) actually enables numerical estimation, or solution, for the unknown, nominal switching frequency, f_x, required, and this nominal frequency is generally placed midway within the VCO output frequency range, f_{min} to f_{max}.

Once the nominal switching frequency is determined, the power stage's gain, that is, output change over frequency move, is easily given as

$$G_f = \frac{d}{df_s}\left(f_s \int_0^{1/f_s} v_r(t)dt\right) \tag{9.8}$$

This is just the low-frequency gain yet to include output filter.

9.7 DESIGN PROCEDURE

Unlike the hard switching PWM, which enjoys a well-documented, straightforward design process yielding accurate duty-cycle determination and magnetic device selection, the resonant converters do not have such advantage, since the power stage waveforms involved in the operation are complex functions of circuit components, and the waveform timing is not necessarily in sync with the external switch ON/OFF instants. The latter is made even more intractable with the existence of multiple diodes that switch and alter circuit structure, autonomously depending on the internal states.

So, we face a challenge and have to approach the task from a different angle.

From (9.5), the peak voltage across the resonant capacitor, and the same reverse blocking voltage D_2 has to sustain, $v_s(1 - \cos\omega t)$. In other words, it can reach twice the magnitude of the drive source, v_s. If we select a diode with a peak reverse blocking voltage of 200 V, a reasonable specification, the secondary drive voltage, v_s, shall be 100 V, and, considering operating margin, we pick 90 V. In the meantime, we also want to use a small inductor and pick 6.8 μH for L_r.

Next, we consider it a sensible choice making the sum of (9.4) equal to about half of the switching period. Therefore,

$$\frac{1}{2(T_{d0}+T_{d1}+T_{d2})}=f(I_o,L_r,C_r) \tag{9.9}$$

The choice allows dynamic modulation when phase four, the freewheel duration, shrinks or expands. (9.9) gives us a good way to decide what resonant capacitor value to choose and the switching frequency to which it corresponds. Figure 9.7 plots the frequency as a function of C_r with fixed load I_o = 3.5 A, L_r, and v_s.

If a switching frequency of about 100 kHz is selected, the required resonant capacitor shall be about 0.05 μF. C_r = 0.047 μF, a standard value, which is therefore selected. We are done, finally, designing the resonant tank.

Next, the isolation transformer turn ratio is evaluated. Given 440 V primary supply and 80 turns winding, a secondary winding of 16 turns yields 88 V, an acceptable number.

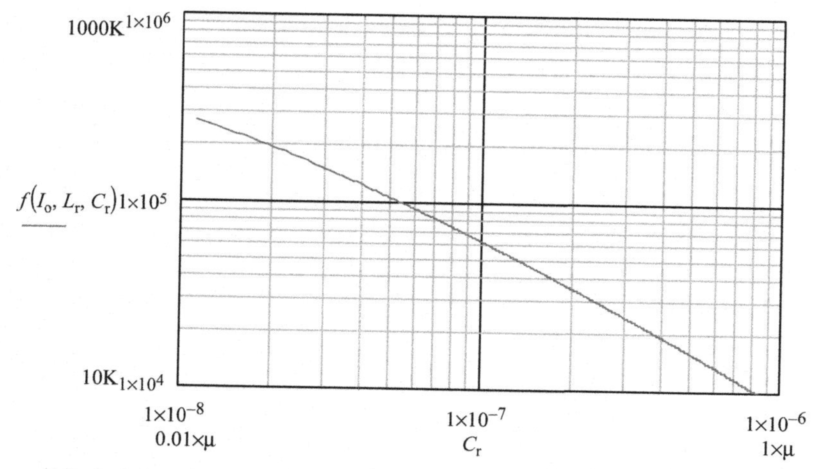

Figure 9.7 ***Switching Frequency Versus Resonant Capacitor, Given I_o, L_r, and v_s in Logarithm Scale.***

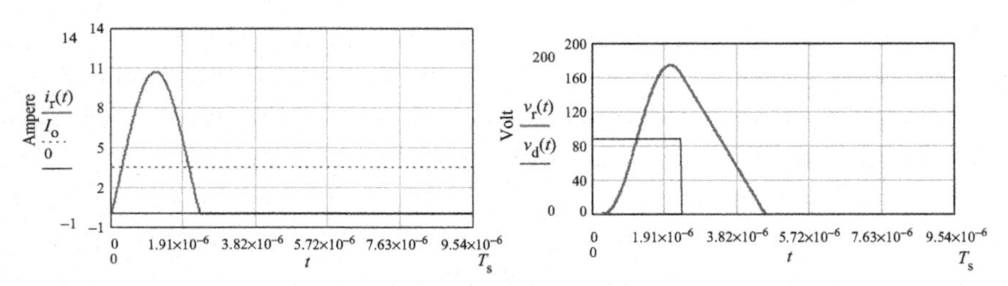

Figure 9.8 ***Expected Current of*** L_r***, Voltage of*** C_r***, and Drive Voltage.***

At this point, the expected current and voltage of the resonant tank is shown in Figure 9.8.

9.8 CLOSE-LOOP UNDER STEADY STATE

Following an identical step outlined by (6.4–6.7), the effective error voltage feeding the VCO is given by

$$v_{ef} = \frac{A\left(V_{ref} - n\left(a e^{b(V_o/m)} + c\right)R_{sen}\left(1 + (R_3 / R_2)\right)\right) - V_F}{R_d}\text{CTR}\,R_e \quad (9.10)$$

Taking (9.10) to (9.1), the VCO generates a square wave at frequency f_{vco}. This is the frequency at which the converter's power MOSFET switch operates. Readers are to be cautioned that this switching frequency is NOT the ω, the resonant tank frequency, mentioned in (9.3).

At the operating frequency f_{vco}, (9.7) generates an output, V_o. The voltage output in turns enters the LED exponential model in (9.10) and closes the regulation loop.

9.9 MODULATOR GAIN AND LOOP GAIN

With the nominal switching frequency set at 100 kHz, the VCO range is selected; f_{min} = 85 kHz corresponding to a control input of 1 V, f_{max} = 110 kHz to 4V. The selection gives, together with (9.1) and LTC6900 application information, R_{set} = 18.2K and R_{in} = 249K. And, (9.2) gives $G_{vco} = 7.302 \times 10^3$ Hz/V.

In the meantime, (9.6) and (9.8) give the power stage gain at low frequency $G_f = 3.886 \times 10^{-4}$ V/Hz.

Other gain blocks needed for establishing the modulator gain are G_{opto}, (6.12); $H(s)$, (6.16); G_{LED}, (6.15); R_{sen}; and $(1+ R_3/R_2)$, and, eventually, the modulator gain $M(s) = G_{opto} G_{vco} G_f H(s) G_{LED} R_{sen}(1+ R_3/R_2)$.

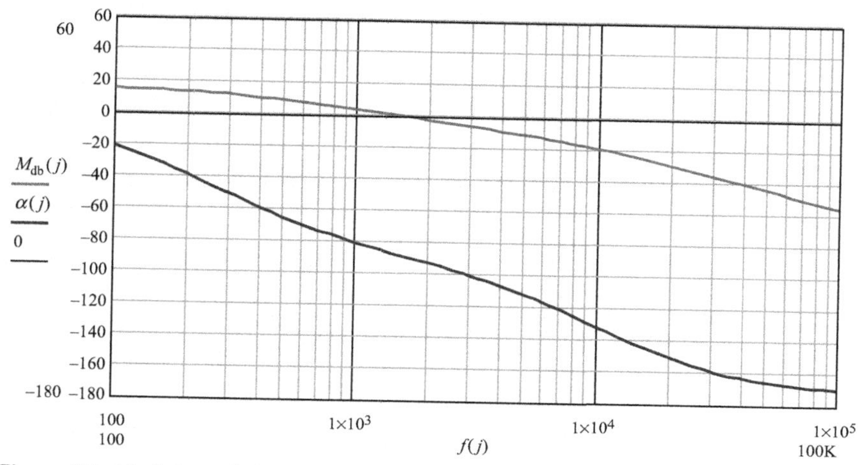

Figure 9.9 ***Modulator Gain.***

With $R_{sen} = 0.01$, $(1 + R_3/R_2) \sim 70$, $R_d \sim 2.7K$, $R_e = 1K$, $L_1 = 470\ \mu H$, $C_1 = 20\ \mu F$, 14 LEDs/string, and six string in parallel, the modulator gain is plotted, Figure 9.9.

At 5 kHz, the modulator has a gain of −11.4 db and phase −112°. A type-II amplifier turns out to be sufficient to meet a 45° phase margin at 5 kHz. Components, (1.14), for the amplifier are obtained: $R_{1a} = 2K$, $C_{1a} = 850$ pF, $R_{2a} = 7.74K$, $C_{2a} = 0.02\ \mu F$. It results in a loop gain, Figure 9.10.

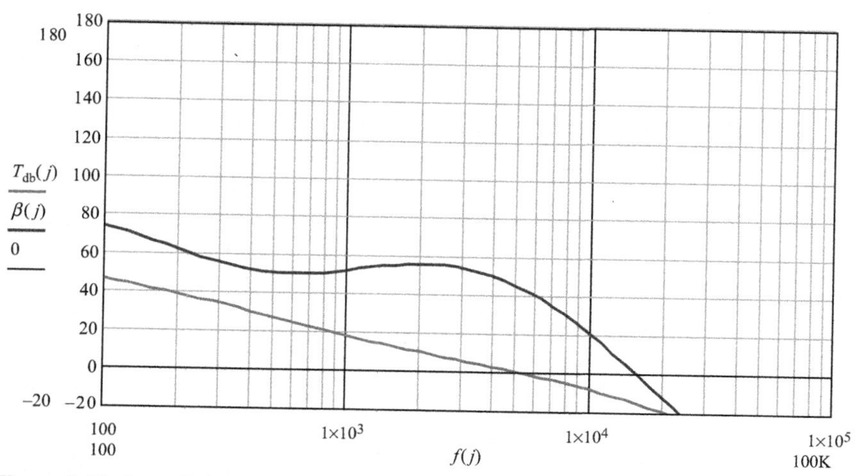

Figure 9.10 ***Loop Gain.***

9.10 PERFORMANCE VERIFICATION WITH SIMULINK

At this point, we are facing a new challenge. That is, in all previous efforts performing time-domain verification, it all begins with sets of differential equations arranged in difference equation form, and then an iterative process ensues to compute series of data points at selected, fixed time steps. Graphical tools embedded in the computing tool then display that datum in visual forms that users have selected.

In the case of the quasiresonant converter, that approach beginning with handwritten equations is no longer feasible, at least for the author with much less programming experience. We therefore bypass the approach. Instead, we shall employ the other, the MATLAB SIMULINK. Figure 9.6 is translated to Figure 9.11.

Since the resonant tank current rings back to zero, there is no need to reset the magnetic core when the power switch turns off. A two-winding transformer with a turn ratio of 80:16 is all it takes to transfer current properly. In Figure 9.6, two diodes were shown to perform ZCS and ZVS. In Figure 9.11, two MOSFETs configured as synchronous rectifiers with associated detectors, zero-current and zero-voltage, perform both critical function with less power consumption. *L*6, 6.8 μH, and *C*54, 0.047 μF, constitutes the resonant tank. *L*5, 470 μH, and *C*32, 20 μF, perform filtering. The filter output voltage is then fed to a block, which emulates Cree LED, given in Figure 6.1, by an exponential function and generates an equivalent current load (3.5 A at steady state) across the filter output. Since a 2.5 V reference voltage is employed in the error amplifier and 3.5 A is expected, a gain block representing both the 0.01-Ω current sensor and additional gain is needed. In other words, the current feedback gain block shall have a magnitude equal to $2.5/3.5 \approx 0.7142857 = 0.01 \times 71.42857$. Next, and in order to save simulation time, the error amplifier was implemented with its equivalent analog transfer function, (1.17).

$$\frac{160.025186 \times 10^{-6} s + 1}{272.709492 \times 10^{-12} s^2 + 43.070336 \times 10^{-6} s + 0} \tag{9.11}$$

The error amplifier output feeds first a saturation block that represents the local power supply limit and then goes to an opto-coupler with gain given by (6.12). Receiving the input from the opto-coupler and constrained by input limit of 0–5 V, the VCO generates a frequency, *f*, based on a design function with selected minimum and maximum frequency. Given R_{in} and

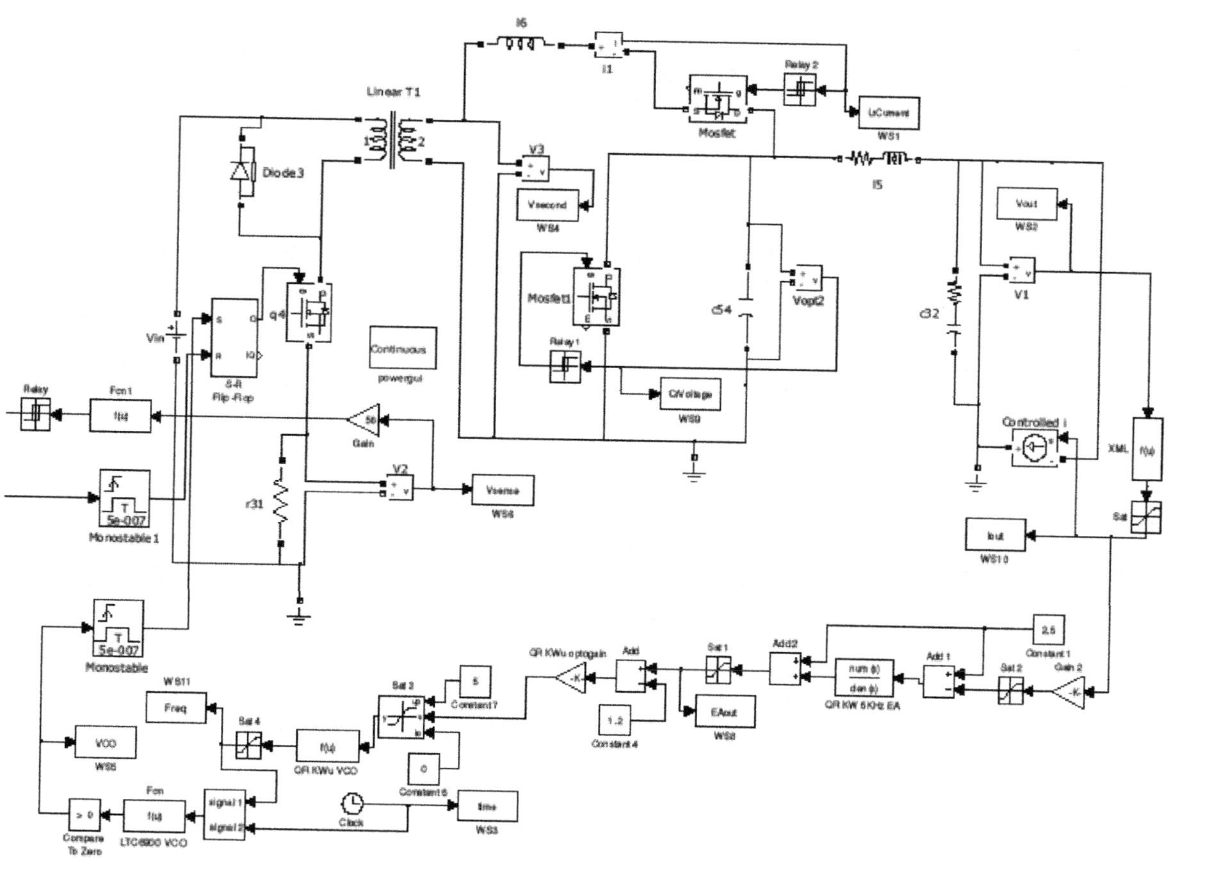

Figure 9.11 *SIMULINK Drawing.*

R_{set} selected in Section 9.9, the VCO output frequency as a function of input control v_x is

$$f(v_x) = 1.17910^5\left(1 + \frac{v_x - 5}{1.1 \times 14.681}\right) \tag{9.12}$$

A second function creates cos(2*πft*) that feeds a comparator and generates further a voltage-controlled square wave with variable frequency. On each rising edge of the square wave, a narrow pulse is generated and set the following reset–set flip-flop (RSFF). The RSFF's Q output subsequently turns on the power MOSFET switch and initiates a quasiresonant cycle. A 0.01-Ω sensing resistor, *R*31, in series with the switch, monitors the switch current by way of a gain and a comparator. When the reflected resonant current rings back to zero, it also triggers a narrow pulse that resets the RSFF and terminates the power switch conduction. Readers should be reminded that the power switch's "ON" duration is identical to the synchronous rectifier in series with the resonant inductor, but is different from the synchronous rectifier across the resonant capacitor.

Taking almost 1 h on an Intel Pentium-4 Dell desktop, Figure 9.11 file is run to cover 1 ms of converter operation. Figure 9.12a–f show the on-transient and the steady states for the resonant inductor and capacitor.

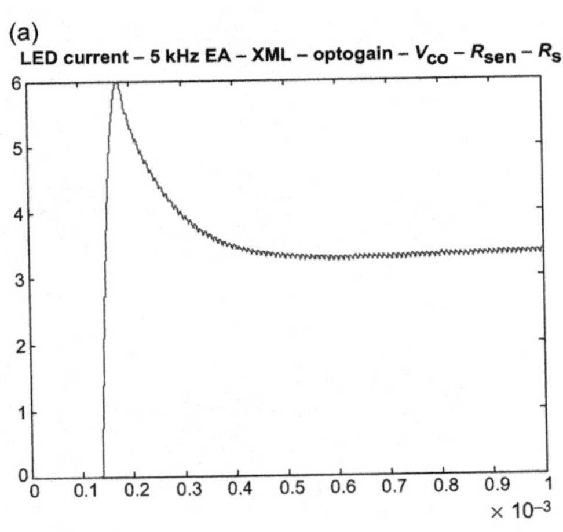

Figure 9.12 Figure 9.11 performance (a) LED array current, (b) converter output voltage, (c) the resonant inductor current, (d) the resonant capacitor voltage, (e) error amplifier output, and (f) VCO frequency.

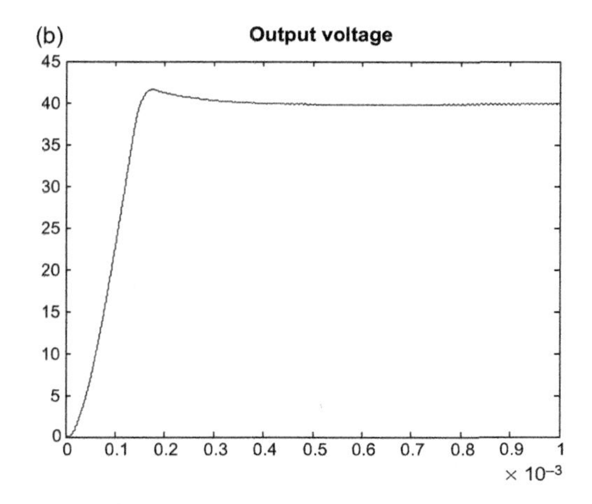

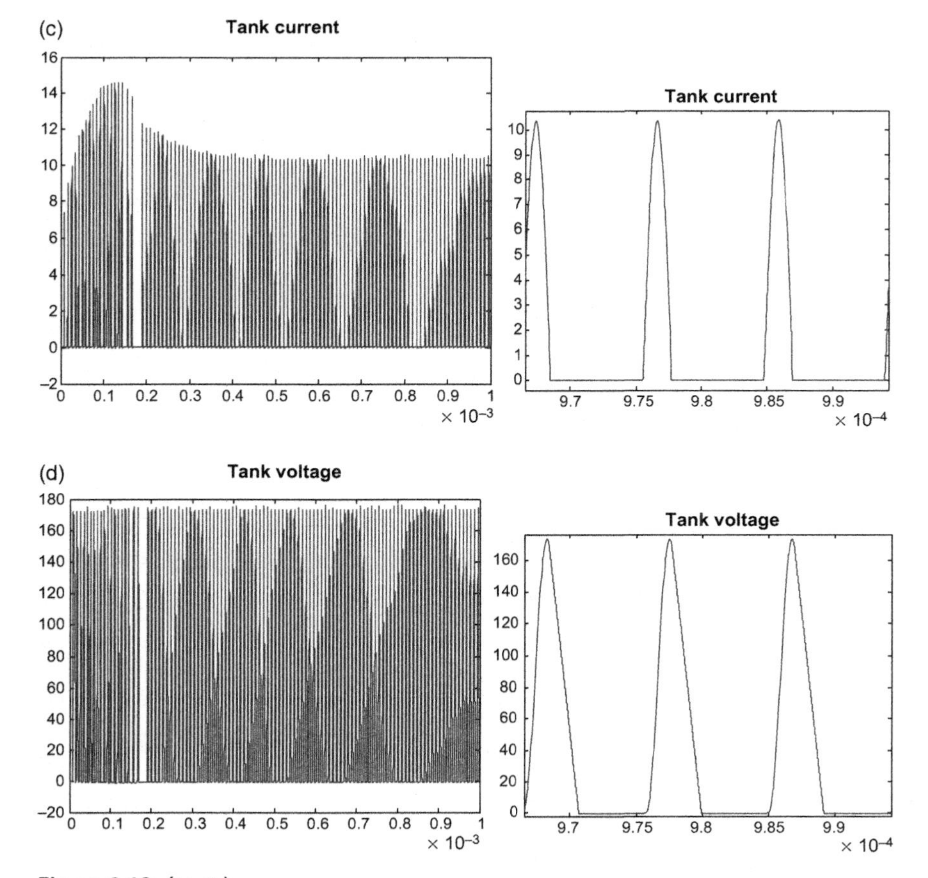

Figure 9.12 ***(cont.)***

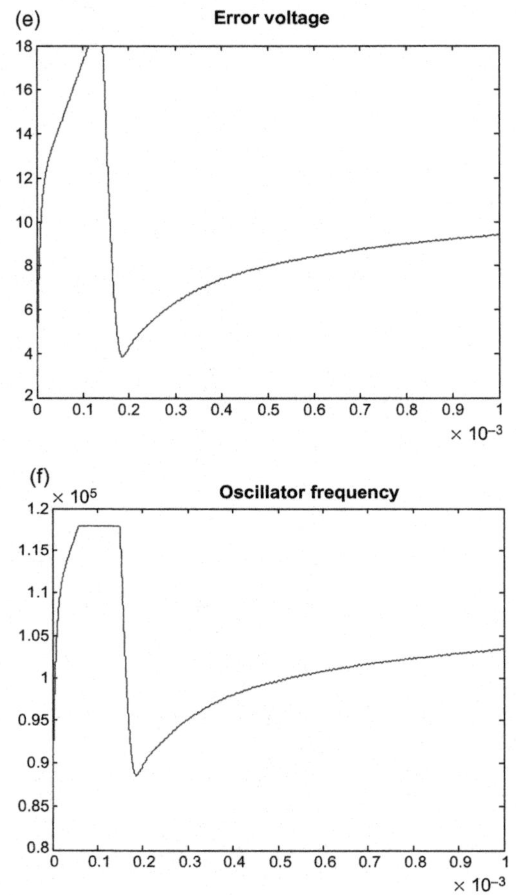

Figure 9.12 ***(cont.)***

Next, the analog transfer function (9.11) is converted to digital with sampling frequency at 2 MHz and bilinear transform constant $C = 4$ MHz, the following $H(z)$ results, and Figure 9.11 becomes Figure 9.13.

$$\frac{141.347448\times10^{-3}+440.952374\times10^{-6}z^{-1}+(-140.906495\times10^{-3})z^{-2}}{1+(-1.92403)z^{-1}+924.030396\times10^{-3}z^{-2}} \tag{9.13}$$

Figure 9.13 yields time-domain performance as shown in Figure 9.14a–e.

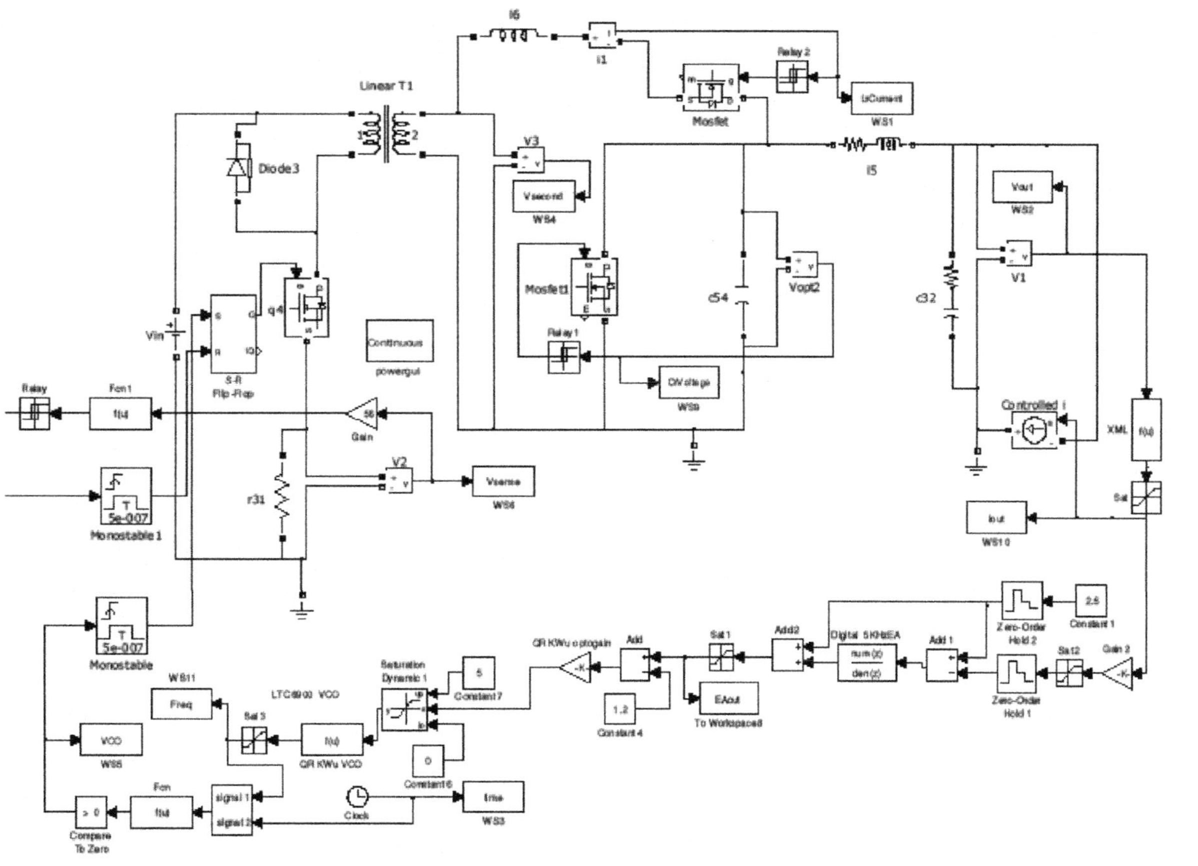

Figure 9.13 *SIMULINK Drawing with Digital Filter.*

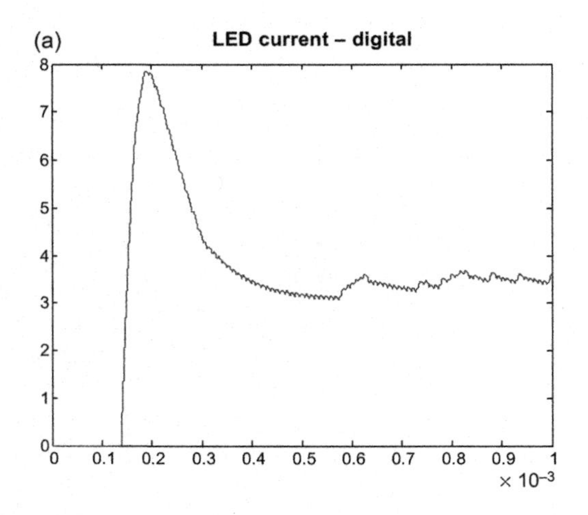

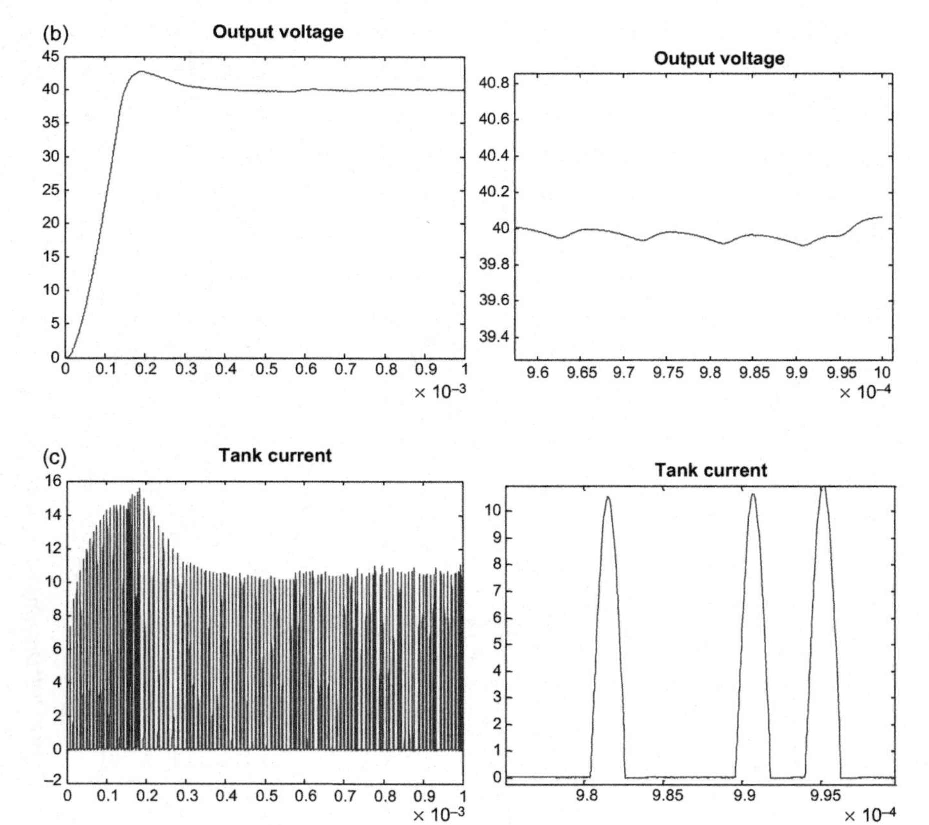

Figure 9.14 Figure 9.13 performance (a) LED array current, (b) converter output voltage, (c) resonant inductor current, (d) resonant capacitor voltage, and (e) VCO frequency.

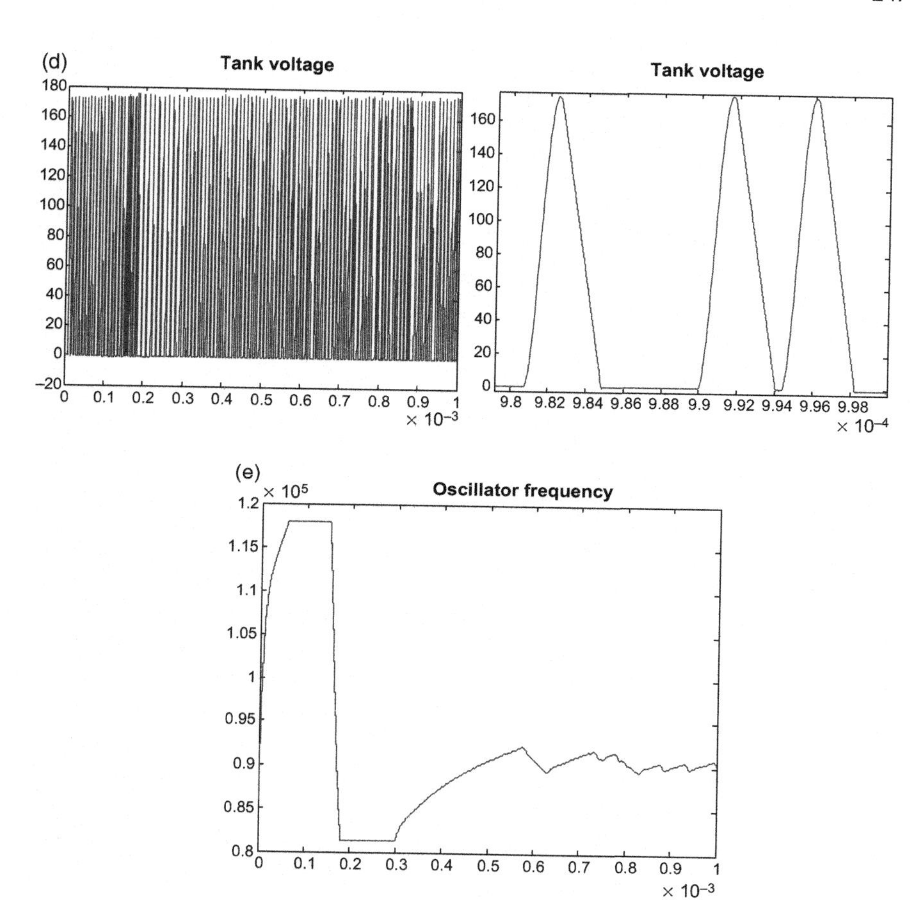

Figure 9.14 ***(cont.)***

CHAPTER 10

Current-Fed Converter

10.1 MERIT OF CURRENT–FED

In Figure 1.1, the source v_s, at the junction of rectifiers D_1 and D_2 feeding the output LC filter, is a pulsating PWM voltage. Being a voltage source, it can feed a large quantity of current, depending on load, if it is not limited. Therefore, if overload occurs, for instance, an output short, a huge amount of current that the input source can provide may pass through the power train. In most cases, this source current exceeds the design rating for components along the train. Obviously, a condition of this nature will destroy the weakest link and apparatus.

A simple solution to alleviate the shortcoming is to certainly replace the feeding voltage source with a current. This can be done by moving the output filter inductor, L, across the isolation transformer, and relocating it to the transformer's primary side. By doing so, the primary transformer is now fed by a current source. This configuration enjoys the property that the peak input current is set and limited by the input voltage, feed choke (inductor), and maximum duty cycle. As long as the relocated and resized inductor does not saturate, input current will not exceed the peak, even if the converter output is overloaded or short.

10.2 A CURRENT-FED CONVERTER

Figure 1.1 represents a push–pull current-fed converter, in which the input choke, L_1, is feeding the center tap of the isolation transformer, T_1. MOSFETs M_2 and M_3 drive the transformer's primary winding at an alternating 50% duty cycle such that the output capacitor and load are constantly coupled to the primary. MOSFET M_1 and diode D_1 perform close-loop PWM such that the current through inductor L_1 meets the output requirement. Provisions, not shown here, can be easily provided to limit the maximum feed current, as shown in Figure 10.1.

The converter is designed to take 400 V input and put out 90 V at 11 A, almost 1 kW. The PWM clock, a sawtooth swinging from 1–4 V, is set at 100 kHz with maximum duty cycle set at 98%, while the push–pull stage is running at an alternating 50 kHz, 50%, L_1 = 7.4 mH, C = 10 μF, and

Power Converters with Digital Filter Feedback Control
http://dx.doi.org/10.1016/B978-0-12-804298-4.00010-1

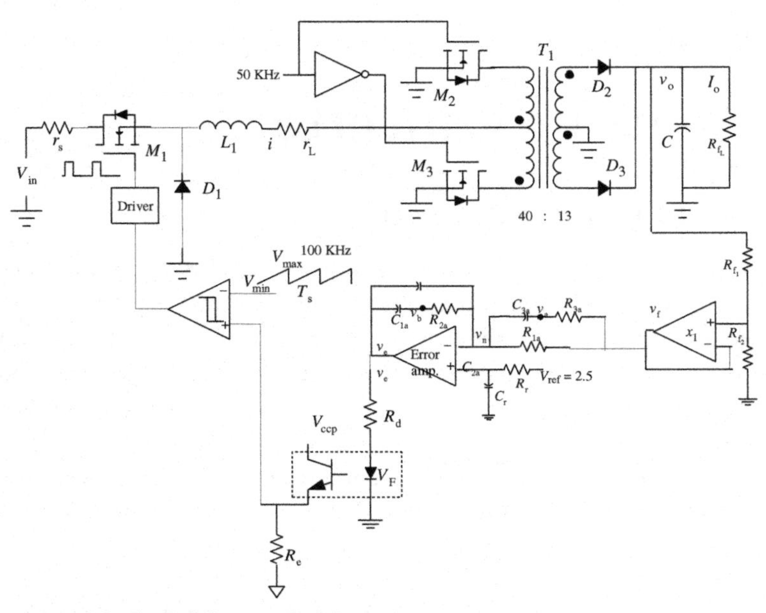

Figure 10.1 ***A Push–Pull Current-Fed Converter.***

$L_k = 10$ nH. Besides, $R_d = 2.29$ kΩ and $R_e = 1.5$ kΩ. Additionally, the optical isolator has a current transfer ratio of about 100% and CTR = 1.

10.3 DERIVATION OF MODULATOR GAIN AND SHAPING LOOP GAIN

We now proceed with the identification of modulator gain. Readers are again reminded of the definition, that is, the modulator gain is a subloop gain excluding the error amplifier.

First, we shall obtain the power stage gain, $G_{yd}(s)$, similar to (A.16). In order to do so, the power stage shall be drawn in its equivalent form, as shown in Figure 10.2.

Given a turn-ratio n, the primary sourcing current is reflected to the secondary side, a manifestation of current-fed operation. Whereas the primary winding is clamped to the secondary voltage multiplied by the turn-ratio. Furthermore, two state variables, the feed choke current, i, and output capacitor voltage, v, are governing the circuit. Both define a state (column) vector $x = (i, v)^T$. The primary loop gives,

$$L_1 \frac{di}{dt} + r_{L_1} i + n v_s = V_{in} \tag{10.1}$$

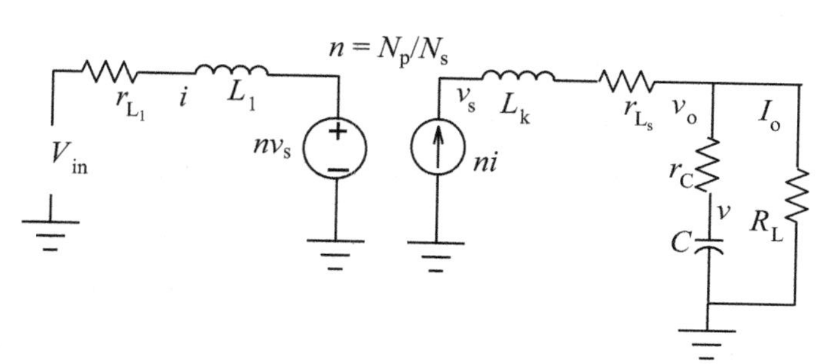

Figure 10.2 ***Power Stage Equivalent Circuit.***

The secondary loop gives,

$$v_s = nL_k \frac{di}{dt} + r_{L_s} ni + v_o \tag{10.2}$$

The output capacitor node gives,

$$C\frac{dv}{dt} = \frac{1}{r_C}\left(v_o - v\right) \tag{10.3}$$

An auxiliary equation yields,

$$v_o = niR_p + k_r v, \quad R_p = \frac{r_C R_L}{r_C + R_L}, \quad k_r = \frac{R_L}{r_C + R_L} \tag{10.4}$$

Then, following the step of Appendix A, state transition matrices are derived.

$$A_1 = \begin{bmatrix} -\dfrac{r_{L1} + n^2(r_{Ls} + R_p)}{L_1 + n^2 L_k} & -\dfrac{nk_r}{L_1 + n^2 L_k} \\ \dfrac{nR_p}{r_C C} & \dfrac{k_r - 1}{r_C C} \end{bmatrix}$$

$$B_1 = \begin{bmatrix} V_{in}/\left(L_1 + n^2 L_k\right) \\ 0 \end{bmatrix}, \quad C_1 = \begin{bmatrix} R_p \\ k_r \end{bmatrix} \tag{10.5}$$

$$A_2 = A_1, \quad B_2 = \begin{bmatrix} -V_D/\left(L_1 + n^2 L_k\right) \\ 0 \end{bmatrix}, \quad C_2 = C_1$$

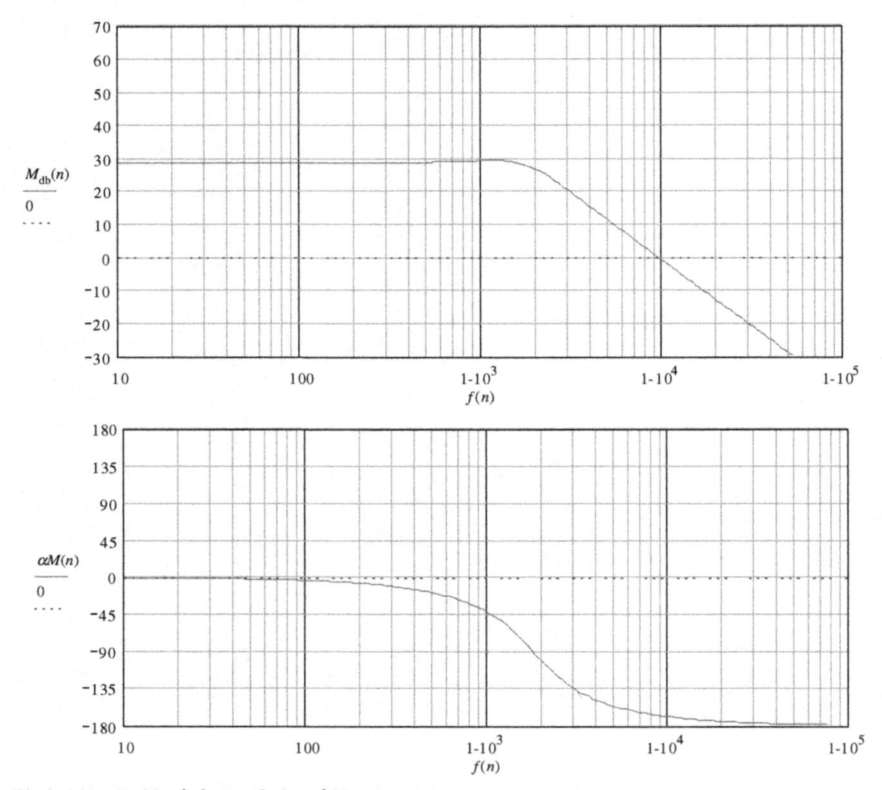

Figure 10.3 ***Modulator Gain of Figure 10.1.***

Equation set (10.5) and (A.7) enables one to find the DC state X and the steady state duty cycle $D = 0.71$. Then, the power stage gain, $G_{yd}(s)$ of (A.16), results. As a result, the modulator gain can be expressed as $M(s) = (1/R_d) \times \text{CTR} \times R_e \times F_m \times G_{yd}(s)$. The current example gives a modulator gain, as shown in Figure 10.3.

At the intended crossover frequency $f_c = 2$ kHz, the modulator has a gain of 27 db and a phase $\theta_M = -99.5°$. We select a type-III error amplifier to compensate the loop with a desired phase margin of $\theta_m = 45°$. The phase boost required, $\alpha_b = \theta_m - (\theta_M + 90°)$, is 54.5°. As a result, the pole–zero separation factor (1.13) equals $k = 2.689$. Then (1.15) gives all the amplifier's components, such as, $R_{1a} = 2\text{K}$, $R_{2a} = 86.7$, $R_{3a} = 1.18\text{K}$, $C_{1a} = 1.5\ \mu\text{F}$,

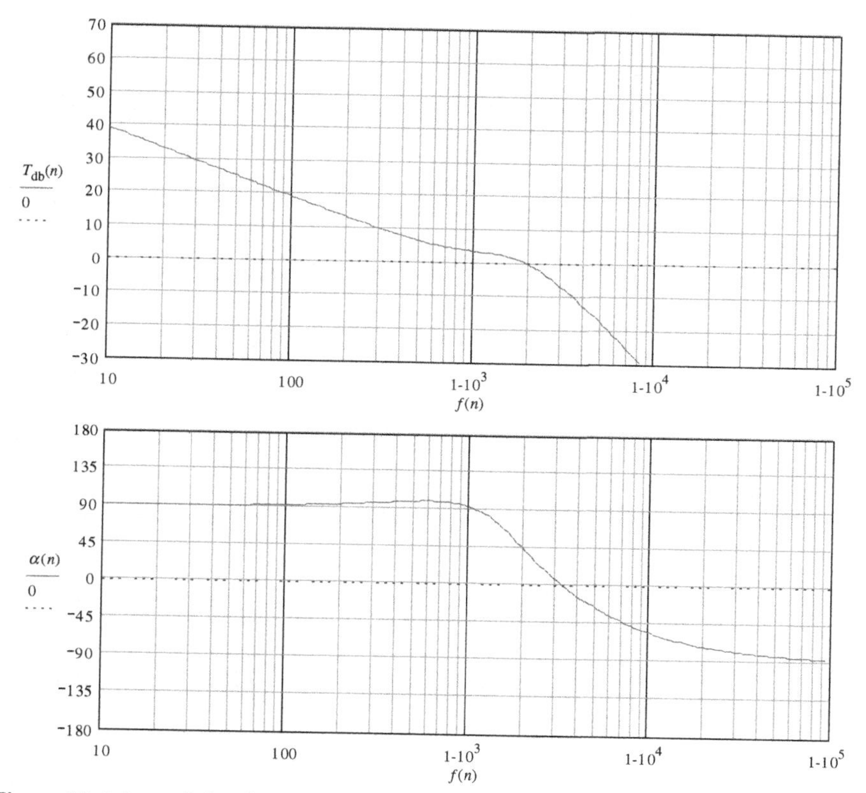

Figure 10.4 ***Loop Gain of Figure 10.1.***

$C_{2a} = 0.9\ \mu F$, and $C_{3a} = 0.04\ \mu F$. Incorporating the error amplifier, overall loop gain is obtained.

As shown in Figure 10.4, the loop gain does cross at 2 kHz with a 45° phase margin.

10.4 TIME DOMAIN PERFORMANCE FOR ANALOG VERSION

In the order of input feed current, output capacitor voltage, error amplifier's local voltages, and error amplifier output, Figure 10.1 operation is expressed and placed in discrete, iterative form (MathCAD),

$$
\begin{pmatrix} i_{p_{j+1}} \\ v_{j+1} \\ v_{a_{j+1}} \\ v_{b_{j+1}} \\ v_{e_{j+1}} \end{pmatrix} =
\begin{bmatrix}
i_{p_j} + \dfrac{\delta t}{L_p + n_t^2 \mathrm{LK}} \begin{bmatrix} -\left[r_p + n_t^2\left(r_k + R_p\right)\right] i_{p_j} - n_t k_r v_j \\ +\mathrm{if}\left(\dfrac{v_{e_j} - 1.2}{R_d} \mathrm{CTR} R_e > s_{w_j}, V_{in} - R_{on} i_{p_j} - \mathrm{VD} \right) \end{bmatrix} \\
v_j + \dfrac{\delta t}{r_c C}\left[n_t R_p i_{p_j} + \left(k_r - 1\right) v_j \right] \\
v_{a_j} + \dfrac{\delta t}{C_{3a}} \dfrac{k_f\left(n_t R_p i_{p_j} + k_r v_j\right) - v_{a_j}}{R_{3a}} \\
v_{b_j} + \delta t \begin{bmatrix} \dfrac{v_{r_j} - v_{b_j}}{R_{2a} C_{2a}} - \dfrac{k_f\left(n_t R_p i_{p_j} + k_r v_j\right) - v_{r_j}}{R_{1a} C_{2a}} \\ - \dfrac{k_f\left(n_t R_p i_{p_j} + k_r v_j\right) - v_{a_j}}{R_{3a} C_{2a}} + \dfrac{v_{r_j} - v_{b_j}}{R_{2a} C_{1a}} \end{bmatrix} \\
\mathrm{if}\left[v_{e_j} < 0, 0, \mathrm{if}\left[v_{e_j} > 12, 12, v_{e_j} + \dfrac{\delta t}{C_{2a}} \begin{bmatrix} \dfrac{v_{r_j} - v_{b_j}}{R_{2a}} - \dfrac{k_f\left(n_t R_p i_{p_j} + k_r v_j\right) - v_{r_j}}{R_{1a}} \\ - \dfrac{k_f\left(n_t R_p i_{p_j} + k_r v_j\right) - v_{a_j}}{R_{3a}} \end{bmatrix} \right] \right]
\end{bmatrix}
$$

$$
v_{o_j} = n_t R_p i_{p_j} + k_r v_j \qquad v_{ef} = \frac{v_{e_j} - 1.2}{R_d} \mathrm{CTR} R_e \tag{10.6}
$$

in which the feedback factor $k_f = 2.5/90$, v_r the command reference and n_t the turn ratio (n in Figure 10.2) 40/13 are assigned. The following, Figure 10.5a–c gives the steady state of key variables.

Figure 10.1 is also simulated in SIMULINK environment, as shown in Figure 10.6a.

10.5 I-FED CONVERTER WITH DIGITAL CONTROL

The analog error amplifier identified in Section 10.3, following the modulator gain evaluation, is transformed to the z-domain with a bilinear transformation constant C = 2 MHz (converter switching frequency 100 kHz, sampling frequency 1 MHz, transform constant = twice of sampling). The

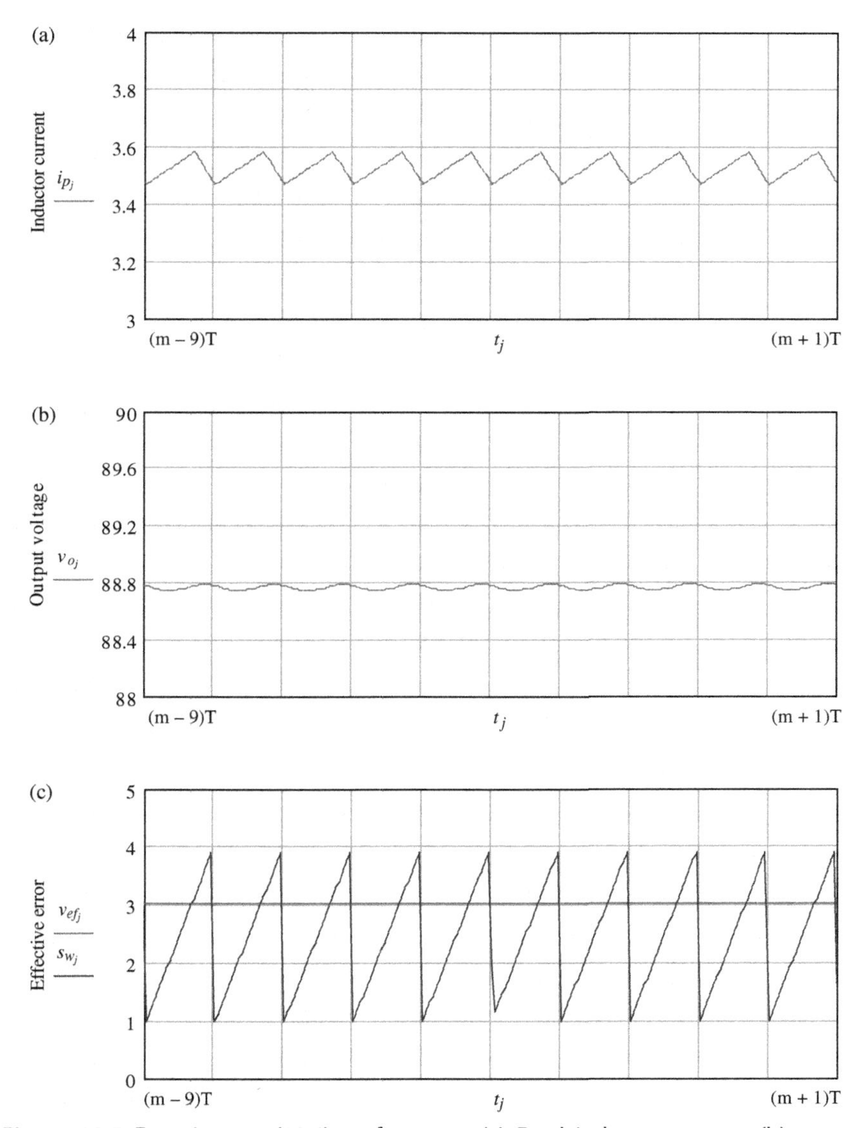

Figure 10.5 Equation set (10.6) performance (a) Feed inductor current, (b) output voltage, and (c) effective error voltage and clock.

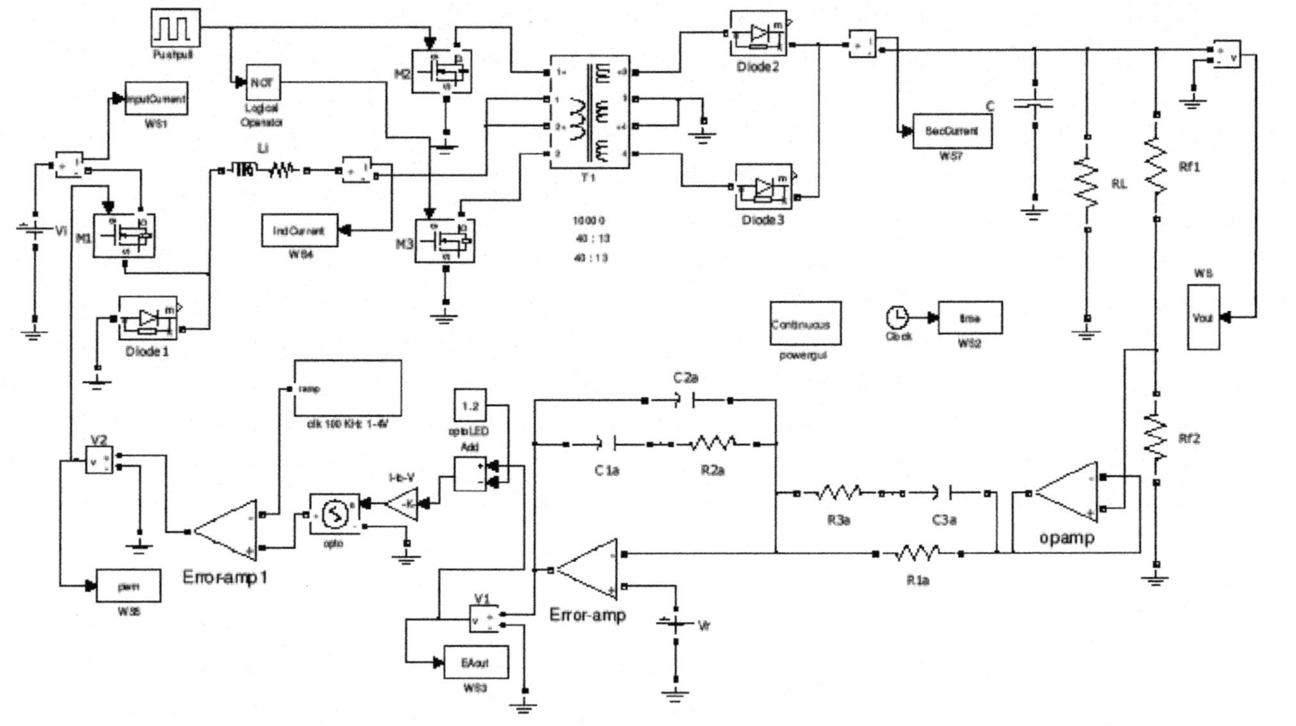

Figure 10.6 ***SIMULINK Schematic for Figure 10.1*** (a) Inductor current, (b) current feeding output capacitor and load, (c) output voltage, and (d) input current.

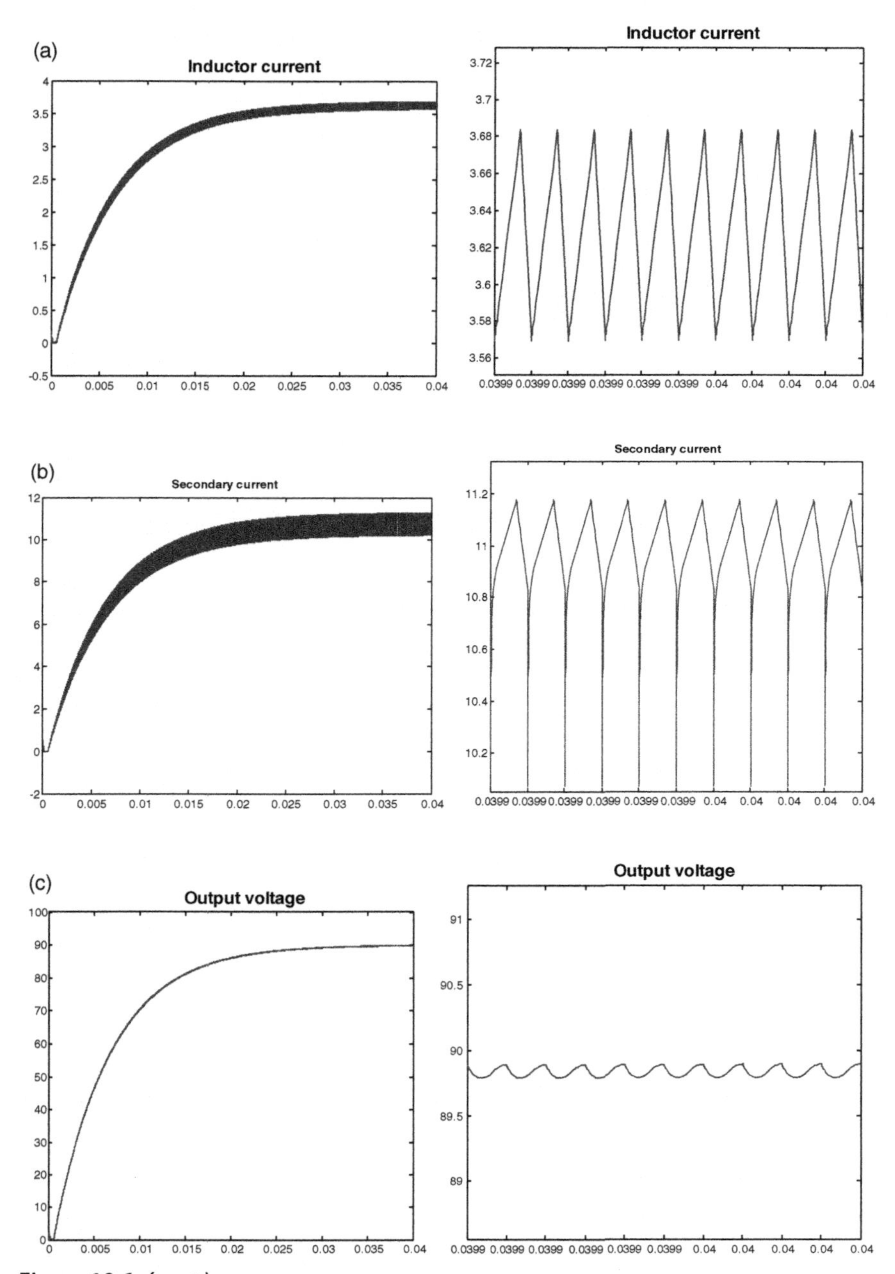

Figure 10.6 ***(cont.)***

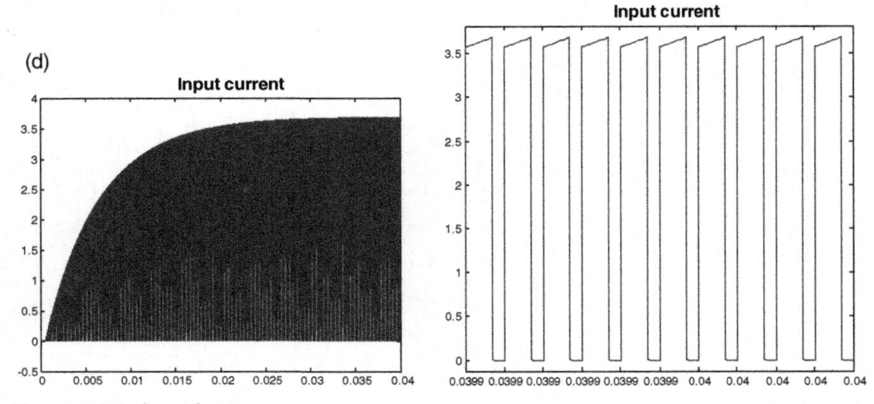

Figure 10.6 ***(cont.)***

resulting z-domain digital filter function is in the form of (1.23) with these coefficients,

$$a_0 = 7.451 \times 10^{-4} \quad a_1 = -7.337 \times 10^{-4} \quad a_2 = -7.451 \times 10^{-4} \quad a_3 = 7.338 \times 10^{-4}$$

$$b_1 = -2.959 \qquad b_2 = 2.919 \qquad b_3 = -0.96$$

Figure 10.7a–d shows the SIMULINK schematic with digital filter and performance in time domain.

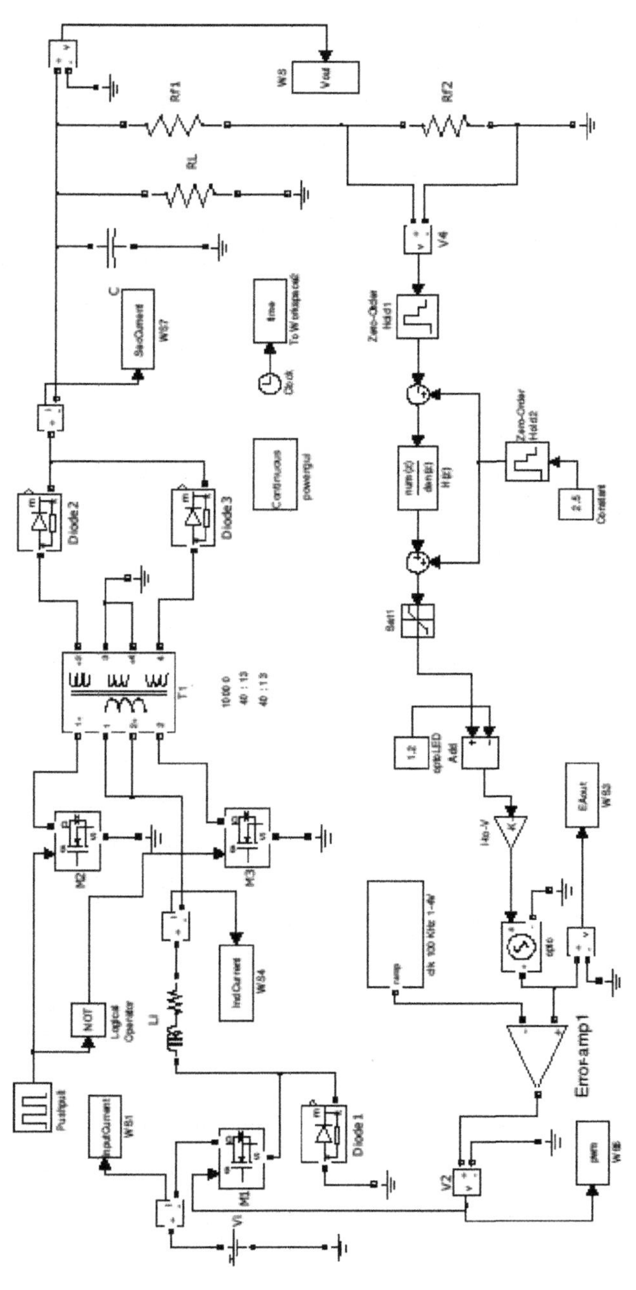

Figure 10.7 ***SIMULINK Schematic with Digital Filter.*** (a) Inductor current, (b) secondary current, (c) output voltage, and (d) input current.

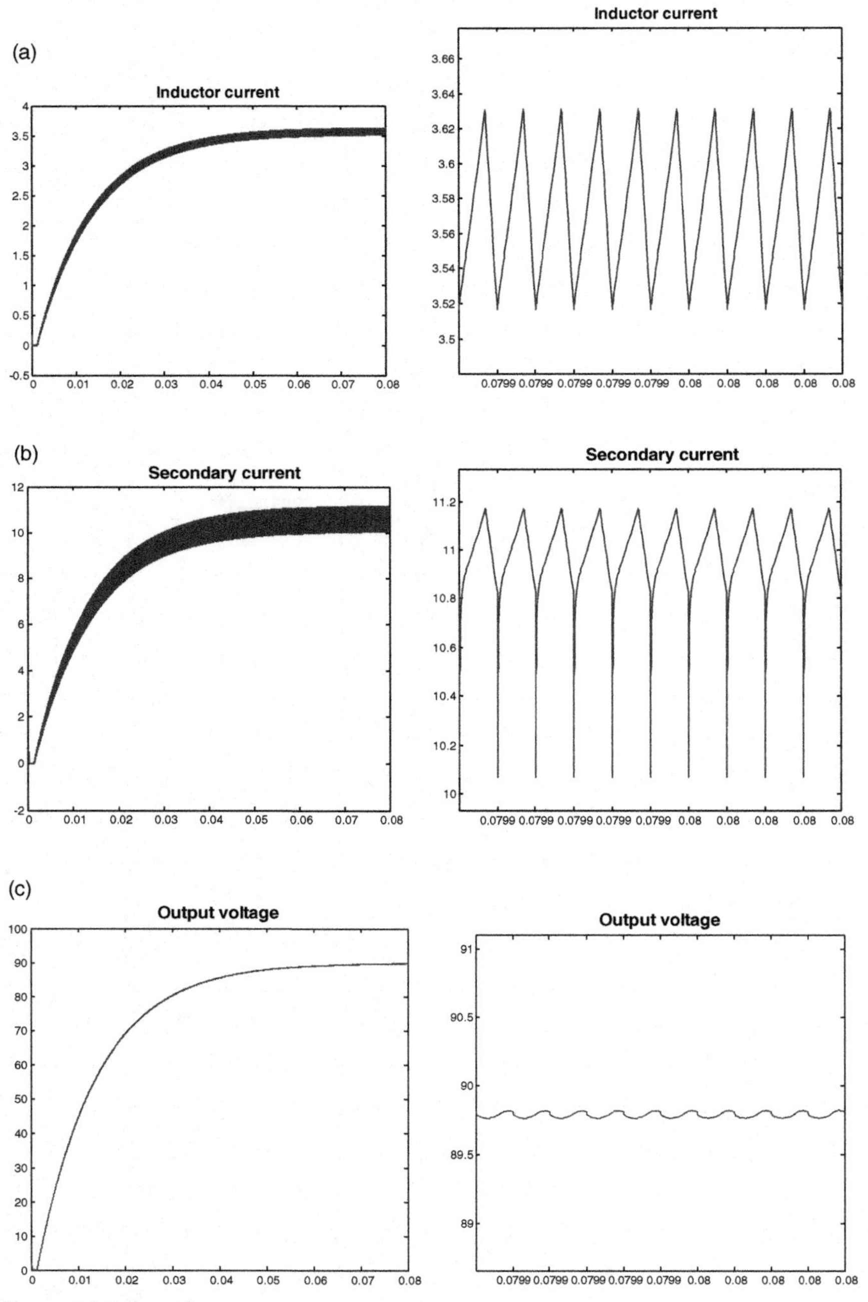

Figure 10.7 ***(cont.)***

(d)

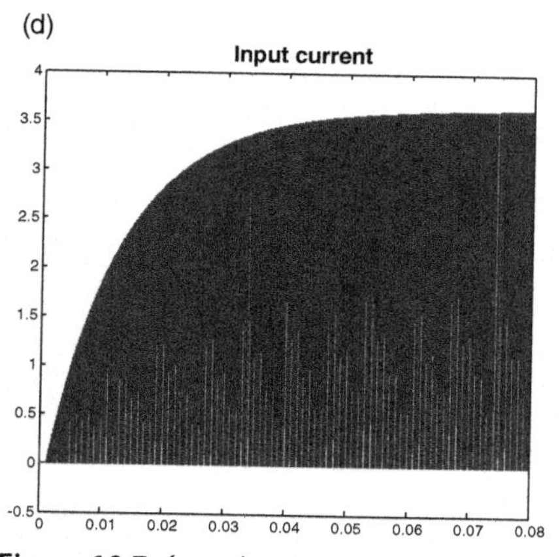

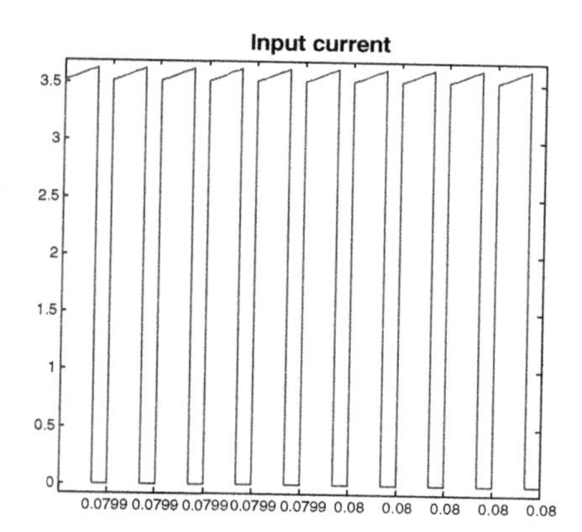

Figure 10.7 ***(cont.)***

CHAPTER 11

Implementing Digital Feedback

Alex Krasner

Implementing digital feedback requires proper development of the control algorithm in the digital domain, as well as proper integration of hardware around the controller chip. Analog ICs and other power controller ICs require the right "house-keeping" circuitry around its inputs and outputs for optimal functionality. These requirements are predetermined by the manufacturer. When developing a custom digital feedback controller, the same principles apply: signals must be adequately conditioned to interface with the user's control algorithm, supply voltages must be accurately set to ensure proper start up and functionality of the digital controller, and appropriate noise countermeasures must be taken to prevent errors in the algorithm. Additionally to the external concerns, the internal structure of the digital control must also be properly configured. This includes timing constraints, numerical representations, and error prevention in the algorithm.

11.1 DATA INPUT AND SIGNAL CONDITIONING

Feedback signal have to be digitized in order to be processed by a digital control algorithm. This is done by selecting the appropriate analog-to-digital converter (A/D). Three parameters are key for selecting the proper A/D: sampling rate, bit resolution, and dynamic range of the analog input. The sampling rate of the chosen A/D should match the data rate that the control algorithm is designed for. Depending on the A/D various ranges of input, clocks are need for its functionality. A/Ds that have serial outputs (SPI or I2C) can require clocks much greater than the sampling frequency. This is because A/Ds require conversion and acquisition stages to collect analog input and convert it to a digital bit format. The specific requirements of the A/D depend on the internal architecture (http://cds.linear.com/docs/en/design-note/dn1013f.pdf).

Four main architectures of A/D exist: sigma delta, delta sigma, SAR, and pipelined. These trade off sampling frequency for resolution. The common A/D architecture to use in a power supply application would be a SAR A/D, due to its appreciable resolution and high sampling rate capabilities, Figure 11.1.

Power Converters with Digital Filter Feedback Control
http://dx.doi.org/10.1016/B978-0-12-804298-4.00011-3

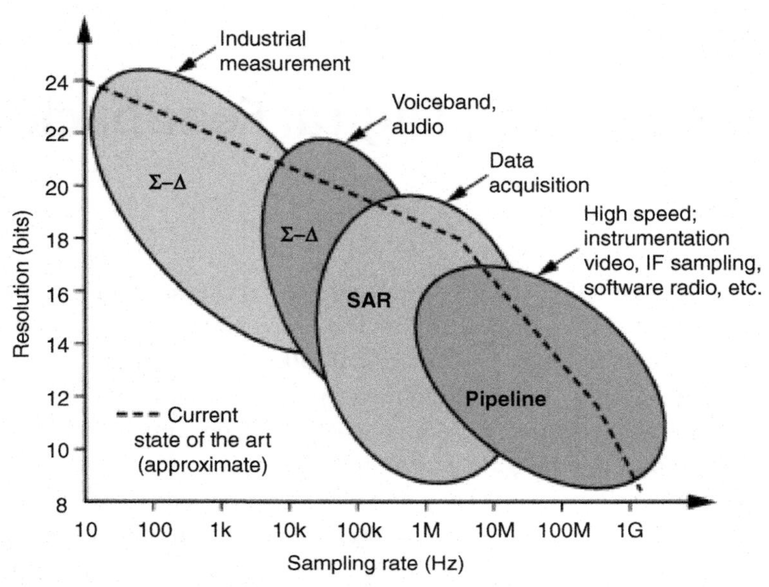

Figure 11.1 ***Main Classes of A/D Architectures.*** Bit resolution and sampling rate are inversely related. For digital power supply controllers, SAR A/Ds provide the best speed to resolution combination. http://www.analog.com/library/analogdialogue/archives/39-06/architecture.html; http://ww1.microchip.com/downloads/en/AppNotes/91078a.pdf.

SAR A/Ds offer high accuracy for sampling rates less than or equal to 5 MSPS, while also ensuring minimal to no pipeline delay.

PLLs (phase lock loops) internal to the chip are used to derive fast clocks for serial A/Ds. The main drawback to a PLL is its ability to produce jitter, that is, transient phase shifts. Skewed clock edges will propagate directly to the ADC, thereby causing bit errors on the output. Depending on the frequency of the analog signal (slew rate), the error of the A/D output may be significant; faster slew rate signals are affected more by jitter than slower signals. Most MCUs and FPGAs are precise enough to limit the amount of jitter on the PLL output; however, it is the designers' responsibility to account for this error.

The signal to noise ratio (SNR) caused by jitter is displayed in the following equation:

$$\mathrm{SNR}\left(d\,\mathrm{BFS}\right) = -20\log\left(2\pi f_{\mathrm{in}}\sigma\right)$$

where σ represents the clock jitter in seconds, and f_{in} is the input signal's frequency.

The dynamic range and bit resolution of the A/D are chosen based on the quality of the signal that is required. Normally, high resolution A/Ds with fast sampling rates sacrifice analog input range. It is important to strike a balance between these parameters. An A/D's signal resolution is calculated by the following formula: $\Delta V = \frac{V\text{pp}}{2^B}$; where Vpp is the analog input range and B represents bit resolution.

An A/D with 12-bit resolution and 4.096 V dynamic range will have a signal resolution of 1 mV. The least significant bit (LSB) of the A/D output corresponds to a 1 mV signal. In power supply applications, 1 mV is far below the noise level on the PCB; therefore, it is guaranteed that some of the LSBs will contain noisy data. It is the PCB designers' priority to assess the noise characteristics of the layout around the A/Ds. Noise minimization can be done by providing local grounds, partitioned ground planes, and bypass capacitors.

A critical question to ask is, "where to place the A/D?" Suppose the feedback path that is to be digitized is the voltage feedback. In an analog circuit, this feedback would go to a type-II or III error amplifier, as it is compared to some references (2.5 V). The user has two choices as seen in Figure 11.2.

In option (a), the A/D is placed before the 2.5 V reference; the reference and the error amplifier are set in the digital domain. In option (b), the A/D is placed after being compared to an analog 2.5V reference. In option (a), the A/D is digitizing the signal $k_f V_{out}$; this signal must be within the dynamic range of A/D. Ideally, the nominal value of $k_f V_{out}$ should be in the center of the dynamic range to allow for the widest V_{out} range. Coincidentally, the nominal $k_f V_{out}$ value should match V_{ref}. Therefore, if V_{ref} is set to 2.5 V, then the appropriate A/D range should be 0–5V analog input. However, it may be difficult to find an A/D with such a wide range and relatively fast speed. The designer should use the next best available dynamic range (4.096V or at worst, 3.3V). Additionally, since the expected result is always greater than zero, a single-ended A/D can be used.

On the other hand, option (b) digitizes the voltage after comparison to V_{ref}: $(k_f V_{out} - V_{ref})$. In this case, the A/D receives a differential signal. If $k_f V_{out} > V_{ref}$, then the analog input is greater than zero; otherwise, it is <0V.

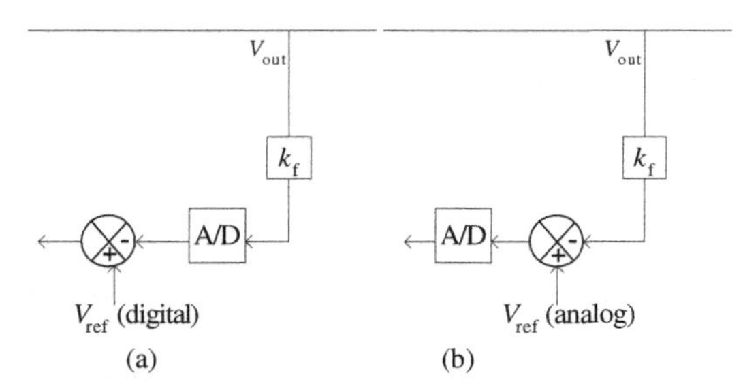

Figure 11.2 ***Placing A/D Converter in Power Converter with Digital Filter Feedback.***

For this purpose, one must use a differential A/D. However, the dynamic range of the A/D (V_{pp}) still remains the same as in option (a). Additionally, the designer must supply an analog V_{ref}. The main advantage of a differential A/D is common mode noise rejection. With single-ended A/D, external circuitry is required to filter out environmental and high-frequency noise.

Both options are therefore valid; for power supply applications, single-ended A/Ds are adequate since signal conditioning circuits usually provide decent noise rejection. The two options given earlier did not cover one more situation: suppose the designer decides to use single-ended A/Ds, but wants to measure an AC signal (such as the auxiliary winding of a transformer). The A/D's analog input is positive, while the desired signal is AC. In this situation, the signal needs to be positively biased to fit within the dynamic range of the A/D. A possible circuit is displayed in Figure 11.3.

Here, an AC decoupling capacitor is used to remove unwanted DC drift and low frequency noise. A resistor divider is used to bias the resulting signal. The DC value of the bias should be set to the middle of the analog range of single-ended A/D to maximize the allowable amplitude of the AC signal. Once the digitized signal enters into the MCU's or FPGA's domain, then the DC bias can be removed.

Figure 11.3 also contains a unity gain op-amp, also known as a buffer. Most A/D require low impedance inputs to preserve the quality of the

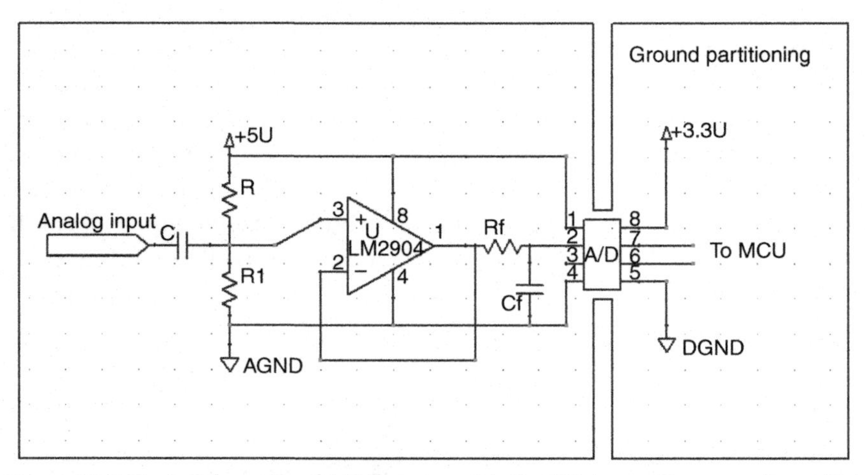

Figure 11.3 ***Basic Schematic of A/D Structure.*** *C* is the AC decoupling capacitor, which removes DC bias. Resistors *R* and *R*1 provide a DC bias to shift the AC signal into the dynamic range of the A/D. R_f and C_f provide passive low passive filtering of the A/D analog input. LM2904 is used as a unity gain buffer. Digital (DGND) and analog (AGND) grounds are partitioned to reduce noise and ground loops.

analog signal. A buffer is the ideal method to separate the external analog environment from the sensitive digital domain, and prevent impedance mismatch (A/D generally require low impedance input). However, when buffering, it is important to note the frequency response of the op-amp used. If a standard LM2904 is used, then high frequency signals (>5 kHz) will be attenuated, resulting in unwanted signals.

Figure 11.4 displays the frequency plot for an LM2904. Using op-amps with wider bandwidths will remove attenuation of signal, but may increase the cost of the design. Therefore, only those signals that are high frequency should use the specialized op-amps; for slow or DC signals, LM2904 is a suitable op-amp.

It is good practice to low pass filter the analog signal to $f_s/2$ to prevent aliasing (f_s = sampling rate). This can be done by using a passive RC low-pass circuit or modifying the buffer op-amp to also act like a low-pass filter. However, antialiasing in the analog domain is not the only option. It is also possible to antialias in the digital domain using a digital low pass filter. To do this, the A/D should oversample the signal and then use a digital filter to down convert and antialias. The resultant signal will be in the correct data rate, but the filter is realized in the digital domain.

Various filters can be used both in the analog and digital domain. The most common filters include the Butterworth filter, Chebyshev, and Bessel filters. These differ in their roll-off characteristics and phase responses; therefore, choosing the appropriate filter is not an easy task and depends on the nature of the signal. The main goal of a filter is to pass wanted frequencies

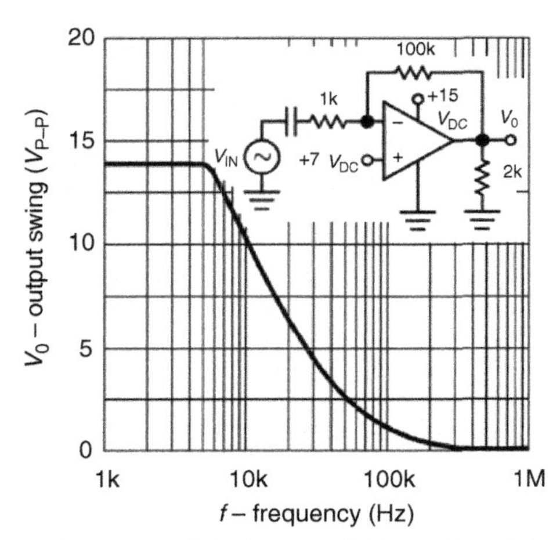

Figure 11.4 ***Frequency Response of the LM2904 (TI Data Sheet).*** Signal attenuation begins at 5 kHz. This is not desirable when digitizing fast signals.

without distortion while maximally attenuating unwanted frequencies. Unfortunately, that is not possible without a compromise: the trade-off is always between magnitude ripple versus the sharpness of frequency cutoff. Filters with sharp cutoff frequencies tend to have ripple around the cutoff.

The Butterworth filter is the simplest low-pass filter to use; it provides steady roll-off without ripple and moderate phase distortion. It is known as the "maximally flat" filter, as it provides the best unity magnitude response for the passband. However, the cutoff rate is slow. It is possible to combine several filters in series (i.e., increase the filter order) to sharpen the cutoff. This method will amplify phase distortion and may introduce ripple. The Butterworth filter has a roll off of -20 dB/decade $\times$ n (filter order). The Chebyshev filter offers a much steeper roll-off as seen in Figure 11.5, but

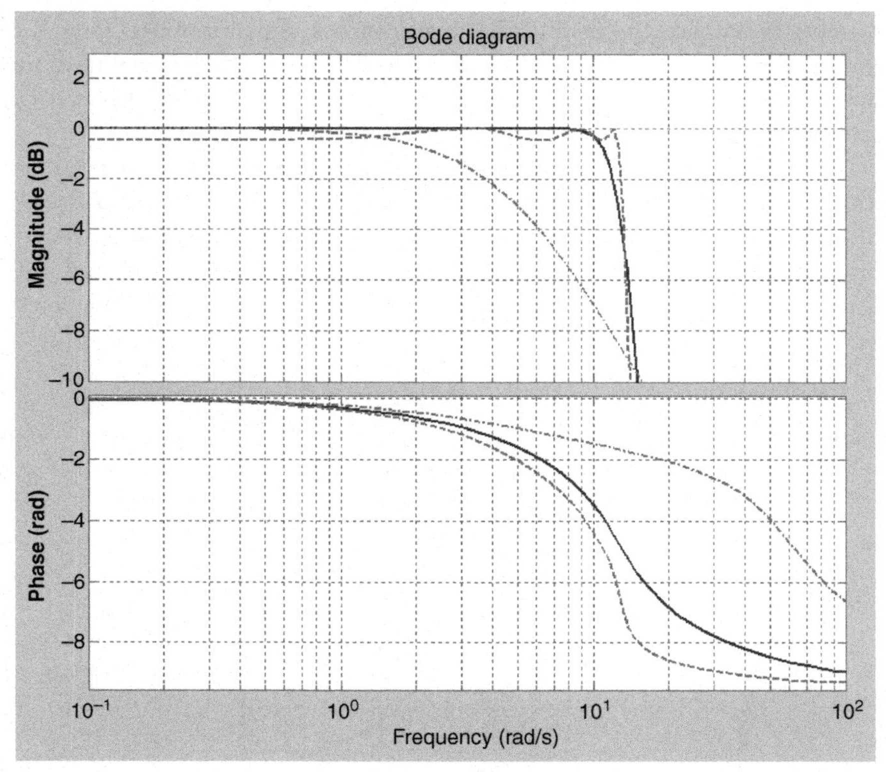

Figure 11.5 ***Comparison of Butterworth (solid), Chebyshev (dash), and Bessel (dot-dash) Filter Responses.*** Butterworth filter has a maximally flat magnitude response and modest roll-off. Chebyshev has fast roll-off but ripple in the magnitude response. Bessel has a gradual roll-off, but the smoothest phase response.

has considerable ripple in the passband (and stopband for high order filters), and a nonlinear phase response.

Finally, the third typically-used filter is the Bessel filter. Unlike the previous filters, Bessel filter's roll-off is considerably more gradual, but it comes with the advantage of having minimal ripple and almost linear phase response. Because of this, the Bessel filter is preferred when signal shape is very important, since all frequencies are delayed equally, therefore preventing signal distortion. The designer must decide what is the optimal filter for his/her application by considering roll-off, phase distortion, and magnitude ripple of these filters.

11.2 DIGITAL REPRESENTATION

When building analog feedback systems, it is generally easy to determine what values (voltages) will be produced at various stages of the feedback. These are known as "real-world" values because they have a physical meaning associated with them (voltage). In the digital domain, values exist only as combinations of bits. It is therefore important to establish a paradigm to interpret the bit values. The previously mentioned A/Ds will output (either serially or in parallel) bit representations of the analog input based on their dynamic range. Suppose the A/D is 10 bits with 0–3.3 V dynamic range; the output "0000000000" will represent 0 V, and "1111111111" will represent 3.3 V, but to the MCU, these outputs are 0 and 1023, respectively. If a different A/D was used with 10 bits but 4 V maximum, then 1,023 would represent 4 V, not 3.3 V. Clearly, the designer must specify the parameters of the A/D within the digital algorithm in order to interpret the value of the A/D output. Let's assume that the voltage to be converted is 2.5 V, using a 10-bit A/D with a 3.3 V range. Then, the output of the A/D would be 776, or "1100001000." Ideally, the digital algorithm should be made as close to real world as possible, so the designer is faced with the task to make the previously mentioned 10-bit value look like "2.5." To do this, the digital algorithm must preprocess the signal by applying scaling and/or bit shifting:

$$\frac{1023}{3.3} = 310 \approx 1.21 \times 2^{\wedge}8$$

The previous expression represents the scaling factor to be used to convert the raw A/D output into a real-world "voltage." It is represented in the scientific representation in base 2 because bit shifts and binary point shifts are easier for digital hardware to accomplish than multiplication and division.

The previous expression states that to convert 1023 into 3.3, one must first shift the binary point the right eight positions and then divide by 1.21:

$$1111111111 \rightarrow 11.11111111 = 3.99609...$$
$$3.99609 / 1.21 = 3.3$$

The extra factor of 1.21 is unfortunate as it takes up computational resources that can be better suited elsewhere. Perhaps, another solution would be to utilize a different A/D with more favorable parameters: 12-bit A/D with a 4V range. In this case, 4V = "111111111111" = "4095"

$$4095 / 4 = 1023.75 \approx 1.0 \times 2 \wedge 10.$$

Here the scaling factor reduces nicely to a 10-position binary point shift without any other nonbase-2 factors. Thus, to represent 4V from 4095, the algorithm must simply shift the binary point 10 times to the right:

$$111111111111 \rightarrow 11.1111111111 = 3.99902 \approx 4$$

A voltage value of 2.5V would similarly be converted:

$$2.5\,\mathrm{V} = 101000000000 \rightarrow 10.1000000000 = 2.5$$

Therefore, the analog voltage value is represented in the digital domain accurately. The previous representation of a real world value using bits and binary points is known as "fixed-point representation." Arithmetic performed in MCUs and FPGAs is based on fixed-point values, which follow different rules than real-world values used in math. MCUs and FPGAs can also utilize "floating point representations," where numbers are represented as a sign, exponent, and mantissa; however, calculations in floating point are computationally intensive and therefore slower without the proper on-chip resources. Floating point representations (based on IEEE 754 standard) can describe real world values with great precision and with an enormous range. 32-bit processors use 1 sign bit, 8 exponent bits, and 23 mantissa bits, thereby achieving resolutions of 2^−127 × 2^−23 and having ranges from −2^128 to 2^128. 64-bit representations will greatly exceed these characteristics (1 signed bit, 11 exponent bits, 52 mantissa bits). The drawback is that representing values in floating point requires large computational power and may not be suitable (or even needed) for control applications.

From here on we will be using Matlab's fixed-point notation: ufix(N,F) and sfix(N,F), where ufix represents an unsigned fixed-point value with N total bit width and F fractional bits, and sfix represents a signed fixed-point value. ufix(N,F) ranges from 0 to $2^N - 2^{-F}$ with a resolution of 2^{-F}, while sfix(N,F) ranges from $-2^{(N-1)}$ to $2^{(N-1)} - 2^{-F}$ with a resolution of 2^{-F}. In sfix values, the MSB (most significant bit) represents the sign of the value.

Fixed-point representation is computationally simple but requires the designer to be very mindful of bit widths. When adding fixed-point values, it is important to increase the bit width by 1 to allow for the new maximum value:

$$\text{ufix}(N,F)+\text{ufix}(N,F)=\text{ufix}(N+1,F)$$

When subtracting, the bit width should increase by 1 as well to accommodate for the sign change:

$$\text{ufix}(N,F)-\text{ufix}(N,F)=\text{sfix}(N+1,F)$$

For multiplication, the resultant bit width and fractional bit length both increase by the sum of each operand (an additional bit is need when multiplying with a signed value):

$$\text{ufix}(N1,F1)\,\text{ufix}(N2,F2)=\text{ufix}(N1+N2,\ F1+F2)$$

This is easy to prove algebraically if the maximal values of each operand are multiplied:

$$\left(2^{N1}-2^{-F1}\right)\left(2^{N2}-2^{-F2}\right)=2^{N1+N2}-2^{N1-F2}-2^{N2-F1}+2^{-(F1+F2)}$$

From the previous expression, it is evident that both the maximal bit width increases *and* the fractional resolution must increase as well.

Division is not advised in fixed-point applications. It is computationally intensive in both floating point and fixed-point representation and, therefore, should be avoided as much as possible. If division by a constant is required, then it is better to precalculate the inverse of the constant and then use fixed-point multiplication. Since fixed-point representation is in base 2, base 10 fractions are difficult to convert:

$$0.1_{10}=.0001100110011001100110_2$$

The previous example displays that 0.1 in base 10 is difficult to express in base 2; therefore, some rounding error will exist. When designing the digital algorithm, the engineer must determine the tolerable amount of error to limit the bit widths used. Large bit width requires more resources and computational power; it is prudent to know the necessary bit width for each signal and calculation. This can be a meticulous process. To miss assigned bit widths can lead to disastrous effects. Overflow in a bit value, that is, a value that exceeds ufix(*N*,*F*) or sfix(*N*,*F*) maximum can either be saturated or wrapped around. Wrap-around is dangerous for a design, since it will propagate large errors to all following operations. Saturation error is more desirable, since the saturated value is closer to the real value. The designer is responsible in preventing wrap-around and saturation errors.

11.3 IMPLEMENTING DIGITAL FILTERS

Two discrete filter topologies exist: FIR (finite impulse response) and IIR (infinite impulse response) filters (http://www.quickfiltertech.com/files/Digital%20Filtering%20Alternatives%20for%20Embedded%20Designs.pdf). FIR filters are inherently stable because they have no feedback paths, only feedforward. Therefore, the transfer function of an FIR filter contains no poles:

$$\frac{Y(z)}{X(z)} = \frac{b_0 + b_1 z^{-1} + b_2 z^{-2} + \ldots + b_N z^{-N}}{1}$$

The FIR filter is commonly used in digital applications, due to its stability. Additionally, since digital logic must use fixed-point representation, the FIR in especially favored, since it will remain stable despite any rounding errors. However, FIRs suffer from several drawbacks: they require lots of coefficients to achieve sharp frequency cutoff (therefore, lots of RAM needs to be allocated for storing values), due to the large amount of coefficients and delays, FIRs have large latency, and they do not provide any phase boost (i.e., constant group delay). In power supply applications where the voltage or current controlled feedback requires a significant phase boost (type-II or III compensator), the FIR filter cannot be used.

IIR filters, on the other hand, have the opposite problems. IIR filters contain both feedforward and feedback paths, and therefore are predisposed to stability problems. IIR filters contain both zeros and poles; therefore, rigorous analysis (via root locus, zero–pole plots, and Nyquist plots) is required to ensure stability:

$$\frac{Y(z)}{X(z)} = \frac{\sum_{i=0}^{N} b_i z^{-i}}{1 - \sum_{j=1}^{M} a_j z^{-j}}$$

Fixed-point values and arithmetic can cause IIR filters to become severely unstable, due to rounding errors during multiplication stages with coefficients. However, if designed properly, IIR filters can achieve sharp frequency cutoffs using only a handful of coefficients (therefore little delay), and provide the necessary phase response needed for power supply stability. Usually to ensure stability of the IIR filter, fixed-point bit widths need to be very large, and thus computational space saved by reducing coefficients is used up for wide bit-width calculations. The tradeoff between FIRs and IIRs is evident; however, for the feedback error amplifier in power supplies, IIRs are preferred.

The most popular IIR topology is the second-order biquad shown in Figure 11.6.

Second-order biquad filters are of the form:

$$H(z) = \frac{b_0 + b_1 z^{-1} + b_2 z^{-2}}{1 - a_1 z^{-1} - a_2 z^{-2}}$$

Complex IIR filters can be reduced into cascaded second-order biquads to improve overall stability and allow for easier debugging. The Matlab command *tf2sos()* can be used to convert any IIR filter (given *b* and *a* coefficients) into a single or multistage biquad cascade. Each biquad can then be stabilized by analyzing it in the Nyquist plot and zero–pole plot. The two poles of the biquad filter must both lie within the zero–pole unit circle.

Aside from properly setting up bit widths to ensure stability, the sampling period must also be chosen correctly. Since phase response is the desired characteristic of the IIR filter, the sampling period must be chosen to match the phase response required. When discretizing a continuous transfer function, the phase response will deviate greatly near the $f_s/2$. Figure 11.7 shows a continuous phase response and those of discrete IIR with various sampling rates.

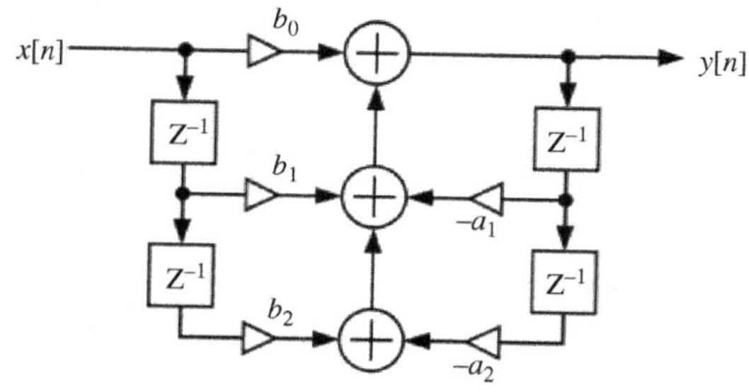

Figure 11.6 ***Second Order Biquad IIR Filter Structure.***

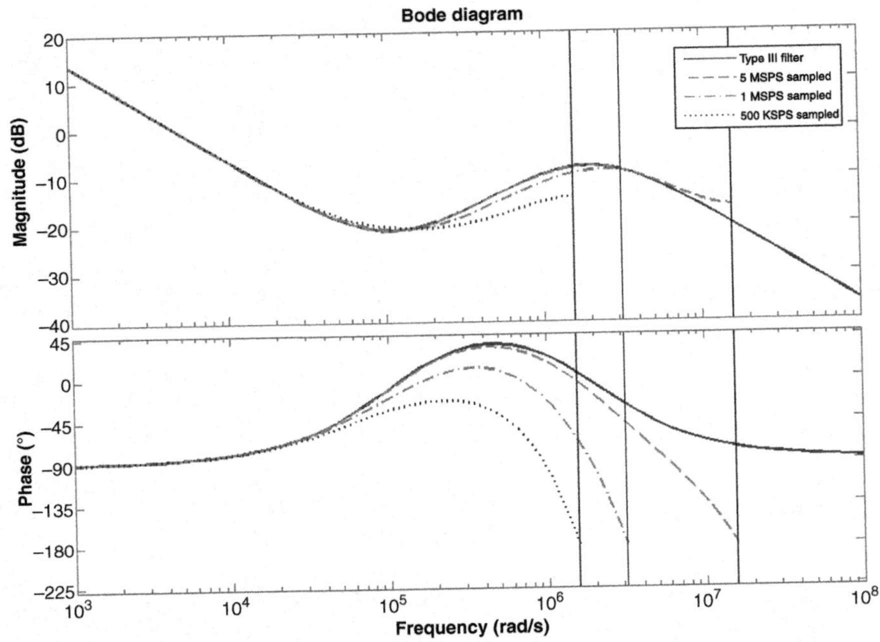

Figure 11.7 ***Distortion of Magnitude and Phase Responses when Digitizing Filters.*** **Higher sampling rates preserve the fidelity of the analog filter.**

It is clear to see that at frequencies closer to the $f_s/2$, the phase response sharply rolls off to $-180°$.

Figure 11.7 shows another intriguing property of quantization. Unlike the magnitude response, which remains relatively true to its analog response, the phase response is altered greatly based on sampling rate. This means that the data must be oversampled at much more than the Nyquist requirement to achieve the desired response. In this figure, 5 MSPS sampling approaches the analog phase response the best. If the designer cannot sample the data at those rates, then interpolation and upsampling of data must occur prior to using the IIR.

Since IIR filters have feedback loops, computational time is another crucial factor. In the biquad filter shown in Figure 11.6, two feedback paths are present: $-a_1y[n-1]$ and $-a_2y[n-2]$. The delays in this structure relate to the "data rate" or sampling period. When implementing digital IIRs on MCUs or FPGAs, the main clock that synchronizes all other clocks (such as the sampling clock) is known as the system clock and can usually run from several MHz to GHz, and is greater than the data rate (although not always). All computation is done relative to the system clock; therefore, the multiplication processes in the IIR paths are relative to the system clock. In order for the IIR to run properly, these computations must finish before the next sample comes in. Let's say, for example, that the system clock is 100 MHz, and the data rate is 10 MSPS; the IIR computation has only 10 clocks cycles (100 MHz/10 MSPS) to produce a result before the next sample arrives. Wide bit-width values can take longer to process because they will require more than one multiplier block (standard 18 bits) to operate. It is crucial to ensure, via simulation, that the computational processes in the IIR filter occur faster than the data rate. If that is not possible, then the data rate should be decreased to allow for more time for calculations, or the clock rate should be increased. The latter option is not practical, since DSP/ multiplier blocks usually run at preset frequencies (≈200 MHz).

MCUs are popular for implementing digital designs. MCUs can have very high-rate system clocks and can support floating-point arithmetic. But MCUs can only do one operation at a time; thus, every multiplication must be buffered in RAM and then recalled for accumulation. MCUs are designed to optimize this operation by use of DMAs (direct memory access controllers) that allow for fast memory reading and writing. The designer, once again, must be mindful of the inherent latencies of background tasks that can disrupt the performance of the IIR.

FPGAs, on the other hand, allow for simultaneous processing of parallel operations. Therefore, all parallel multiplications can be processed together and then accumulated. FPGA clock speeds are comparable but no greater than MCU speeds; therefore, the parallelized nature of the FPGA should be exploited to overcome this. Since all operations occur in parallel, one may question race conditions. Proper FPGA design is made to be synchronous, that is, all processes occur relative to a system clock edge, and all values are held in registers until the next clock edge. This ensures that all interdependent paths have a known state at each given clock cycle. Propagation delays of bits between registers are minimal but important to consider when implementing FPGA designs. Unlike in MCUs, where the digital design is ultimately software, FPGA designs are digital hardware; therefore, physical constraints such as voltage, temperature, and delay can affect performance. By creating synchronous designs, most of these concerns are mitigated, and the performance of the designs becomes invariant. The designer is responsible for simulating the digital design to ensure that all timing conditions are met and that the IIR filter remains stable at all edge conditions (voltage and temperature).

In most cases, the system clock will be much faster than the data rate, thus leaving many clocks' cycles of operation time between new data samples. In this, resources can be saved by *folding* the IIR filter. Folding is a method of reusing the same multiplier and accumulator for multiple calculations and buffering results. In the second order biquad filter, five multiplications must be done, followed by an accumulation of all five results. If the design is folded, all five multiplications can be performed by the same multiplier sequentially, while the accumulator adds up each new result. In this case, one-fifth of the multipliers is required (at minimum: large bit widths may require more multipliers to be used). IIRs can be folded as long as the loop delay properties are conserved; otherwise, the IIR will become unstable and produce erroneous results.

A basic example of an IIR that can be implemented is an integrator. In the z-domain an integrator takes the form:

$$Y(z) = TX(z) + Y(z)z^{-1}$$

Hence,

$$H(z) = \frac{Y(z)}{X(z)} = \frac{T}{1 - z^{-1}}$$

T, the sampling period, can be a very small fractional value if the sampling rate is high. Figure 11.8 displays the graphical representation of the digital integrator.

In fixed-point notation, the value, *T*, may require a large amount of bits to minimize rounding error. Suppose the data sampling rate is 10 kHz (i.e.,

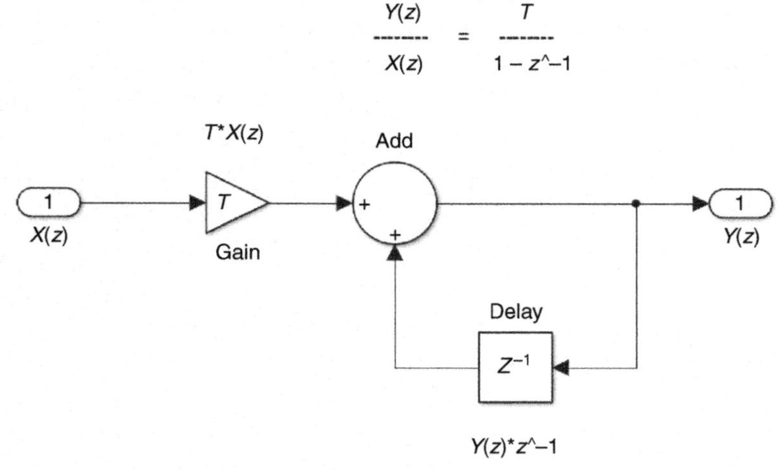

Figure 11.8 ***Digital Integrator of the Form H(z) = T/(1 − z^− 1).***

the sampling period of 0.0001 s). Representing 0.0001 accurately in binary is difficult; using a decimal to binary conversion tool 0.0001 can be accurately converted using 66 bits. This is a huge bit width to use for calculations and will require significant resources to implement. If the application is not too sensitive, a 0.0001 can be rounded to a more binary friendly value, namely 1/8192 (0.000122...). This value can be represented precisely using only 13 bits (2^−13). By accepting some rounding error, it is possible to reduce the calculation bit width by 53 bits, thus greatly saving resources and computational time. As stated before, all feedback paths must complete calculation processes before the next sample arrives. By rounding the sampling period constant, *T*, to the closest power of 2, the computational task is much simpler; therefore, the feedback path can meet timing requirements. Of course, simulations must be done to fully ensure that this compromise does not affect the integrity of the IIR. Figure 11.9 displays the bode plots of the original integrator using $T = 0.0001$, and the rounded integrator, using $T = 1/8192$.

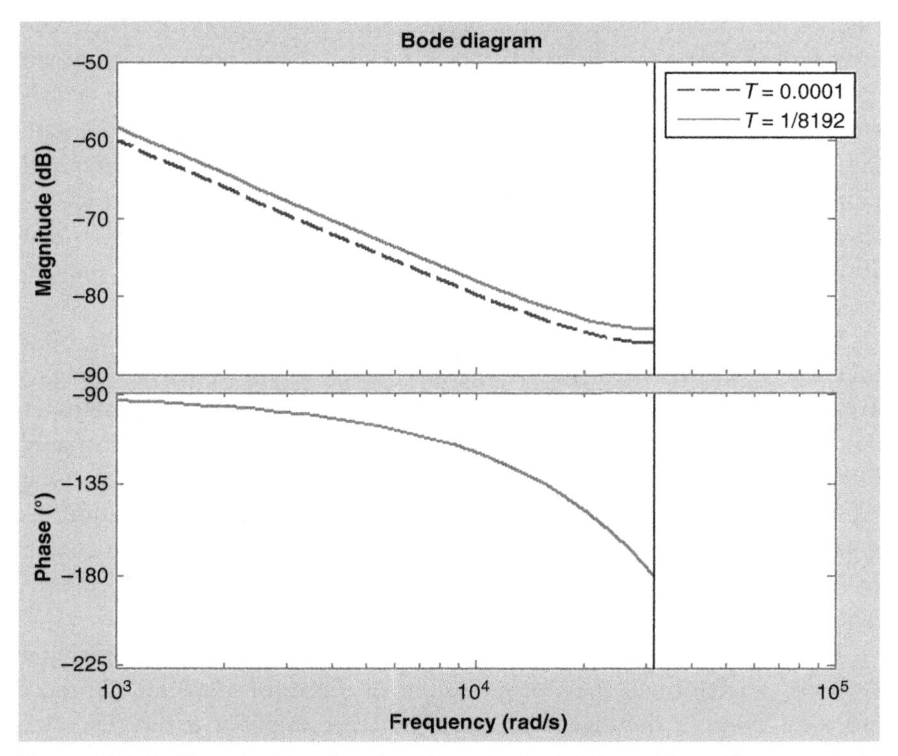

Figure 11.9 By Changing the Time Step Value to 1/8192 From 0.0001. The magnitude response increased by a constant 1.73 db, but the phase response did not change.

In this example, only the magnitude response was affected by 1.7322 dB (= 20log(10000/8192)).

Other IIR filters, such as digital type-II or III filters, will be more sensitive to coefficient rounding and may become unstable. It is important to always plot Nyquist and zero–pole plane plots of altered IIR filters, and to perform simulations to ensure that the IIR filter remains stable.

11.4 DIGITAL PWM

The outputs of the digital controller for a power supply will most likely be a gate drive signal. Since MCUs and FPGAs cannot drive more than 16 mA of current, external gate drivers need to be implemented to convert the logical drive signal into a physical drive signal (1 A or more source/sink) for the MOSFET gate. These gate drivers should have minimal latency to ensure that the digital signal is synced to the external environment.

Just like in the ADC discussion, the gate drive signal will be affected by clock jitter. In synchronous designs, all state changes are triggered on an edge of the system clock, so if the system clock experiences jitter, then the jitter will propagate to derived signals. Clock jitter will normally range no more than a few 100 ps. For power supply applications, gate drive signals will be in the 100 s of kHz range (>1 μs). Will the jitter affect the gate signal enough to cause unwanted gate signals? No. This is because jitter will only represent less than a tenth of a percent of the duty cycle. Even for synchronous switching applications, the required dead time is usually at least 50 ns (100 ps +/−). That value will not cause damage to the FETs and can be considered noise on the gate driveline.

One also has to understand the physical implications of turning on a FET. The function of the gate driver is to pump current to the FET gate to overcome the input capacitance. The stronger the drive current, the faster the MOSFET will turn on, but it is by no means instantaneous. Normally, the turn-on period for a power MOSFET ranges from 10 ns to 50 ns. One hundred extra picoseconds of turn-on time will not have any significant impact on the operation of the FET, since in that time, very little charge will be supplied to the gate. Therefore, jitter is not a significant factor for digital MOSFET drive.

In analog feedback circuits, PWM signals are generated by comparing a sawtooth waveform to the voltage output of the error amplifier. A similar process is done in the digital domain. The error amplifier is the IIR filter described earlier, while the sawtooth waveform is generated by a variable rate

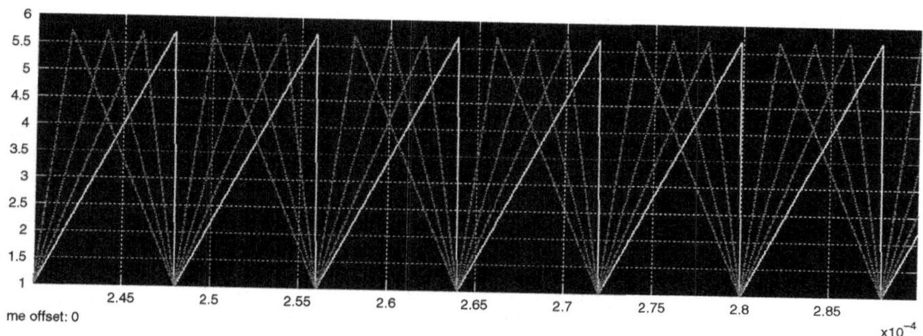

Figure 11.10 ***Digital Sawtooth Waveforms with 98% (solid), 75% (dash), 50% (dot), and 25% (dot-dash) Duty Cycle.***

counter. Digital counters have a great property of wrapping around when overflowing. Note that in the IIR filter discussion, bit overflow is to be avoided at all costs, but in the sawtooth application, the bit overflow can be exploited to generate the waveform. A digital counter increments at a set rate (by 1 or otherwise) each system clock cycle. When the counter reaches its maximal value, the next increment will return it back to zero. In this case, only the rise rate of the sawtooth waveform can be set. However, what if the sawtooth waveform should have a rising rate and a falling rate, as shown in Figure 11.10.

In this application, the counter should not simply wrap around and continue to increment, it should increment to the maximum value at one rate and decrement back to zero at another rate. Digital counters can be set to have a step size greater than 1; thus, each system clock cycle the counter will increment or decrement by the set step size.

The PWM is then created by comparing the current counter state to the IIR output. Using a sawtooth waveform that has a rising duty cycle and a falling duty cycle can be useful for phase shifting the PWM signal, giving the designer ability to do some frequency modulation.

In some power supply applications, a fixed-frequency PWM gate drive is not an optimal control algorithm. Certain topologies require frequency modulated gate signals (synchronous gate drives) at 50% duty cycle. One method of creating a dynamic frequency modulate algorithm in the digital domain is the use of a NCO (numerically controlled oscillator). NCOs work on the principle of using a look-up table and a phase step to cycle through the look up table. The larger the phase step, the faster the frequency (almost like in the sawtooth example: the larger the step size, the greater the rise or fall rate). The lookup table stored 1/4 of a sine wave period (zero to $\pi/2$). Figure 11.11 shows the internal architecture of a digital NCO (MATLAB).

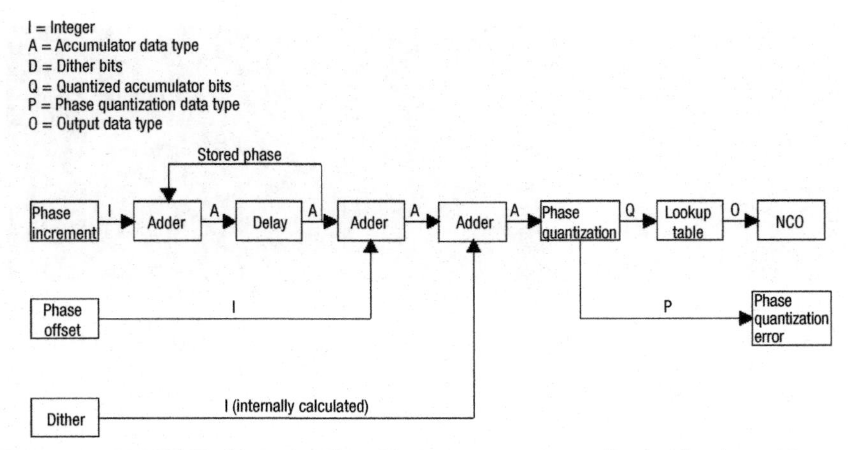

Figure 11.11 ***NCO Architecture.*** **The phase increment is used to read a sinusoid out of the lookup table. Increasing the phase increment will increase the output frequency (MATLAB).**

The phase increment and the output frequency are related by the equation:

$$\phi = F_o \frac{2^N}{F_s}$$

where F_o is the desired output frequency, N is the quantization bit width (more bits means better resolution), and F_s is the data rate. The phase increment can be dynamically changed, thus outputting a frequency-modulated waveform. The NCO outputs a sine waveform; the user can then add threshold comparators (digitally) to convert the sine waveform into a square wave. For example, if the condition is:

$$\begin{cases} \text{if NCO} > 0; \text{OUT} = 1 \\ \text{if NCO} \leq 0; \text{OUT} = 0 \end{cases}$$

This will generate a 50% square wave. By varying these thresholds and conditions, the duty cycle can be varied as well, allowing the designer to create signals with varying frequency and duty cycle. The digital domain allows the designer to create algorithms that would be difficult to realize using analog components.

Once the gate drive signals are derived in logic, external FET drivers can be used to provide the current and voltage required to turn on MOSFETs. For low-side drivers, any FET driver with small propagation delay (<15 ns) can be used. For high side FETs, the gate driver must include a bootstrap circuit so that the gate voltage of the FET exceeds the source voltage. For synchronous FETs, the driver must also include a cross-conduction prevention circuit. It is, of course, the designer's responsibility to prevent turning out both low and high side FETs together in the digital logic; however, to protect from glitches and possible start-up issues, an external anticross conduction circuit is desirable.

11.5 POWERING DIGITAL HARDWARE

None of the above will function if the MCU or FPGA is not powered. Hence, in the start-up sequence of the power supply, DC voltage should be generated to initialize digital hardware. This can be a difficult task, since digital hardware functions on low-voltage DC rails. As technology improves, and transistor geometries get smaller, digital hardware must run on smaller DC voltage rails and requires high current capabilities. Most MCUs, for example, require 3.3 V DC, but have internal regulators to supply 1.2 V to the core. FPGAs usually do not include DC–DC regulators, so the designer must provide several DC rails (3.3, 2.5, 1.8, and/or 1.2 V). This is an arduous task to include so many rails, because (1) it requires many DC–DC bucks or LDOs; (2) noise considerations; and (3) voltage ripple constraints.

Digital core voltages are specified to have <100 mV of ripple; this requires the regulator deriving this voltage to have a significant output capacitance. Additionally, processors that perform complex operations or DSP functions require a significant amount of current. Heavy duty MCUs or FPGAs can require up to 4 A of core current, but for the applications discussed herein, 2 A maximum provides a healthy margin. Since the core is very sensitive, the core voltage rail should have ample TVS and overvoltage protection to ensure that the processor is not damaged.

The I/O voltage to the processor, A/Ds, and other digital hardware is usually 3.3 V. Since this voltage is used by many ICs, it is important to have adequate PCB routing and denoising capacitors by each IC to ensure clean power. The current requirements for the 3.3 V rail can reach up to 1 A, but most likely will be several 100 mA. In either case, it is prudent to use a DC–DC buck regulator to generate this rail, as well as the core voltage rail. Because of this, there will be a lot of high frequency switching to generate these voltages. The designer must ensure a good digital ground plane to minimize this switching noise. It would be detrimental if this noise was to cause voltage oscillations on the rails and contaminate A/D signals.

Gate drivers also required their own power rail, usually 12 V. The best design would be to generate 12 V at a large amount of current (>1 A), and then use DC–DC regulators to derive the low DC voltages (3.3, core, etc). However, generating 12 V is difficult in power supply applications. In a boost topology, the initial PFC voltage is $\sqrt{2}\ V_{rms}$, which at 120 VAC would be 169 VDC. Stepping down 169 V to 12 V is not easy, but even worse, at full operation, the PFC voltage can reach 400 V, thus stepping down to 12 V is even more difficult. It is up to the designer to make the proper DC–DC bucking (perhaps multistage) topology to do so.

The situation seems counterintuitive; one must create a power supply to power a digital processor that would then control a power supply. The initial power supply would be very low power (<10 W) that will start up the digital hardware. Once that occurs, the digital hardware will control the high wattage power supply. This is essentially what "analog" power supply controller ICs do. Internal to these ICs is a power supply section that initializes the ICs. Unlike what the designer must do using discrete buck regulators, these ICs usually include BCD silicon that can directly connect to high voltage DC lines (>600 V). Nonetheless, these ICs consume a similar amount of power as digital hardware.

Besides providing power rails, the designer must also provide proper ground planes. Digital and analog (high power) grounds need to be segregated, but not disconnected. They must connect at the "mecca" ground using the star ground topology. In most power supplies, "mecca" ground would be the ground of the large DC capacitor. Digital and analog grounds are shown be partitioned and connected with ground channels. This prevents unwanted ground loops and noise interference between the two domains. This also reduces radiated and conducted noise and reduces ground plane inductance.

Using digital hardware and digital control, the designer is open to many theoretical possibilities, but is also posed with new challenges. IC manufacturers develop their products to be as user-friendly as possible; this means that many concerns are taken care of. When designing digital hardware, these concerns, which were previously taken for granted, become important.

APPENDIX A
State Space Averaging

While dwelling in one structure in the time frame of dT, the state equation and the output are described by

$$\dot{x} = A_1 x + B_1 E_1, \quad y = C_1^{\mathrm{T}} x, \quad 0 < t < dT \tag{A.1}$$

In the rest-time duration $(1 - d)T$, both equations change since structure varies.

$$\dot{x} = A_2 x + B_2 E_2, \quad y = C_2^{\mathrm{T}} x, \quad dT < t < T \tag{A.2}$$

For switch-mode converters, the power source often remains one and the same; $E_1 = E_2 = E$. Superscript T associated with column matrix C_1 and C_2 stands for "Transposition" and is not to be confused with switch period T.

State space average proposed a view that under low frequency perturbation much less than the switching frequency, the nonlinear portion of the converter system consisting of two periodically alternating structures and described by the two distinctive state equations, (6.2) and (6.3), may be consolidated in one and represented by a weighted sum.

$$\begin{aligned}\dot{x} &= d\left(A_1 x + B_1 E\right) + (1-d)\left(A_2 x + B_2 E\right) \\ &= \left(dA_1 + (1-d)A_2\right)x + \left(dB_1 + (1-d)B_2\right)E \\ y &= dC_1^{\mathrm{T}} x + (1-d)C_2^{\mathrm{T}} x = \left(dC_1^{\mathrm{T}} + (1-d)C_2^{\mathrm{T}}\right)x\end{aligned} \tag{A.3}$$

When subjected to small-signal perturbation, for instance, source fluctuation $(E + \hat{e})$, both the state vector and the duty cycle also exhibit corresponding deviation

$$x = X + \hat{x}, \quad d = D + \hat{d} \tag{A.4}$$

Plugging (6.5) and source disturbance in (6.4), the averaged state equation becomes

Power Converters with Digital Filter Feedback Control
http://dx.doi.org/10.1016/B978-0-12-804298-4.00017-4

$$\dot{X} + \dot{\hat{x}} = \dot{\hat{x}} = (AX + BE) + A\hat{x} + B\hat{e} + [(A_1 - A_2)X + (B_1 - B_2)E]\hat{d} \tag{A.5}$$

in which second order product terms are ignored, and

$$A = DA_1 + (1 - D)A_2, \quad B = DB_1 + (1 - D)B_2 \tag{A.6}$$

Equation (6.6) consists of two parts, steady state DC and small signal AC. The DC part yields

$$\dot{X} = AX + BE = 0, \quad X = -A^{-1}BE \tag{A.7}$$

and the AC part

$$\dot{\hat{x}} = A\hat{x} + [(A_1 - A_2)X + (B_1 - B_2)E]\hat{d} + B\hat{e} \tag{A.8}$$

Taking Laplace transform of (6.9) gives

$$\begin{aligned} \hat{x}(s) &= G_{xd}(s)\hat{d}(s) + G_{xe}(s)\hat{e}(s) \\ G_{xd}(s) &= (sI - A)^{-1}[(A_1 - A_2)X + (B_1 - B_2)E] \\ G_{xe}(s) &= (sI - A)^{-1}B \end{aligned} \tag{A.9}$$

in which I is the unit matrix. That is, (6.10) identifies two critical transfer functions, the duty-to-state $G_{xd}(s)$ gain and the source-to-state $G_{xe}(s)$ gain.

A similar procedure can also be carried out for the output, the second equation of (6.4), by plugging in both duty cycle and output disturbances

$$Y + \hat{y} = [(D + \hat{d})C_1^{\mathrm{T}} + (1 - D - \hat{d})C_2^{\mathrm{T}}](X + \hat{x}) \tag{A.10}$$

Again separating (6.11) into DC and AC parts and recognizing the average,

$$C^{\mathrm{T}} = DC_1^{\mathrm{T}} + (1 - D)C_2^{\mathrm{T}} \tag{A.11}$$

we obtain the DC output

$$Y = C^{\mathrm{T}}X = -C^{\mathrm{T}}A^{-1}BE \tag{A.12}$$

and the AC part

$$\hat{y}(s) = C^{\mathrm{T}}\hat{x}(s) + \left(C_1^{\mathrm{T}} - C_2^{\mathrm{T}}\right)X\hat{d}(s) \tag{A.13}$$

This is, however, not the end yet since (6.10) needs to be incorporated in (6.14).

$$\hat{y}(s) = C^{\mathrm{T}}[G_{\mathrm{xd}}(s)\hat{d}(s) + G_{\mathrm{xe}}(s)\hat{e}(s)] + \left(C_1^{\mathrm{T}} - C_2^{\mathrm{T}}\right)X\hat{d}(s) \qquad \text{(A.14)}$$

Further processing gives

$$\begin{aligned} \hat{y}(s) &= [C^{\mathrm{T}}G_{\mathrm{xd}}(s) + \left(C_1^{\mathrm{T}} - C_2^{\mathrm{T}}\right)X]\hat{d}(s) + C^{\mathrm{T}}G_{\mathrm{xe}}(s)\hat{e}(s) \\ &= G_{\mathrm{yd}}(s)\hat{d}(s) + G_{\mathrm{ye}}(s)\hat{e}(s) \end{aligned} \qquad \text{(A.15)}$$

where

$$\begin{aligned} G_{\mathrm{yd}}(s) &= C^{\mathrm{T}}(sI - A)^{-1}[(A_1 - A_2)X + (B_1 - B_2)E] + \left(C_1^{\mathrm{T}} - C_2^{\mathrm{T}}\right)X \\ G_{\mathrm{ye}}(s) &= C^{\mathrm{T}}(sI - A)^{-1}B \end{aligned} \qquad \text{(A.16)}$$

It is abundantly clear that the DC operating point does impact, and should, the duty-to-output transfer function $G_{\mathrm{yd}}(s)$, since X appears in square in the equation.

APPENDIX B

(1.17) to (1.19) and (1.22) to (1.23) Transform

Equation (1.17) has the form

$$\frac{A_1 s + A_0}{B_2 s^2 + B_1 s + B_0} \tag{B.1}$$

The Laplace operator s shall be replaced by

$$s = C\frac{1 - z^{-1}}{1 + z^{-1}} \tag{B.2}$$

(B.1) becomes

$$\frac{A_1 C\left((1 - z^{-1})/(1 + z^{-1})\right) + A_0}{B_2 C^2\left((1 - z^{-1})/(1 + z^{-1})\right)^2 + B_1 C\left((1 - z^{-1})/(1 + z^{-1})\right) + B_0} \tag{B.3}$$

It can be reduced to

$$\frac{A_1 C(1 - z^{-1})(1 + z^{-1}) + A_0(1 + z^{-1})^2}{B_2 C^2(1 - z^{-1})^2 + B_1 C(1 - z^{-1})(1 + z^{-1}) + B_0(1 + z^{-1})^2} \tag{B.4}$$

There are now two polynomials. Both need to be rearranged in the standard form preferred in digital signal processing. First, the numerator is expanded employing symbolical processing of MathCAD (PTC Mathsoft). It results in

$$A_1 C - A_1\frac{C}{z^2} + A_0 + 2\frac{A_0}{z} + \frac{A_0}{z^2} \tag{B.5}$$

The next step is to collect terms in power of z^{-1}.

$$A_0 + A_1 C + 2A_0 z^{-1} + \left(A_0 - A_1 C\right)z^{-2} \tag{B.6}$$

Then the same steps are carried for the denominator. It gives

$$B_2 C^2 + B_0 + B_1 C + \left(-2B_2 C^2 + 2B_0\right)z^{-1} + \left(-B_1 C + B_2 C^2 + B_0\right)z^{-2} \tag{B.7}$$

Power Converters with Digital Filter Feedback Control
http://dx.doi.org/10.1016/B978-0-12-804298-4.00018-6

Therefore, (B.1) is now in the form of

$$\frac{[A_0 + A_1C + 2A_0z^{-1} + \left(A_0 - A_1C\right)z^{-1}]}{[B_2C^2 + B_0 + B_1C\left(-2B_2C^2 + 2B_0\right)z^{-1} + \left(-B_1C + B_2C^2 + B_0\right)z^{-1}]} \tag{B.8}$$

We then place (B.8) in the following

$$\frac{\left[\dfrac{A_0 + A_1C}{B_2C^2 + B_0 + B_1C} + \dfrac{2A_0}{B_2C^2 + B_0 + B_1C}z^{-1} + \left(\dfrac{A_0 - A_1C}{B_2C^2 + B_0 + B_1C}\right)z^{-2}\right]}{\left[1 + \left(\dfrac{-2B_2C^2 + 2B_0}{B_2C^2 + B_0 + B_1C}\right)z^{-1} + \left(\dfrac{-B_1C + B_2C^2 + B_0}{B_2C^2 + B_0 + B_1C}\right)z^{-2}\right]} \tag{B.9}$$

This is in line with the standard form

$$\frac{(a_0 + a_1z^{-1} + a_2z^{-2})}{(1 + b_1z^{-1} + b_2z^{-2})} \tag{B.10}$$

(B.9) and (B.10) together confirm (1.19).

By the same token, for form (1.22), by plugging the bilinear transform (B.2) and rearranging both the numerator and denominator polynomial, the third order transfer function (1.22) becomes

$$\frac{A_2[C(1-z^{-1})]^2(1+z^{-1}) + A_1C(1-z^{-1})(1+z^{-1})^2 + A_0(1+z^{-1})^3}{B_3[C(1-z^{-1})]^3 + B_2[C(1-z^{-1})]^2(1+z^{-1}) + B_1C(1-z^{-1})(1+z^{-1})^2 + B_0(1+z^{-1})^3} \tag{B.11}$$

Performing again expansion, collecting terms in power of z^{-1}, and placing it in the standard form, the following results

$$\frac{\dfrac{A_2C^2 + A_1C + A_0}{B_3C^3 + B_2C^2 + B_1C + B_0} + \left(\dfrac{-A_2C^2 + A_1C + 3A_0}{B_3C^3 + B_2C^2 + B_1C + B_0}\right)z^{-1} + \left(\dfrac{-A_2C^2 - A_1C + 3A_0}{B_3C^3 + B_2C^2 + B_1C + B_0}\right)z^{-2} + \left(\dfrac{A_2C^2 - A_1C + A_0}{B_3C^3 + B_2C^2 + B_1C + B_0}\right)z^{-3}}{1 + \left(\dfrac{-3B_3C^3 - B_2C^2 + B_1C + 3B_0}{B_3C^3 + B_2C^2 + B_1C + B_0}\right)z^{-1} + \left(\dfrac{3B_3C^3 - B_2C^2 - B_1C + 3B_0}{B_3C^3 + B_2C^2 + B_1C + B_0}\right)z^{-2} + \left(\dfrac{-B_3C^3 + B_2C^2 - B_1C + B_0}{B_3C^3 + B_2C^2 + B_1C + B_0}\right)z^{-3}} \tag{B.12}$$

And, (1.23) is established.

APPENDIX C

Setting Up Difference Equation (1.33)

For this effort, Figure 1.1, with some component renamed, is repeated here to avoid flipping pages and to aid learning efficiency (Figure C.1).

A nomenclature issue shall be spelled out prior to analytical derivation.

In electrical system studies, the wordings of state variables are invoked quite frequently in setting up differential equations describing electrical circuit functions. Accordingly, capacitor voltages and inductive currents in Figure 1.1, repeated earlier, albeit some symbol changes, are considered state variables. Here, capacitor voltage means the voltage across a capacitor; for instance state variable, v_{C16}, of C16, equals $v_n - v_e$. However, electrical engineers are, for practical purposes, more interested in node voltages referred to ground than a difference voltage across a component. Therefore, in this writing and in this appendix in particular, state variables for capacitors are not actively invoked. Instead, we focus directly on node voltages. That being said, inductive currents do not face such a problem. It is what it is.

Now, we proceed with developing (1.33).

First, node voltages and inductive branch currents are identified. There are (1) the input current i_1; (2) the primary (winding terminal) node voltage v_p; (3) the output loop current i; (4) the output node voltage v_o; (5) the output filter damping node voltage v_{cd}; (6) the error amplifier input node voltage v_a; (7) the error amplifier feedback node voltage v_b; (8) the error amplifier output node voltage v_e; (9) the magnetizing current i_m; (10) the input filter damping node voltage v_{id}; and (11) the transformer core reset current i_r. Up to this point, variable subscripts are still enforced. However, in what is to follow, a decision was made to forgo all subscript format, since in MathCAD environment for seeded iterative computation, the variable subscripts make iteration index, a built-in range variable in subscript form defined by the software tool, quite cumbersome. The decision also saves a lot of typing time.

For (1), the input filter loop gives

$$V_{in} = L_1 \frac{di_1}{dt} + r_1 i_1 + v_p \tag{C.1}$$

Power Converters with Digital Filter Feedback Control
http://dx.doi.org/10.1016/B978-0-12-804298-4.00019-8

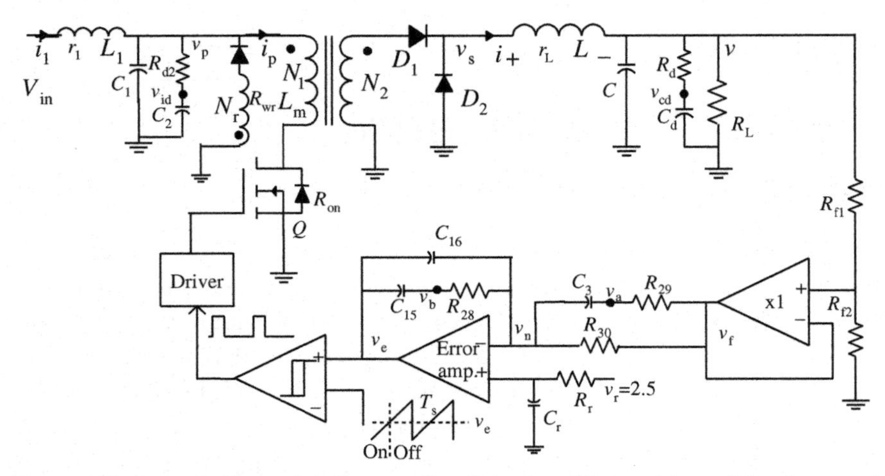

Figure C.1 ***Parts, in Figure 1.1, Surrounding Error Amplifier and Input Filter Renamed to Match (1.33).***

Rewrite it in a different form, (C.1) becomes

$$di_1 = \frac{dt}{L_1}\left[V_{\text{in}} - (r_1 i_1 + v_{\text{p}})\right] \tag{C.2}$$

In finite difference form, given a small time step, δt, its size yet to be determined, and step index j, (C.2) becomes

$$i_{1_{j+1}} - i_{1_j} = \frac{\delta t}{L_1}\left[V_{\text{in}} - (r_1 i_{1_j} + v_{p_j})\right] \tag{C.3}$$

In other words, the input current change within a very small time step is determined by its current value, the time step size, the input inductor value, and the effective difference voltage across the inductor (the term in bracket).

One more step places it in a form suitable for iterative computation, the first in (1.33)

$$i_{1_{j+1}} = i_{1_j} + \frac{\delta t}{L_1}\left[V_{\text{in}} - (r_1 i_{1_j} + v_{p_j})\right] \tag{C.4}$$

Next we go to (2), the same voltage across C_1. This node requires more attention and preparation since it consists of four currents: the input line current, C_1 current, C_2 (or R_{d2}) current, and primary winding current. The primary winding current further comprises of two components, the transformer core magnetizing current and reflected load current. Thus, applying the Kirchhoff's current law to this node, C_1 current is expressed as

$$C_1 \frac{dv_p}{dt} = i_1 - \frac{v_p - v_{id}}{R_{d2}} - i_m - \frac{N_2}{N_1} i \qquad \text{(C.5)}$$

Or, C_1 voltage change is expressed as

$$dv_p = \frac{dt}{C_1}\left(i_1 - \frac{v_p - v_{id}}{R_{d2}} - i_m - \frac{N_2}{N_1} i \right) \qquad \text{(C.6)}$$

In iterative form and considering the fact that the reflected load current, i, is a switched quantity depending on the power switch ON/OFF state, (C.6) as a consequence becomes (C.7), and it is the second equation of (1.33). Here the periodic sawtooth, s_w, carrier is introduced with the error voltage, v_e, in the framework of "*if(condition, true, false)*" statement of MathCAD.

$$v_{p_{j+1}} = v_{p_j} + \frac{\delta t}{C_1}\left(i_{1_j} - \frac{v_{p_j} - v_{id_j}}{R_{d2}} - i_{m_j} - \frac{N_2}{N_1} \text{if}(v_{e_j} > s_{w_j}, i_j, 0) \right) \qquad \text{(C.7)}$$

For the third equation, related to variable (3), the output inductor current and parasitic series resistance r (subscript omitted), Kirchhoff's voltage law gives

$$v_s = ri + L\frac{di}{dt} + v_o, \quad di = \frac{dt}{L}\left[v_s - (ri + v_o) \right] \qquad \text{(C.8)}$$

Written in basic difference form, (C.8) becomes

$$i_{j+1} = i_j + \frac{\delta t}{L}\left[v_{s_j} - (ri_j + v_{o_j}) \right] \qquad \text{(C.9)}$$

However, it is recognized that the voltage source, v_s, feeding the output filter is also a switched quantity. It is actually the rectangular PWM voltage wave swinging between $[(N_2/N_1)$(primary winding voltage) – rectifier drop] and [−rectifier drop], again depending on the switch ON/OFF state. Therefore, we have

$$i_{j+1} = i_j + \frac{\delta t}{L}\left[\text{if}\left(v_{e_j} > s_{w_j}, \frac{N_2}{N_1} v_{\text{primary}} - v_d, -v_d\right) - (ri_j + v_{o_j})\right] \quad \text{(C.10)}$$

(C.10) is not the final form since the primary winding voltage has yet to account for the power switch drop due to its on resistance, R_{on}. This is taken care of by the product of the on resistance and the switch-on current, the sum of reflected load current and the magnetizing current.

$$i_{j+1} = i_j + \frac{\delta t}{L}\left[\text{if}\left(v_{e_j} > s_{w_j}, \frac{N_2}{N_1}\left(v_{p_j} - R_{on}\left(\frac{N2}{N1} i_j + i_{m_j}\right)\right) - v_d, -v_d\right) - (ri_j + v_{o_j})\right] \quad \text{(C.11)}$$

In addition, because of the continuous conduction mode operation and the unidirectional nature of both output rectifier diodes, the output inductor current never runs dry, that is, less than zero, or reverse. The effect is also taken care of by an overriding nested "*if*" statement. With this additional constraint, the third equation takes the final form as shown in (1.33).

Next is the output voltage, actually the voltage across the output filter capacitor C. The output node equation, based on KCL, gives

$$C\frac{dv_o}{dt} = i - \frac{v_o}{RL} - \frac{v_o - v_{cd}}{R_d}, \quad dv_o = \frac{dt}{C}\left(i - \frac{v_o}{RL} - \frac{v_o - v_{cd}}{R_d}\right) \quad \text{(C.12)}$$

In finite difference form, (C.12) leads to the fourth equation in (1.33).

Variable number (5) is quite easy to handle. The output filter damping capacitor current equals

$$C_d\frac{dv_{cd}}{dt} = \frac{v_o - v_{cd}}{R_d}, \quad dv_{cd} = \frac{dt}{C_d}\left(\frac{v_o - v_{cd}}{R_d}\right) \quad \text{(C.13)}$$

Fifth equation in (1.33) presents (C.13) in iterative finite difference form.

We proceed next with variable (6), the voltage at junction of $C3$ and $R29$. Setting up this equation requires a little bit extra reasoning. The current through $C3$ is given by

$$C3\frac{d(v_a - v_n)}{dt} = \frac{v_f - v_a}{R29} \quad \text{(C.14)}$$

where v_f, the feedback voltage, is equal to $k_f v_o$, the product of output voltage and feedback factor. Furthermore, if high-gain (open loop) error amplifier is used, the inverting node voltage, v_n tracks the command reference voltage at the noninverting node; $v_n = v_r$. Since the reference voltage is constant, its derivative equals zero. Therefore (C.14) reduces to (C.15).

$$C3\frac{dv_a}{dt} = \frac{k_f v_o - v_a}{R29}, \quad dv_a = \frac{dt}{C3}\frac{k_f v_o - v_a}{R29} \tag{C.15}$$

By the same token, (C.15) turns into sixth equation in (1.33).

At this point, the processing sequence shall be disrupted. Instead of (7), variable (8), equivalent to $C16$ voltage, will be handled now. The current through that capacitor is

$$\begin{aligned} C16\frac{d(v_n - v_e)}{dt} &= \frac{k_f v_o - v_a}{R29} + \frac{k_f v_o - v_n}{R30} - \frac{v_n - v_b}{R28} \\ \frac{dv_e}{dt} &= \frac{1}{C16}\left(\frac{v_r - v_b}{R28} - \frac{k_f v_o - v_r}{R30} - \frac{k_f v_o - v_a}{R29}\right) \end{aligned} \tag{C.16}$$

Again, following the same process of finite difference and also taking into consideration the error amplifier's local supply at 15 V, eighth equation in (1.33) with nested "*if*" takes its place.

In the same neighborhood of the error amplifier, the local feedback voltage, v_b, at the junction of $C15$ and $R28$, is the next to be treated. Current of $C15$ equals

$$C15\frac{d(v_b - v_e)}{dt} = \frac{v_n - v_b}{R28}, \quad \frac{dv_b}{dt} - \frac{dv_e}{dt} = \frac{v_r - v_b}{C15R28} \tag{C.17}$$

Here, a little bit of complication arises. There are two dynamic derivatives, dv_b/dt and dv_e/dt, embedded in the equation. The second derivative must be removed. (C.16)'s second line can do exactly that and was left in that form for the anticipated need.

$$\begin{aligned} \frac{dv_b}{dt} &= \frac{v_r - v_b}{C15R28} + \frac{dv_e}{dt} \\ \frac{dv_b}{dt} &= \frac{v_r - v_b}{C15R28} + \frac{1}{C16}\left(\frac{v_r - v_b}{R28} - \frac{k_f v_o - v_r}{R30} - \frac{k_f v_o - v_a}{R29}\right) \end{aligned} \tag{C.18}$$

This form of course yields the seventh equation in (1.33).

Similar to the output damping capacitor, the input damping capacitor, $C2$, is governed by

$$C2\frac{dv_{\mathrm{id}}}{dt}=\frac{v_{\mathrm{p}}-v_{\mathrm{id}}}{R_{\mathrm{d}2}},\quad \frac{dv_{\mathrm{id}}}{dt}=\frac{dt}{C2}\frac{v_{\mathrm{p}}-v_{\mathrm{id}}}{R_{\mathrm{d}2}} \tag{C.19}$$

This leads to tenth equation in (1.33).

The treatment of (9), the magnetizing current, takes a bit more consideration. The key is to recognize that the effective voltage across the primary winding is less than the primary node voltage.

$$\begin{aligned} L_{\mathrm{m}}\frac{di_{\mathrm{m}}}{dt} &= v_{\mathrm{p}}-\left(\frac{N_2}{N_1}i+i_{\mathrm{m}}\right)R_{\mathrm{on}} \\ di_{\mathrm{m}} &= \frac{dt}{L_{\mathrm{m}}}\left[v_{\mathrm{p}}-\left(\frac{N_2}{N_1}i+i_{\mathrm{m}}\right)R_{\mathrm{on}}\right] \end{aligned} \tag{C.20}$$

Moreover, the magnetizing current is a switched quantity subjected to the power switch ON/OFF condition. When the switch is turned ON, the current ramps up. When the switch is OFF, the magnetic core current is diverted to the reset winding and becomes zero. With the switching condition included, (C.20) becomes the ninth equation in (1.33).

The last variable (11), during the time frame the power switch is turned OFF, the transformer core resets and the reset winding conducts. By doing so, the reset winding clamps across the primary winding terminal.

$$\begin{aligned} L_{\mathrm{m}}\frac{di_{\mathrm{r}}}{dt}+R_{\mathrm{w,r}}i_{\mathrm{r}}+v_{\mathrm{p}} &= 0 \\ di_{\mathrm{r}} &= \frac{dt}{L_{\mathrm{m}}}\left[-v_{\mathrm{p}}-R_{\mathrm{w,r}}i_{\mathrm{r}}\right] \end{aligned} \tag{C.21}$$

It is also a switched quantity and is of course subjected to the same switching state of the power switch, except in reverse logic. That is, when the power switch is turned ON, it is a zero current. When the power switch is turned OFF, the reset winding steers the transformer core magnetizing current and returns it to the line input source. In analytical terms, the peak magnetizing current serves as the initial condition (current) for the reset winding when it conducts.

This concludes the long derivation for (1.33). For (2.7) and similar equation sets of rest of the chapters, there is no need to duplicate the effort given earlier. Readers are encouraged to do it on their own.

APPENDIX D

Flyback Converter DCM Operation

The flyback converter in steady-state DCM operation is understood to go through three phases as shown in Figure D.1. In Phase I, the power switch turns on, and the primary winding current ramps up, starting from zero. It reaches a peak and stores energy in the magnetic core when the switch is commanded off. The output capacitor supports the load alone during this phase. In Phase II, the power switch turns off. The secondary winding kicks in, steering the stored energy to support the load and also replenish the output capacitor. When the secondary winding current ramps down to zero, Phase III commence until power switch is commanded on.

During Phase I, the primary winding loop is governed by

$$\frac{di_{\mathrm{p}}}{dt} + \frac{1}{\tau_{\mathrm{p}}} i_{\mathrm{p}} = \frac{V_{\mathrm{bus}}}{L_{\mathrm{p}}}, \quad \tau_{\mathrm{p}} = \frac{L_{\mathrm{p}}}{r_{\mathrm{p}}} \tag{D.1}$$

With the ramping-up current starting from a zero initial condition, (D.1) has an exponential solution

$$i_{\mathrm{p}}(t) = \frac{V_{\mathrm{bus}}}{r_{\mathrm{p}}} \left(1 - \mathrm{e}^{-(t/\tau_{\mathrm{p}})}\right)\left(u(t) - u(t - D_1 T)\right) \tag{D.2}$$

The output capacitor is governed by

$$\frac{dv}{dt} + \frac{1}{\tau_{\mathrm{c}}} v = 0, \quad \tau_{\mathrm{c}} = (r_{\mathrm{c}} + R_{\mathrm{L}})C \tag{D.3}$$

A crucial point must be made here. That is the initial voltage of the output capacitor, C, is NOT zero. It has a starting value, V_{01}, yet unknown. As such, (D.3)'s solution, valid only for Phase I, is given in the following with the unknown, initial voltage included.

$$v_1(t) = V_{01}\, \mathrm{e}^{-(t/\tau_{\mathrm{p}})} \left(u(t) - u(t - D_1 T)\right), v_{01}(t) = k_{\mathrm{R}} v_1(t), k_{\mathrm{R}} = \frac{R_{\mathrm{L}}}{r_{\mathrm{c}} + R_{\mathrm{L}}} \tag{D.4}$$

Power Converters with Digital Filter Feedback Control
http://dx.doi.org/10.1016/B978-0-12-804298-4.00020-4

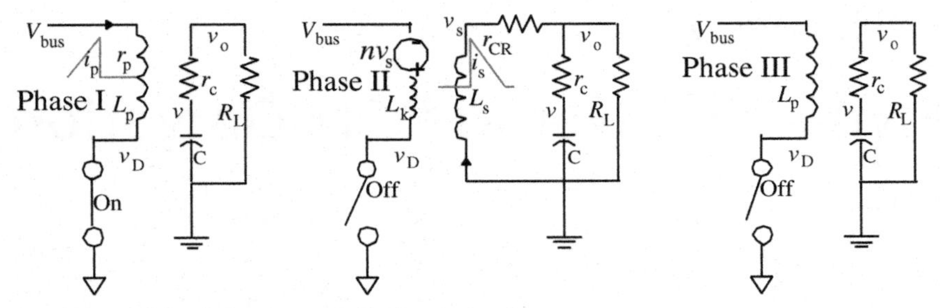

Figure D.1 ***Flyback Converter DCM Operation Sequence.***

A word of caution is appropriate here: output voltage ≠ capacitor voltage.

During Phase II, there are two governing equations: a loop and a node.

$$\frac{di_s}{dt} + \frac{1}{\tau_s} i_s + \frac{k_R}{L_s} v = 0, \quad \tau_s = \frac{L_s}{r_{CR} + R_p}, \quad R_p = \frac{r_C R_L}{r_C + R_L}$$
$$-\frac{R_L}{\tau_c} i_s + \frac{dv}{dt} + \frac{1}{\tau_c} v = 0 \tag{D.5}$$

Taking the Laplace transform of (D.5), both equations in the transformed domain are

$$\left(s + \frac{1}{\tau_S}\right) I_s(s) + \frac{k_R}{L_s} V(s) = I_s, \quad -\frac{R_L}{\tau_c} I_s(s) + \left(s + \frac{1}{\tau_c}\right) V(s) = V_{02} \tag{D.6}$$

in which the secondary winding's initial condition, I_s, and the capacitor voltage, V_{02}, are also included.

Kramer's rule enable us to solve $I_s(s)$ and $V(s)$ in terms of I_s and V_{02}

$$D(s) = s^2 + \left(\tfrac{1}{\tau_C} + \tfrac{1}{\tau_S}\right) s + \tfrac{k_R R_L}{L_S \tau_C} + \tfrac{1}{\tau_S \tau_C}$$
$$F_1(s) = \tfrac{s+(1/\tau_c)}{D(s)}, \quad F_2(s) = \tfrac{k_R}{L_s D(s)}, \quad F_3(s) = \tfrac{R_L}{\tau_C D(s)}, \quad F_4(s) = \tfrac{s+(1/\tau_c)}{D(s)} \tag{D.7}$$
$$I_s(s) = F_1(s) I_s - F_2(s) V_{02}, \quad V_2(s) = F_3(s) I_s + F_4(s) V_{02}$$

A voltage subscript, 2, is introduced for the capacitor voltage to signify that it is valid for Phase II only.

Now, taking the inverse Laplace transform for (D.7) and considering time delay, the loop current and the capacitor voltage are expressed as

$$\begin{aligned}
&f_1(t) = \ell^{-1}\left[F_1(s)\right], \quad f_2(t) = \ell^{-1}\left[F_2(s)\right] \\
&i_s(t) = \left[I_s f_1(t - D_1T) - V_{02} f_2(t - D_1T)\right]\left[u(t - D_1T) - u(t - (D_1 + D_2)T)\right] \\
&f_3(t) = \ell^{-1}\left[F_3(s)\right], \quad f_4(t) = \ell^{-1}\left[F_4(s)\right] \\
&v_2(t) = \left[I_s f_3(t - D_1T) + V_{02} f_4(t - D_1T)\right]\left[u(t - D_1T) - u(t - (D_1 + D_2)T)\right] \\
&v_{o2}(t) = R_p i_s(t) + k_R v_2(t)
\end{aligned} \tag{D.8}$$

During Phase III, the output capacitor is again supporting the load alone. Its voltage profile looks similar to (D.4), except unknown initial voltage V_{03}.

$$\begin{aligned}
&v_3(t) = V_{03}\, e^{-\left[(t-(D_1+D_2)T)/\tau_p\right]}\left(u(t - (D_1 + D_2)T) - u(t - T)\right), \\
&v_{o3}(t) = k_R v_3(t)
\end{aligned} \tag{D.9}$$

The output capacitor voltage is understood to be continuous. Therefore,

$$V_{01}e^{-(D_1T/\tau_p)} = V_{02}, \quad I_s f_3(D_2T) + V_{02} f_4(D_2T) = V_{03}, \quad V_{03}e^{-(T-(D_1+D_2)T/\tau_p)} = V_{01} \tag{D.10}$$

Successive substitution of initial voltages enables us to solve first V_{01}, the cyclic starting voltage of the output capacitor in Phase I.

$$V_{01} = \frac{I_s f_3(D_2T)}{e^{\left[(T-(D_1+D_2)T)/\tau_p\right]} - e^{-(D_1T/\tau_p)} f_4(D_2T)} \tag{D.11}$$

The second equation of (D.8) and the first of (D.10) give the secondary winding current

$$i_s(t) = I_s f_1(t - D_1T) - V_{01}\, e^{-(D_1T/\tau_p)} f_2(t - D_1T) \tag{D.12}$$

Plug in (D.11) and isolate the secondary peak current

$$i_s(t) = I_s\left[f_1(t - D_1T) - \frac{f_3(D_2T)}{e^{\left[(T-(D_1+D_2)T)/\tau_p\right]} - e^{-(D_1T/\tau_p)} f_4(D_2T)}\, e^{-(D_1T/\tau_p)} f_2(t - D_1T)\right] \tag{D.13}$$

The secondary winding current must reach zero at $t = (D_1 + D_2)T$.

$$i_s\left[(D_1 + D_2)T\right] =$$

$$I_s\left[f_1(D_2T) - \frac{f_3(D_2T)}{e^{\left[(T-(D_1+D_2)T)/\tau_p\right]} - e^{-(D_1T/\tau_p)}\, f_4(D_2T)}\, e^{-(D_1T/\tau_p)}\, f_2(D_2T)\right] = 0 \tag{D.14}$$

Therefore,

$$f_1(D_2T) - \frac{f_3(D_2T)}{e^{\left[(T-(D_1+D_2)T)/\tau_p\right]} - e^{-(D_1T/\tau_p)}\, f_4(D_2T)}\, e^{-(D_1T/\tau_p)}\, f_2(D_2T) = 0 \tag{D.15}$$

Given desired, or selected, D_1 and D_2, (D.15) enables one to find a more accurate inductor value for the secondary winding.

REFERENCES

[1] Cuk S, Middlebrook RD. Modeling, analysis and design of switching converters. 1978. NASA CR-135174; TRW A72042-RHBE; TRW D04803-CFCM.
[2] Cuk S, Middlebrook RD. Advances in switch-mode power conversion, vols. 1, 2, and 3. TESLAco; 1983.
[3] Wu KC. Switch-mode power converters: design and analysis. Elsevier, Academic Press; 2005.
[4] Wu KC. Pulse width modulated DC–DC converters. Springer; 1997.
[5] Linear Technology VCO LTC6900 application note.

Power Converters with Digital Filter Feedback Control
http://dx.doi.org/10.1016/B978-0-12-804298-4.00021-6

INDEX

G

H

I

K

L

Q

R

S

T

U

V

Printed in the United States
By Bookmasters